Diverse Quantization Phenomena in Layered Materials

T0136396

Diverse Quantization Phenomena in Layered Materials

Edited By

Chiun-Yan Lin, Ching-Hong Ho, Jhao-Ying Wu, Thi-Nga Do,
Po-Hsin Shih, Shih-Yang Lin, and Ming-Fa Lin

CRC Press
Taylor & Francis Group
Boca Raton London New York

CRC Press is an imprint of the
Taylor & Francis Group, an **informa** business

CRC Press
Taylor & Francis Group
6000 Broken Sound Parkway NW, Suite 300
Boca Raton, FL 33487-2742

First issued in paperback 2021

ISBN-13: 978-0-367-42028-4 (hbk)
ISBN-13: 978-1-03-208268-4 (pbk)

Library of Congress Cataloging-in-Publication Data

Names: Lin, Chiun-Yan, editor.
Title: Diverse quantization phenomena in layered materials / [edited by]
    Chiun-Yan Lin [and 6 others].
Description: First edition. | Boca Raton, FL : CRC Press/Taylor & Francis
    Group, 2020. | Includes bibliographical references and index. | Summary:
    "This book offers a comprehensive overview of diverse quantization
    phenomena in layered materials, covering current mainstream experimental
    and theoretical research studies. It presents essential properties of
    layered materials and provides a wealth of figures. Aimed at readers
    working in materials science, physics, and engineering this book should
    be useful for potential applications in energy storage, electronic
    devices, and optoelectronic devices"– Provided by publisher.
Identifiers: LCCN 2019037804 | ISBN 9780367420284 (hardback) | ISBN
    9781003004981 (ebook)
Subjects: LCSH: Layer structure (Solids)–Mathematical models. | Electronic
    structure–Mathematical models. | Geometric quantization. |
    Electromagnetic testing.
Classification: LCC QD921 .D58 2020 | DDC 530.4/11–dc23
LC record available at https://lccn.loc.gov/2019037804

Visit the Taylor & Francis Web site at
http://www.taylorandfrancis.com

and the CRC Press Web site at
http://www.crcpress.com

Contents

Preface

Layered materials have attracted lots of research interests in the fields of condensed-matter physics, physical chemistry, and materials science due to the nature of its particular 2D quasi-particle properties. Furthermore, they promise a broad range of applications in energy storage, electronic, optoelectronic, and photonic devices. 2D materials, such as graphene, silicene, germanene, and topological insulators, exhibit rich electronic and optical properties under external electric and magnetic fields. Especially, Landau quantization of the quasi-particles in a magnetic field is a fundamental and contemporary research topic for many scientific and applied fields, a phenomenon being strictly distinct from what is usually seen in the conventional 2D electron gas systems. We devote to give a general overview of the diverse quantization phenomena in layered materials, covering both of up-to-date mainstream experimental and theoretical research studies.

The diverse quantization phenomena in 2D condensed-matter systems, being due to a uniform perpendicular magnetic field and the geometry-created lattice symmetries, are the focuses of this book. We first give an introduction of the up-to-date research of 2D layered materials, such as graphene, silicene, germanene, and topological insulators, and illustrate the quantization phenomena in the framework of the generalized tight-binding model. The Landau-level spectra are obtained for those systems. Various magneto-optical selection rules, unusual quantum Hall conductivities, and single- and many-particle magneto-Coulomb excitations, based on the characteristics of the Landau-level spectra and the wave functions, can be further calculated under influences of external electric fields and charge dopings.

The generalized tight-binding model with an effective diagonalization method can be used to efficiently obtain the well-depicted energy spectrum and wave functions over a wide energy range, exceeding the accuracy of the effective-mass model that expands the low-energy Hamiltonian equation near the K point with perturbations of fields and some interlayer atomic interactions. The interplay between external fields and the geometric configuration determines the relationships between the components of wave functions in different sublattices.

This book presents the full progress about this topic and provides intuitive physical pictures with the importance and significance in the fields of experimental and theoretical research. The commonly used synthesis methods of those 2D materials are illustrated in the introduction (Chapter 1). Also included are the comparisons between the calculated results and the experimental measurements, and some novel features not yet reported. Chapter 2 discusses experimental measurements on magnetic quantization. The theoretical modeling for the studying systems is addressed in Chapter 3. The quantization phenomena cover diversified magneto-electronic properties, various magneto-optical selection rules, unusual quantum Hall conductivities, and single- and many-particle magneto-Coulomb excitations. The rich and unique behaviors are clearly revealed in few-layer graphene systems with the distinct stacking configurations (Chapters 4 and 9), the stacking-modulated structures (Chapter 5), and the silicon-doped lattices (Chapter 8), bilayer silicene (Chapters 6 and 7)/germanene (Chapter 10) systems with the bottom-top and bottom-bottom buckling structures, monolayer and bilayer phosphorene systems, and quantum topological insulators (Chapter 11). The generalized tight-binding model, the static and dynamic Kubo formulas, and the random-phase approximation are developed/modified to thoroughly explore the fundamental properties and propose the concise physical pictures. The different high-resolution experimental measurements are discussed in detail, and they are consistent with the theoretical predictions.

Authors

Chiun-Yan Lin obtained his PhD in 2014 in physics from the National Cheng Kung University (NCKU), Taiwan. Since 2014, he has been a postdoctoral researcher in the department of physics at NCKU. His main scientific interests are in the field of condensed matter physics, modeling and simulation of nanomaterials. Most of his research is focused on the electronic and optical properties of two dimensional nanomaterials.
cylin@mail.ncku.edu.tw

Jhao-Ying Wu received his PhD in 2009 in physics from the National Cheng Kung University (Tainan, Taiwan). After that, he was a postdoctoral fellow until 2016. He became professor in the National Kaohsiung University of Science and Technology. His main scientific interests focus on theoretical condensed matter physics, including the electronic and optical properties of low-dimensional systems, Coulomb excitations, and quantum transport.
yarst5@gmail.com

Ming-Fa Lin is a distinguished professor in the Department of Physics, National Cheng Kung University, Taiwan. He received his PhD in physics in 1993 from the National Tsing-Hua University, Taiwan. His main scientific interests focus on essential properties of carbon related materials and low-dimensional systems.
mflin@mail.ncku.edu.tw

Ching-Hong Ho is a postdoctoral researcher in the Department of Physics, National Cheng Kung University, Taiwan, where he received his PhD in 2011. His research area of interest is theoretical condensed matter physics with a focus on topological aspects in the past years.
hohohosho@gmail.com

Shih-Yang Lin received his PhD in physics in 2015 from the National Cheng Kung University (NCKU), Taiwan. Since 2015, he has been a postdoctoral researcher at NCKU. His research interests include low-dimensional group IV materials and first-principle calculations.
sylin.1985@gmail.com

Thi-Nga Do is a theoretical physicist whose scientific interests mainly focus on the field of solid state physics. She received her PhD degree in physics from National Cheng Kung University in 2017. Since then, Dr. Do has been working as a postdoctoral researcher at different affiliations, National Cheng Kung University, National Kaohsiung Normal University, and Academia Sinica, Taiwan. She is currently a researcher at Ton Duc Thang University, Vietnam.
dothinga@tdtu.edu.vn

Po-Hsin Shih is a postdoctoral researcher at National Cheng Kung University, Taiwan. He received his PhD in physics from the same university in 2018. He has been working on modeling of fundamental physics of nanomaterials, including electronic, optical, and transport properties.
phshih@phys.ncku.edu.tw

1 Introduction

Shih-Yang Lin,[e] Thi-Nga Do,[c,d] Chiun-Yan Lin,[a]
Jhao-Ying Wu,[b] Po-Hsin Shih,[a]
Ching-Hong Ho,[b] Ming-Fa Lin[a,f,g]

[a] Department of Physics, National Cheng Kung University, Tainan 701, Taiwan
[b] Center of General Studies, National Kaohsiung University of Science and Technology, Kaohsiung 811, Taiwan
[c] Laboratory of Magnetism and Magnetic Materials, Advanced Institute of Materials Science, Ton Duc Thang University, Ho Chi Minh City, Vietnam
[d] Faculty of Applied Sciences, Ton Duc Thang University, Ho Chi Minh City, Vietnam
[e] Department of Physics, National Chung Cheng University, Chiayi 621, Taiwan
[f] Quantum Topology Center, National Cheng Kung University, Tainan 701, Taiwan
[g] Hierarchical Green-Energy Materials Research Center, National Cheng Kung University, Tainan, Taiwan

CONTENTS

How to diversify the quantization phenomena is one of the mainstream issues in physics. Up to now, they can be achieved by the intrinsic lattice symmetries [1], the distinct dimensions [2, 3], the diminished scales [4], the various stacking configurations [1, 2], the mechanical strains [5], the uniform and/or modulated magnetic fields [1, 2, 6], and the electric fields [2, 3, 7]. The focuses of this book are the magnetically quantized behaviors of the emergent layered materials, in which they will be clearly revealed in the electronic properties, optical absorption spectra, quantum Hall transports, and magneto-Coulomb excitations. Such 2D condensed-matter systems are very suitable for exploring the rich and unique essential properties, such as the twisted bilayer graphene systems [8–10], the stacking-modulated bilayer graphene ones [11, 12], the sliding bilayer graphenes [3, 13], the few-layer graphenes with the AAA [2, 3], ABA [2, 3], ABC [2, 3] and AAB stackings [3], the silicon-doped graphene systems [1], the AA- and AB-stacked bilayer silicene/germanene systems with the bottom-top and bottom-bottom bucking structures [14, 15], the monolayer and bilayer phosphorene systems [24, 25], and the quantum topological insulators [TIs] [16]. To thoroughly present and comprehend the diverse magnetic quantization, the previous theoretical modes need to be further developed and modified. The significant magnetic Hamiltonians, which cover the single- or multi-orbital chemical bondings, the intralayer & interlayer hopping integrals, the important spin-orbital couplings, and the external field will be efficiently solved by the generalized tight-binding model [1,2]. Its direct combinations with the dynamic Kubo formula is reliable in calculating the magneto-optical excitation spectra and analyzing the selection rules under the gradient approximation [3]. Another linking with the static one can investigate the unusual quantum conductivities [17]. Moreover, the random-phase approximation [RPA], which is available for the electronic excitations in all the condensed-matter systems, requires the exact modifications being closely related to the layered structures [15, 18, 19]. The detailed comparisons between the theoretical predictions and the high-resolution experimental measurements are also made in this work.

The magnetic quantization can be greatly diversified by modulating/changing the various geometric structures of condensed-matter systems and the external field forms. The diverse phenomena have been predicted/observed in the mainstream or emergent materials. For example, the hexagonal symmetry, stacking configuration play critical roles in graphene-related honeycomb lattices. Under a uniform perpendicular magnetic field $[B_z\hat{z}]$, the low-lying Landau-level energies of monolayer graphene are proportional to the square root of nB_z [n quantum number] [20], the sliding bilayer graphene systems exhibit three kinds of Landau Levels [the well-behaved, perturbed, and undefined Landau levels] [13], the twisted ones present the fractal energy spectra during the variation of the magnetic-field strength [21], the AAA-stacked graphene systems only display the quasi-monolayer Landau levels with the different initial energies for the distinct groups [the vertical Dirac-cone structure], the trilayer ABA stacking shows the unusual superposition of monolayer- and bilayer-like Landau levels [20], the irregular Landau-level energy spectra and the frequent anti-crossing & crossings come to exist in the ABC-stacked graphene systems [22], the strongly, significantly, and weakly k_z-dependent Landau subbands, respectively, appear in the AA-, AB-, and ABC-stacked 3D graphites [2]. The low-energy quasi-Landau-levels survive in the 1D graphene nanotubes [4], and the dispersionless Landau levels are absent in carbon nanotubes/carbon tori [23]. Such diverse Landau levels are only created by the significant single-$2p_z$ orbital hybridizations of carbon atoms. Furthermore, the multi-orbital chemical bondings and/or the non-negligible spin-orbital interactions, which can create new categories of Landau levels, are frequently revealed in the emergent layered materials, such as few-layer silicene [14,26], germanene [27–29], stanene [30,31], bismuthene [32], antimonene [33], GaAs [34], and MoS_2-related systems [35, 36]. The above-mentioned Landau levels are drastically changed by a uniform perpendicular electric field (the layer-dependent Coulomb potentials), e.g., the induced Landau-level splittings, crossings, and anti-crossings. When an opposite external field, a uniform perpendicular magnetic field accompanied with a spatially modulated magnetic/electric one, is present in 2D systems, the nonuniform magnetic quantization will lead to the abnormal energy dispersions without the very high state degeneracy and the irregular wave functions [6]. The previous theoretical studies are focused on monolayer graphene, and the similar methods could be generalized to 2D emergent materials. The up-to-date calculations on the main features of Landau levels and Landau subbands are covered in the following chapters. Some magnetic issues, being associated with the novel geometries and the adatom chemisorptions, are worthy of the systematic studies, e.g., the Landau-level characteristics in the 1D folded [37], curved [4,38], and scrolled [39] graphene nanoribbons [4], amorphous and defect-enriched graphene systems [40,41], adatom-adsorbed graphene systems [42].

Various types of experimental equipment can accurately examine and then identify the diverse magnetic quantization phenomena in layered materials. Their measurements cover the magneto-electronic energy spectra & Landau level/Landau subband wave functions [1–3], magneto-optical absorption spectra [3], Hall conductivities [17], magnetic-field-dependent specific heat [43, 44], and magneoplasmon modes [45]. First, the energy-related and energy-fixed measurements of scanning tunneling spectroscopy [STS; details in Sec. 2.1] are, respectively, available in exploring the main features of Landau-level energy spectra [46,47] and probability distributions [48]. Most of the experimental verifications are conducted on the former, while the opposite is true for the latter, e.g., the various B_z dependencies for few-layer graphene systems with different stacking configurations [1–3]. The direction identifications of the number of zero points and distribution symmetries are worthy of the systematic studies under the great enhancement in STS measurement techniques [46, 47]. Second, the transmission [49], absorption [50], reflection [51], and Rayleigh [52] optical spectroscopies, accompanied by an external magnetic field, are very useful in examining the special structures and selection rules of electronic excitation spectra [details in Sec. 3.2]. Up to now, the high-resolution measurements have confirmed and the symmetric absorption peaks and the specific magneto-selection rule of $\Delta n = \pm 1$ [n quantum number] for the well-behaved Landau levels in monolayer-like graphene systems [49]. However, the different excitation categories and the extra selection rules, which are, respectively, due to the multi-group Landau levels and the abnormal spatial distributions, require the further experimental test. Third, the Hall transport measurements are frequently utilized to investigate the normal and unusual quantum conductivities arising from the static scatterings of Landau-level states. Specifically, the experimental and theoretical studies are consistent with each other for monolayer [53], AB [54], and ABC [55] stacking graphene systems. For example, magnetic transport measurements have verified the unconventional half-integer Hall conductivity $\sigma_{xy} = (m + 1/2)4e^2/h$, in which m is an integer and the factor of 4 stands for the spin- and sublattice-induced Landau-level state degeneracy. This novel quantization is attributed to the quantum anomaly of $n = 0$ Landau levels corresponding to the Dirac point. Fourth, a delicate calorimeter, which is sensitive to the temperature and magnetic-field strength, can accurately identify the very high Landau-level degeneracy, a weak but significant Zeeman splitting, and few Landau levels nearest to the Fermi level. Such measurements on graphene-related materials are absent up to date. However, they have been done a 3D electron gas in GaAs-GaALAs [56, 57], showing the non-monotonous T- and B_z-dependences. Finally, in general, the electronic Coulomb excitations could be measured by the electron energy loss spectroscopy [EELS] and the hard-x ray inelastic scatterings. The EELS measurements are very difficult to the collective excitations in the presence of a magnetic field, since the incident electron beam is easily perturbed by such field. The light scattering experiments are very suitable for the full exploration of magnetoplasmon modes, e.g., the clear verifications on the magnetoplasmon modes due to electron gases in the doped semiconductor compounds [58].

Up to now, there exist the effective-mass approximation [59–62] and the generalized tight-binding model [1–3], being frequently utilized in exploring the rich magnetic quantization of layered materials. It should be noticed that the first-principles method can solve the electronic properties [63, 64], but not the magneto-electronic ones, as a result of the enlarged unit cell by the vector potential [details discussed later in Sec. 3.1]. Such calculations could provide the reliable energy bands at the high symmetry points (e.g., the K and Γ point) and thus the reliable hopping integrals related to the single- or multi-orbital hybridizations in chemical bonds by using the Wanner functions [65]. As to the low-energy approximation, this method is suitable in dealing with the magneto-electronic properties, if the layered systems have the simple and monotonous valence and conduction bands near the Fermi level, e.g., monolayer graphene [53], silicene [66], and phosphorene [67,68], and the bilayer AA- and AB-stacked graphene systems [59,60]. The zero-field and magnetic Hamiltonian matrices have the same dimension, so the main features of Landau levels could be solved very quickly. However, it might create the high barriers in calculating the other essential physical properties [detailed comments in Sec. 3.1], e.g., magneto-optical absorption spectra [3], unusual Hall conductivities [67,68], and magnetoplasmons [69,70]. On the other side, the generalized tight-binding model covers all the intrinsic interactions, the intralayer and interlayer hopping integrals due to the single- or multi-orbital hybridizations and the spin-orbital couplings; that is, such model does not ignore the significant atomic and spin interactions, e.g., the complicated non-vertical interlayer hopping integrals in the trilayer ABC-stacked graphene [22]. Such interactions and the magnetic field [the composite field] are included in the numerical calculations simultaneously. The concise physical pictures could be proposed to clarify how many groups of the different sublattice-dominated Landau levels can be classified [1–3]. Furthermore, the magneto-electronic energy spectra and the Landau-level wave functions are very useful in the further understanding of the other physical properties.

It is well known that the geometric symmetries are one of the critical factors in determining the fundamental physical, chemical, and material properties. The condensed-matter systems, which are purely made up of carbon atoms, include the significant sp^3 [71–74], sp^2 [71, 73, 74], and sp chemical bondings [71, 75], being, respectively, revealed as the 3D bulk, 2D surface, and 1D line structures, such as diamond, graphene, and carbon chain. Apparently, the diversified orbital hybridizations in C—C bonds originate from the four half occupied orbitals ($2s, 2p_x, 2p_y, 2p_z$). Among carbon allotropes, the open/closed surface geometries might appear in the planar/nonplanar forms, covering graphites [76], layered graphenes [77], graphene nanoribbons [37] and quantum dots [78] [3D-0D]/carbon nanotubes [79], C_{60}-related fullerenes [80], and carbon onions [81] [1D-0D]. These unusual materials possess the unique honeycomb lattices with the hexagonal symmetries; therefore, the arrangements of normal hexagons are expected to play important roles in diversifying the various essential properties. A perfect 2D graphene crystal, with three nearest neighbors, is responsible for the isotropic Dirac-cone band structures [77]. Very interestingly, the achiral or chiral arrangement on a cylindrical nanotube surface/two open boundaries of nanoribbon has been clearly identified to dominate the metallic or semiconducting behaviors across the Fermi level [Refs. [82, 83]/Refs. [84]]. The similar diverse phenomena are verified/predicted to appear in bilayer graphene systems through the twisting [8, 9] or sliding effects [13]. The former behave as a Moire superlattice, with a lot of carbon atoms in an enlarged unit cell of bilayer graphene, while the latter only have four ones being identical to those in AA or AB stackings [details in Sec. 9.1]. The distinct geometric structures of twisted bilayer graphenes are accurately confirmed through the high-resolution scanning tunneling microscopy [STM] measurements [details in Sec. 2.1], e.g., the twisted angles between two honeycomb lattices of $\theta = 1.4°, 3.5°, 6.4°, 9.6°$, and $1.79°$ [85]. In addition, the spatially resolved Raman spectroscopy also provides the geometric information, a bilayer system with a mixture of AB stacking and twisted structure [86–88]. Simultaneously, the STS examinations clearly identify three kinds of van Hove singularities in density of states, covering the V-shape across the Fermi level, shoulders, and logarithmically divergent peaks at the negative and positive energies. The measured results are consistent with the theoretical calculations of the first-principles method [89, 90] and tight-binding model [91, 92]. Such special structures indicate the linear and parabolic energy dispersions, as verified from the angle-resolved photo-emission spectroscopy [93]. The main features of geometric and electronic properties are directly reflected in optical absorption spectra [94] and quantum Hall transports [95]. Very interestingly, the specific bilayer system, with a magic angle of $\theta \sim 1.1°$, is thoroughly examined and verified to exhibit the superconducting characteristics [almost vanishing resistance and diamagnetism] at the critical temperature of $T_c = 1.7$ K [96]. However, the systematic theoretical investigations on the essential properties are absent up to date. For example, the magneto-electronic states, being closely related to the unusual optical selection rules and Hall conductivities, need to be well characterized through the magnetic wave functions with the normal or irregular spatial oscillations. This will be completely explored by the generalized tight-binding model in Chap. 4.

In addition to the variations among the well-behaved stacking configurations [1–3], a creation of the nonuniform sublattices will lead to the dramatic changes in the fundamental physical properties, especially for the thorough transformation from the 2D to quasi-1D behaviors. The modulated geometries in few-layer graphene systems have been successfully synthesized using the STM tips [97], mechanical exfoliation [98], and chemical vapor deposition [99]. From the theoretical point of view, the stacking symmetries of bilayer graphene systems can be drastically altered by the significant sliding [13], twisting [92], and generation of domain walls [12]. When only the low-lying electronic states due to the C-$2p_z$ orbitals are taken into account, the first and second systems are predicted to exhibit two pairs and many pairs of valence and conduction bands along the high symmetry points in the hexagonal Brillouin zone, respectively. Energy bands have the regular dispersion relations with the 2D wave vectors. Furthermore, their band-edges states possess the 2D-specified structures in van Hove singularities of density of states [12] and the similar ones in absorption spectral structures [12]. The magnetic quantization in the first [second] system creates the highly degenerate Landau states with the well-behaved, perturbed, or undefined oscillation modes (the normal ones). Apparently, the mandatory modulations of stacking configuration, which are due to the changes of bond lengths, have generated a large unit cell and the nonuniform intralayer and interlayer hopping integrals. Even in the zero-field case, the dependence of essential properties on the wave vector along the modulated direction is expected to be negligible. That is, the quasi-1D phenomena come to exist in certain special cases, e.g., plenty of 1D energy subbands with the unusual dispersion relations, and many significant band-edge states and absorption structures [12]. A uniform magnetic field strongly affects the commensurate periods and the vector-potential-induced Peierls phase on the neighboring atomic interactions. As a result, the nonuniform magnetic quantization can create a plenty of the low-degenerate Landau subbands with the sufficiently wide oscillating energy widths and the strong couplings [6], being in sharp contrast with the high-degenerate Landau levels. It is worthy of making a systematic comparison for the main features of Landau subbands in stacking-modulated bilayer graphene systems [12], graphene nanoribbons [4], and Bernal and rhombohedral graphites [RG] [2].

Obviously, the 3D silicon crystals, with the sp^3 chemical bondings, have shown the critical roles in the basic sciences [100], applied engineering [101, 102], and high-technique semiconductor applications [103]. From the experimental and theoretical viewpoints, the reduction in dimension should be a very effective way in diversifying the various physical phenomena. The researches on layered silicene systems are getting into one of the main-stream topics in 2D emergent materials. In general, the epitaxial growth is frequently utilized to synthesize monolayer and bilayer silicene systems on the different substrates, e.g., the former grown on Ag(111) [104, 105], Ir(111) [106], and ZrB$_2$(0001) [107], with the 4×4, $\sqrt{3} \times \sqrt{3}$, and 2×2 unit cells, respectively. The STM [details in Sec. 2.1] and low-energy electron diffraction can identify a buckled single-layer honeycomb lattice, as well as an enlarged unit cell arising from the significant atomic interactions between the host and guest atoms. Such geometry clearly indicates the dominating sp^2 bonding and its competition with the weak, but important sp^3 one. The phenomenon of orbital hybridizations is further reflected in the low-lying valence and conduction bands. The former, which possesses a slightly separated Dirac-cone structure (approximately a zero-gap semiconductor; the monolayer-graphene-like behavior), is confirmed through the high-resolution angle-resolved photoemission spectroscopy [ARPES; [105]]. However, it is very difficult to successfully generate the pure bilayer silicene systems through the various experimental methods. The previous study shows that there exist three types of bilayer silicenes after treating the calcium-intercalated monolayer silicene (CaSi$_2$) around a BF$_4^-$-based ionic liquid. That is to say, the bilayer silicenes might consist of the four-, five-, and six-membered silicon rings with three nearest neighbors [104], where the third ones correspond to the AB-bottom-top stacking configuration (AB-bt). In addition, the experimental evidences of AA stackings are absent up to now. Such images are accurately achieved by a high-angle annular dark field scanning transmission electron microscopy (HAADF-STEM). Moreover, the measured absorption spectrum suggests an indirect optical gap of 1.08 eV (or a threshold excitation frequency). Energy gaps are quite different from those in monolayer and bilayer materials; therefore, the stacking configurations, the interlayer hopping integrals, and the spin-orbital couplings play a critical role in the fundamental physical properties. Most importantly, the multi-metastable geometric structures come to exist simultaneously, clearly illustrating the very active chemical environments due to the four half-occupied orbitals of silicon atoms ($3s$, $3p_x$, $3p_y$, $3p_z$). Few-layer silicene systems are expected to become the 2D materials in the next generation of electronic devices [108], while their structural instabilities need to be overcome for the highly potential applications.

Also, many theoretical predictions on 2D-layered silicene materials present very interesting results, especially for the diversified physical phenomena. Generally speaking, their essential properties are thoroughly investigated by the first-principles calculations, *ab initio* molecular dynamics, the tight-binding model, and the effective-mass approximation. The optimal geometric structures are accurately presented through the first and second methods. For example, monolayer silicene has a buckled honeycomb lattice, with a significant height difference between A and B sub-lattice [104], and bilayer ones possess the distinct members in silicon rings and the various stacking configurations [104], such as the metastable configurations of AA-bt, AA-bb, AB-bt, and AB-bb in bilayer materials. Specifically, the Vienna *ab initio* simulation package is quite powerful in predicting electronic properties, covering band structures, spatial charge distributions, atom-, orbital-, and spin-projected density of states, magnetic moments, and spin distribution configuration. For example, the chemisorption-diversified fundamental properties are clearly revealed in the hydrogenated and oxidized silicene systems [111, 112]. For pristine bilayer silicenes, their band properties across the Fermi level are predicted to be very sensitive to the stacking configurations, e.g., a semimetal with some free carrier densities [113], and a finite indirect-gap/direct-gap semiconductor [$\sim, 0.5 - 1.0$ eV [14, 104], being in sharp contrast with a very narrow gap of monolayer system [$\sim, 5$ meV; [105]]. On the other hand, the third and fourth methods are quite efficient in studying the diverse physical phenomena after getting a set of reliable parameters for all the intrinsic interactions. It should be noticed that an outstanding fitting in non-monolayer silicene systems, which is done by a detailed comparison with the first-principles calculations, will become a giant barrier. The more complex interlayer hopping integrals and enhanced spin-orbital couplings, due to the rich bucklings and stackings, are very difficult to obtain even for bilayer silicenes. Up to date, the effective-mass can only deal with the fundamental properties for monolayer silicene under the external electric and magnetic fields, covering the in-orbital-induced slight separation of Dirac points, gate-voltage-created state splitting [14], and the magnetic-field-generated degenerate Landau levels and specific selection rule [14]. There are no excellent parameters for two pairs of low-lying valence and conduction bands in AA- and AB-stacked systems. However, the first pair nearest to the Fermi level is well fitted for the non-monotonous energy dispersions and the irregular/normal valleys. Apparently, the diversified Coulomb excitations and magnetic quantizations are effectively solved by the generalized tight-binding model, e.g., the diverse (momentum, frequency)-phase diagrams in monolayer silicene under the composite effects of spin-orbital interactions, electric potentials, and magnetic fields, the inter-Landau-level and electron-gas-like magnetoplasmon modes with the rather strong electron-hole dampings for the same system [114], and the spin- and valley-split Landau levels of bilayer silicenes in the frequent non-crossings, crossings, and anti-crossings [14]. The systematic investigations will be made on the fundamental properties of AA- and AB-stacked bilayer systems.

The chemical modifications, as revealed in the theoretical and experimental researches [115–117], are one of the most efficient methods in dramatically changing the essential properties of condensed-matter systems. They are very effective for the layered materials, since such systems can provide the outstanding chemical environments. For example, there are a lot of dangling C-$2p_z$ orbitals on the planar structures of few-layer graphene systems. Recently, the adatom chemisorptions [115, 116] and the guest-adatom substitutions [117] in the emergent 2D materials are easily achieved during the experimental growths. Apparently, how to generate the uniform adatom/guest-atom distributions in experimental laboratories is the studying focus/issue. Up to now, many graphene-related binary and ternary compounds have been successfully synthesized using various methods. Among them, the silicon carbide compounds, with the sp^2 or sp^3 multi-orbital chemical bondings, frequently appear in the distinct dimensions, e.g., the 3D bulk compounds [118–120], the 2D systems [121], and the 1D nanotube ones [122, 123]. Similarly, there also exist the dimension-dependent $B_x C_y N_z$ compounds. The 3D silicon carbide systems possess the sp^3 multi-orbital hybridizations, mainly owing to the four half-occupied orbitals of $[3s, 3p_x, 3p_y, 3p_z]$ and & $[2s, 2p_x, 2p_y, 2p_z]$ orbitals. Such chemical bondings are similar to those in diamond [124], so the SiC materials have the outstanding mechanical properties, as observed in the consistence between the calculated results [125, 126] and experimental measurements [127]. Such SiC-related materials have presented the high potentials in applications, such as the energy storage [128], and the semiconducting optical devices [129]. Recently, the planar silicon-carbon compounds have been synthesized in the experimental laboratories [130]. The theoretical predictions under the first-principles calculations clearly show that these 2D compounds display the sp^2 and π chemical bondings [131]. The latter will be responsible for the low-energy essential physical properties. Specifically, the π bondings, which mainly originate from the $2p_z$-$2p_z$ hybridization in C—C bond and the $2p_z$-$3p_z$ mixing in C-Si one, is expected to be significantly distorted by the Si-substitutions. These are very sensitive to the concentrations and distributions of the Si-guest atoms, and so do the other physical, material, and chemical phenomena [131]. It is very interesting to fully comprehend the close relations between the substitution-enriched chemical bondings and the magnetic quantization behaviors. The nonuniform ionization potentials, C—C & Si—C bond lengths and hooping integrals are deduced to be the critical factors in determining how many kinds of Landau levels and magneto-optical selection rules can be created during the variation of the Si-substituted geometric structures [1–3].

For the quantum transport properties, the 2D graphene-related systems are very suitable for fully exploring the diverse magnetic quantization phenomena. Apparently, such layered materials sharply contrast with the doped semiconductors/2D electron gases revealing the integral and fractional Hall effects [132]. The previous investigations clearly show that honeycomb lattices [53], few layers [54, 55], stacking configurations [55], planar/rippled/curved/folded structures [91], negligible spin-orbital interactions [133], dimensions [134], nanodefects [135], chemical absorptions [136], and guest-atom substitutions [117] are responsible for the rich band structures and thus the unusual magneto-electronic and transport properties. The close and complex relations between the critical factors and the diversified Hall conductivities are the studying focus of Chap. 9. Up to date, the experimental measurements and theoretical predictions (the tight-binding model and the effective-mass approximation) have delicately identified the rich characteristics of quantum Hall conductivities in monolayer [53], bilayer [54] & trilayer AB stackings [55], and trilayer ABC configuration [55] under a low Fermi level and magnetic-field strength. For example, there are one, two & six, and three Landau levels at/across the Fermi energy, leading to the anomalous structures there in terms of height, width, and sequence of quantum plateaus [53–55]. This clearly illustrates the significant roles due to the well-defined modes, state degeneracies, and normal energy spectra. Obviously, the more complicated transport phenomena, which might be created by the non-well-behaved magneto-electronic states and the frequently crossing & anti-crossing behaviors, deserve a systematic investigation. The sliding bilayer graphenes and AAB trilayer stacking will be chosen for a model study, in which the former have the structural transformation between two high-symmetry configurations [13, 137], and the low-symmetry latter presents the very complex interlayer hopping integrals [13, 137]. These systems are predicted to exhibit the diversified energy bands and magneto-electronic properties, such as three kinds of Landau levels (the well-behaved, perturbed and undefined ones) with the complex B_z-dependent energy spectra. To propose the completely physical pictures, the quantum Hall effects are conducted on the trilayer AAA, ABA, ABC and AAB stacking. On the theoretical side, the linear static Kubo formula is directly combined with the generalized tight-binding model [discussed later in Sec. 3.3.1], based on the model consistence between the gradient approximation in evaluating the electric dipole moment and the sublattice-defined subenvelope functions. Such direct association is expected to be very efficient in solving the open transport issues due to the abnormal Landau-level wave functions and energy spectra [13, 137]. That is to say, the developed theoretical framework is suitable and reliable for any kinds of Landau levels, e.g., the quantum transports arising from the well-behaved, perturbed, and undefined oscillation modes. In addition, the effective-mass perturbation method might be quite difficult in exploring the irregular magneto-electronic states and even the higher-energy/strong-field transport properties of the well-behaved Landau levels, mainly owing to the complicated effects coupled by the intrinsic interactions and the magnetic field.

The magneto-electronic specific heats of conventional quantum Hall systems are very suitable for fully understanding the diverse magnetic quantization phenomena. They have been extensively investigated both theoretically [138] and experimentally [56, 139, 140]. For example, a lot of researches on GaAs-GaAlAs [56, 140] show that the thermal properties are very sensitive to the temperature and magnetic-field strength. The temperature-dependent specific heat reveals an obvious peak structure at a low critical temperature [56, 140]. Furthermore, the magnetic-field dependence exhibits an oscillation behavior [56, 140]. These two kinds of thermal behaviors directly reflect the main features of Landau levels, such as the filling factor and energy spacing. Also, the high-resolution calorimeter measurements are conducted on the graphene-related materials, such as 1D carbon nanotubes [141] and 3D graphite systems [142]. The former clearly display a linear temperature-dependence in $C[T]$, and so do for the latter. These results are deduced to be closely related to the 1D and 3D metals. This book will illustrate the rich thermal excitation phenomena on monolayer graphene, e.g., their unique dependences on the Zeeman splitting energy, doping level of conduction electrons/valence holes, temperature, and magnetic field.

The 2D emergent materials are ideal condensed-matter systems in thoroughly exploring the many-particle Coulomb excitations under the magnetic quantization. The strong competition in magneto-excitation spectra, that is due to the longitudinal electron-electron interactions and the transverse cyclotron forces, is a very interesting research focus. Recently, the various experimental methods are delicately developed to successfully synthesize a lot of layered materials, e.g., few-layer germanene, silicene, and tinene being, respectively, grown on Pt(111), Au(111), & Al(111) [143–145], Ag(111), Ir(111), & ZrB_2 [104–107], and Bi_2Te_3 surfaces. Apparently, the obvious buckled structures, the important orbital hybridizations, the significant spin-orbital interactions, and the external electric and magnetic fields would play critical role in the essential physical properties, especially for their strong effects in creating the diverse magneto-electronic excitation. The previous theoretical studies, which are focused on monolayer graphene and bilayer AA and AB stackings [1–3] clearly show the rich and unique phenomena: a lot of inter-Landau-level single-particle and collective excitations, the unusual magnetoplasmon modes with the non-monotonic momentum dispersion relations, the very strong dependences on the temperature, doping level and interlater atomic interactions, the stacking-enriched (momentum, frequency)-phase diagrams, and the 2D electron-gas behavior only under a sufficiently high free carrier density. On the experimental side [details in Sec. 2.4], the high-resolution EELS is very powerful in the full examinations of the carbon-orbital- [146], dimension- [147], layer- [148], stacking- [148], doping- [149], and temperature-diversified plasmon modes [150]. Up to now, most of the measurements are consistent with the calculated results [18]. A thorough investigation on monolayer germanene is expected to present the distinct magneto-electronic excitation phenomena, since its band structure might be quite differ from those of graphene and silicene systems [15]. The composite effects arising from the significant spin-orbital coupling, free carrier doping, static Coulomb on-site energies, and magnetic field would play critical roles in diversifying magnetoplasmons.

The final topic is concerned about 3D layered systems of topological matter in view of the extensively and intensively researched topological phases of condensed matter [151–153]. Specifically, it is about matter that can host 2D Dirac fermions. In this realm, the phases with topological order cannot be characterized by symmetry alone. Moreover, the relevant phase transitions do occur without spontaneous symmetry breaking, beyond the scope of Landau's theory [154]. The first realization of such phases is the discovery of the integer quantum Hall effect [QHE] [155], which was followed soon by a topological interpretation [156–158]. Later on, the distinct, half-integer QHE was also found in graphene [159, 160], which has almost spin degeneracy described by $SU[2]$ symmetry, say, spinless. This discovery realized previous theoretical predictions [161, 162]. It has been well understood that the anomalous half-integer QHE originates from the chiral zero-mode Landau level, which is protected in the presence of 2D massless Dirac fermions around the Dirac points [DPs]. In graphene, the DPs exist only if the spinless condition is assumed. It is, however, worthy to consider such systems, which can be realized by means of cold atoms in optical lattices. As described in Sec. 3.3, the Hall conductivity and the associated

Landau levels are of the relativistic type in graphene. The very characteristic lies in that there exists a topologically robust zero-mode Landau level. This zero-mode Landau level is protected by a continuous chiral symmetry [CS] of the effective Dirac Hamiltonian [163–165], where the continuous CS results from sublattice symmetry (SLS) of the lattice Hamiltonian within the minimal model for spinless graphene. Since massless Dirac fermions are broadly present in condensed matter with various symmetries, it is of interest to see what about the topological characteristics in other systems with 2D massless Dirac fermions [166–169]. In particular, 3D layered systems of topological matter is central to Chap. 11. Specifically, the symmetry protection of the zero-mode Landau level attributed to 2D massless Dirac fermions is elaborated.

In addition to the interesting topics in this work, the open issuers, being clearly illustrated through a final chapter [details in Chap. 13], are proposed to promote the understanding of magnetic quantizations and explore the other diverse phenomena in 2D emergent materials. Certain mainstream researches might be very difficult to thoroughly investigate their main features according to the current methods/the present calculations, mainly owing to the complicated geometric structures with the nonuniform/quasi-random chemical environments, or a lot of important intrinsic interactions (such as the significant couplings of multi-orbital hybridizations and spin-orbital interactions). For example, a single-layer graphene could present the vacancy and Stone-Wales defects [170]; that is, a defective and amorphous system possesses hexagons, pentagons, and heptagons. It is deduced that non-homogenous sp^2-bonding network can create the great modifications on the basic magnetic properties, e.g., the strong dependences of energy spectra and wave functions on the concentrations and distribution configurations of various defects, and the close relationships of the defect-created sublattices and the vector-potential-dependent Peierls phases. Whether the highly degenerate Landau levels appear in the seriously deformed/curved surfaces is one of the mainstream topics. The folded graphene nanoribbons [37], curved ones [171], carbon nanoscrolls [39], carbon toroids [23], and carbon onions [81] are very suitable for the model studies on the curvature effects (the misorientation of carbon-$2p_z$ orbitals and hybridization of π and σ electrons), critical roles of the position-dependent hopping integrals, the strong competition of the quantum-confinement-induced standing wave functions with the magnetically localized ones, and the diamagnetic, paramagnetic, ferromagnetic, and anti-ferromagnetic behaviors [23, 37, 39, 81, 171]. The more complex magnetic quantization phenomena will frequently come to exist in the adatom-chemisorption/substitution layered materials (the chemically modified systems). The previous first-principles calculations show that the halogenated [172], hydrogenated, and oxidized graphenes/silicenes [111, 112] exhibit the main characteristics of Moire superlattices in the presence of multi-orbital chemical bondings. The superposition of guest and host sublattices should be the critical mechanism responsible for the localized magneto-electronic states; therefore, plenty of non-equivalent sublattices can take part in the highly irregular magnetic quantizations. Apparently, the above-mentioned geometry- and chemical-modification-dominated novel materials could serve as the typical models in fully exploring the diversified optical properties [3], quantum Hall transports [17], and Coulomb excitations [18] under any external fields. The studying focuses might cover the different magneto-optical selection rules of Landau levels or subbands due to the combined effects of dimensions, edge structure, curvatures and external fields, the integer or non-integer Hall conductivities with the regular/abnormal dependences on the magnetic-field strength and the Fermi energy [53], and the (momentum. frequency)-phase diagrams of the inter-Landau-level magnetoplasmon modes and electron-hole excitations [69]. Also noticed that the theoretical frameworks of the different physical/chemical/material properties, belonging to the single- and many-particle models, might need to make the great modifications for fully understanding their diverse phenomena. Among them, the suitable Hamiltonians with any number of layers, which clearly show the dramatic transformation in multi-orbital hybridizations from few-layer sp^2-bonding silicenes to 3D diamond-like silicon, are one of the challenges.

This book covers Chaps. 1–10 for fully exploring the diverse magnetic quantization phenomena. Chapter 2 discusses the various high-resolution experimental measurements, including STM, STS, optical absorption reflection/transmission spectroscopies, transport/thermal instruments, and light inelastic scattering spectroscopy. The theoretical frameworks, the generalized tight-binding model, its combinations with the static and dynamic Kubo formulas, and the modified random-phase approximation, are systematically developed in Chap. 3. They are, respectively, responsible for the magnetically quantized electronic states, the unusual Hall conductivities, the magneto-optical selection rules, and the magneto-electronic excitations. Chapter 4 will thoroughly investigate the diversified magneto-electronic states and optical properties of bilayer graphene systems due to the very strong cooperations/competitions between the twist-induced Moiré superlattice and the enlarged unit cell by the vector potential. The modulation of stacking configuration, which is created by the domain walls in one graphene layer, is clearly illustrated for the AB-stacked bilayer graphene in Chap. 5. Its significant effects on the electronic and optical properties are studied in detail, and furthermore, the detailed comparisons with the sliding and twisted bilayer graphene systems, graphene nanoribbons, and Bernal and RG are also made. Chapter 6 fully explores the rich and unique phenomena in an AA-bt bilayer silicene with the very strong interlayer hopping integrals and the negligible spin-orbital couplings; furthermore, the significant differences two- and single-layer systems are made in detail. The composite effects, due to the intrinsic atomic interactions and the magnetic & electric fields, on the essential properties are the studying focuses through the proposed viewpoint: the valley structures in electronic energy spectra will play critical roles in the fundamental physical properties, such as the diverse magnetic quantizations due to the different valleys in monolayer and bilayer silicene systems. The layer-dependent spin-orbital interactions, as reported in Chap. 7, will come to exist in AB-stacked silicene systems. Apparently, the more complicated effects arising from stacking configuration and buckling structure are expected to diversify the magnetic quantization. Chapter 8 clearly illustrate the dramatically chemical modifications of the fundamental properties by using the silicon guest-atom substitution in monolayer graphene. Four kinds of Landau levels and magneto-optical selection rules will be classified according to the main features of magneto-electronic properties and absorption transition channels. Few-layer graphenes in Chap. 9, with the distinct layer numbers and stacking configurations, can present the rich and unique quantum transport properties, the unusual Hall conductivities and magneto-heat capacity. The available scattering events are clarified by delicately analyzing the dipole matrix elements associated with the initial and final Landau-level-state spatial distributions. The magneto-electronic specific heat of monolayer graphene provides another magnetic quantization phenomenon, especially for its rich dependences on the Dirac cone/the unusual Landau-level energy spectrum, Zeeman splitting effect, temperature and magnetic-field strength. The superposition of the transverse magnetic force and the longitudinal Coulomb one leads to the abnormal electronic excitations, where germanene, silicene, and graphene will exhibit the diverse magnetoplasmon modes induced by the significant/negligible spin-orbital couplings, as shown in Chap. 10. Chapter 11 covers the topological characterization of the Landau levels in 3D layered systems of topological matter. The focus is put, in a topological viewpoint, on the existence and stability of 2D massless Dirac fermions when an external perpendicular magnetic field is applied. As is well known, it is 2D massless Dirac fermions that provide the symmetry protection of a chiral zero-mode Landau level, which, in turn, dictates the half-integer QHE. The half-integer QHE is dictated by the topologically robust zero-mode Landau level in the relativistic Landau level spectrum. Two systems are considered after Sec. 11.1 which introduces some relevant topological aspects. Section 11.2 aims at 3D layered spinless nodal-line topological semimetals [TSMs] in an elaborate extent, exemplified by spinless RG, a 3D stack consisting of spinless graphene layers in ABC configuration. In this system, 2D massless Dirac fermions are hosted along the nodal lines in the bulk. In Sec. 11.3, a discussion is held on spinful, time reversal invariant 3D strong TIs, which can be layered systems such as Bi_2Se_3 and Bi_2Te_3. These systems are insulating in the bulk, while possessing an odd number of Dirac cones on the surface, where 2D massless Dirac fermions are hosted. how to derive such an effective Hamiltonian from the lattice Hamiltonian is discussed. Finally, Chaps. 12 and 13, respectively, present the concluding remarks, and future perspectives & open issues.

REFERENCES

1. Chen S C, Wu J Y, Lin C Y, and Lin M F 2017 Theory of magnetoelectric properties of 2D systems *IOP Concise Physics*. San Raefel, CA, USA: Morgan & Claypool Publishers.
2. Lin C Y, Chen R B, Ho Y H, and Lin M F 2018 *Electronic and Optical Properties of Graphite-Related Systems*. Boca Raton, FL: CRC Press.
3. Lin C Y, Do T N, Huang Y K, and Lin M F 2017 Electronic and optical properties of graphene in magnetic and electric fields *IOP Concise Physics*. San Raefel, CA, USA: Morgan & Claypool Publishers.
4. Chung H C, Chang C P, Lin C Y, and Lin M F 2016 Electronic and optical properties of graphene nanoribbons in external fields *Phys. Chem. Chem. Phys.* **18** 7573.
5. Wong J H, Wu B R, and Lin M F 2012 Strain effect on the electronic properties of single layer and bilayer graphene *J. Phys. Chem. C* **116** 8271–8277.
6. Ou Y C, Sheu J K, Chiu Y H, Chen R B, and Lin M F 2011 Influence of modulated fields on the Landau level properties of graphene *Phys. Rev. B* **83** 195405.
7. Ou Y C, Chiu Y H, Lu J M, Su W P, and Lin M F 2013 Electric modulation effect on magneto-optical spectrum of monolayer graphene *Comput. Phys. Commun.* **184** 1821.

8. Cherkez V, Trambly de Laissardiere G, Mallet P, and Veuillen J Y 2015 Van Hove singularities in doped twisted graphene bilayers studied by scanning tunneling spectroscopy *Phys. Rev. B* **91** 155428.

9. Huang S, Kim K, Efimkin D K, Lovorn T, Taniguchi T, Watanabe K, et al. 2018 Topologically protected helical states in minimally twisted bilayer graphene *Phys. Rev. Lett.* **121** 037702.

10. Fan Z, Yan J, Zhi L, Zhang Q, Wei T, Feng J, et al. 2010 A three-dimensional carbon nanotube/graphene sandwich and its application as electrode in supercapacitors *Adv. Mater.* **22** 3723.

11. Vaezi A, Liang Y, Ngai D H, Yang L, and Kim E A 2013 Topological edge states at a tilt boundary in gated multilayer graphene *Phys. Rev. X* **3** 021018.

12. Huang B L, Chuu C P, and Lin M F 2019 Asymmetry-enriched electronic and optical properties of bilayer graphene *Sci. Rep.* **9** 859.

13. Huang Y K, Chen S C, Ho Y H, Lin C Y, and Lin M F 2014 Feature-rich magnetic quantization in sliding bilayer graphenes *Sci. Rep.* **4** 7509.

14. Do T N, Shih P H, Gumbs G, Huang D, Chiu C W, and Lin M F 2018 Diverse magnetic quantization in bilayer silicene *Phys. Rev. B* **97** 125416.

15. Shih P H, Chiu Y H, Wu J Y, Shyu F L, and Lin M F 2017 Coulomb excitations of monolayer germanene *Sci. Rep.* **7** 40600.

16. Hasan M Z and Kane C L 2010 Colloquium: Topological insulators *Rev. Mod. Phys.* **82** 3045.

17. Do T N, Chang C P, Shih P H, and Lin M F 2017 Stacking-enriched magneto-transport properties of few-layer graphenes *Phys. Chem. Chem. Phys.* **19** 29525.

18. Ho J H, Lu C L, Hwang C C, Chang C P, and Lin M F 2006 Coulomb excitations in AA- and AB-stacked bilayer graphites *Phys. Rev. B* **74** 085406.

19. Lin C Y, Lee M H, and Lin M F 2018 Coulomb excitations in trilayer ABC-stacked graphene *Phys. Rev. B Rapid Communication* **98** 041408.

20. Lin C Y, Wu J Y, Ou Y J, Chiu Y H, and Lin M F 2015 Magneto-electronic properties of multilayer graphenes *Phys. Chem. Chem. Phys.* **17** 26008.

21. Wang Z F, Liu F, and Chou M Y 2012 Fractal Landau-level spectra in twisted bilayer graphene *Nano Lett.* **12** 3833.

22. Lin C Y, Wu J Y, Chiu Y H, and Lin M F 2014 Stacking-dependent magneto-electronic properties in multilayer graphenes *Phys. Rev. B* **90** 205434.

23. Tsai C C, Shyu F L, Chiu C W, Chang C P, Chen R B, and Lin M F Magnetization of armchair carbon tori *Phys. Rev. B* **70** 075411.

24. Wu J Y, Chen S C, Gumbs G, and Lin M F 2017 Field-created diverse quantizations in monolayer and bilayer black phosphorus *Phys. Rev. B* **95** 115411.

25. Wu J Y, Chen S C, Do T N, Su W P, Gumbs G, and Lin M F 2018 The diverse magneto-optical selection rules in bilayer black phosphorus *Sci. Rep.* **8** 13303.

26. Wu J Y, Chen S C, Gumbs G, and Lin M F 2016 Feature-rich electronic excitations of silicene in external fields *Phys. Rev. B* **94** 205427.

27. Tabert C J and Nicol E J 2013 Magneto-optical conductivity of silicene and other buckled honeycomb lattices *Phys. Rev. B* **88** 085434.

28. Tabert C J and Nicol E J 2013 Valley-spin polarization in the magneto-optical response of silicene and other similar 2D crystals *Phys. Rev. Lett.* **110** 197402.

29. Tabert C J, Carbotte J P, and Nicol E J 2015 Magnetic properties of Dirac fermions in a buckled honeycomb lattice *Phys. Rev. B* **91** 035423.

30. Chen S C, Wu J Y, Lin C Y, and Lin M F 2017 Theory of magnetoelectric properties of 2D systems *IOP Concise Physics*. San Raefel, CA, USA: Morgan & Claypool Publishers.

31. Chen S C, Wu C L, Wu J Y, and Lin M F 2016 Magnetic quantization of sp3 bonding in monolayer gray tin *Phys. Rev. B* **94** 045410.

32. Chen S C, Wu J Y and Lin M F 2018 Feature-rich magneto-electronic properties of bismuthene *New Jour. Phys. Fast Track Communication* **20** 062001.

33. Yu J, Katsnelson M I, and Yuan S 2018 Tunable electronic and magneto-optical properties of monolayer arsenene: from GW0 approximation to large-scale tight-binding propagation simulations *Phys. Rev. B* **98** 115117.

34. Chung H C, Chiu C W, and Lin M F 2019 Spin-polarized magneto-electronic properties in buckled monolayer GaAs *Sci. Rep.* **9** 2332.

35. Ho Y H, Su W P, and Lin M F 2015 Hofstadter spectra for d-orbital electrons: a case study on MoS2 *RSC Adv.* **5** 20858.

36. Ho Y H, Wang Y H, and Chen H Y 2014 Magnetoelectronic and optical properties of a MoS_2 monolayer *Phys. Rev. B* **89** 55316.

37. Vo T H, Shekhirev M, Kunkel D A, Morton M D, Berglund E, Kong L, et al. 2014 Large-scale solution synthesis of narrow graphene nanoribbons *Nat. Comm.* **5** 3189.

38. Lin C Y, Chen S C, Wu J Y, and Lin M F 2012 Curvature effects on magnetoelectronic properties of nanographene ribbons *J. Phys. Soc. Jpn.* **81** 064719.

39. Patra N, Wang B, and Kral P 2009 Nanodroplet activated and guided folding of graphene nanostructures *Nano Lett.* **9** 3766.

40. Yazyev O V and Helm L 2007 Defect-induced magnetism in graphene *Phys. Rev. B* **75** 125408.

41. Kotakoski J, Krasheninnikov A V, Kaiser U, and Meyer J C 2011 From point defects in graphene to two-dimensional amorphous carbon *Phys. Rev. Lett.* **106** 105505.

42. Lin S Y, Tran N T T, Chang S L, Su W P, and Lin M F 2018 *Structure- and Adatom-Enriched Essential Properties of Graphene Nanoribbons*. Boca Raton, FL: CRC Press.

43. Lin M F and Shung K W-K 1996 The electronic specific heat of single-walled carbon nanotubes *Phys. Rev. B* **54** 2896.

44. Chiu C W, Lin M F, and Shyu F L 2001 Electronic specific heat of nanographite ribbons *Physica E* **11** 356.

45. Wu J Y, Chen S C, Roslyak O, Gumbs G, and Lin M F 2011 Plasma excitations in graphene: their spectral intensity and temperature dependence in magnetic field *ACS Nano* **5** 1026.

46. Matsui T, Kambara H, Niimi Y, Tagami K, Tsukada M, and Fukuyama H 2005 STS observations of Landau levels at graphite surfaces *Phys. Rev. Lett.* **94** 226403.

47. Li G and Andrei E Y 2007 Observation of Landau levels of Dirac fermions in graphite *Nat. Phys.* **3** 623.

48. Fu Y S, Kawamura M, Igarashi K, Takagi H, Hanaguri T, and Sasagawa T 2014 Imaging the two-component nature of Dirac-Landau levels in the topological surface state of Bi_2Se_3 *Nat. Phys.* **10** 815.

49. Orlita M, Faugeras C, Martinez G, Maude D K, Sadowski M L, and Potemski M 2008 Dirac fermions at the H Point of graphite: magnetotransmission studies *Phys. Rev. Lett.* **100** 136403.

50. Keppler H, Dubrovinsky L S, Narygina O, and Kantor I 2008 Optical absorption and radiative thermal conductivity of silicate perovskite to gigapascals *Science* **322** 1529.

51. Toyt W W and Dresselhaus M S 1977 Minority carriers in graphite and the H-point magnetoreflection spectra *Phys. Rev. B* **15** 4077.

52. Casiraghi C, Hartschuh A, Lidorikis E, Qian H, Harutyunyan H, Gokus T, et al. 2007 Rayleigh imaging of graphene and graphene layers *Nano Lett.* **7** 2711.

53. Novoselov K S, Jiang Z, Zhang Y, Morozov S V, Stormer H L, Zeitler U, et al. 2007 Room-temperature quantum Hall effect in graphene *Science* **315** 1379.

54. Zhang Y, Tan Y W, Stormer H L, and Kim P 2005 Experimental observation of the quantum Hall effect and Berry's phase in graphene *Nature* **438** 201.

55. Zhang L, Zhang Y, Camacho J, Khodas M, and Zaliznyak I 2011 The experimental observation of quantum Hall effect of l=3 chiral quasiparticles in trilayer graphene *Nature* **7** 953.

56. Gornik E, Lassnig R, Strasser G, Stormer H L, Gossard A C, and Wiegmann W 1985 Specific heat of two-dimensional electrons in GaAs-GaAlAs multilayers *Phys. Rev. Lett.* **54** 1820.

57. Mendez E E, Esaki L, and Wang W I 1986 Resonant magnetotunneling in GaAlAs-GaAs-GaAlAs heterostructures *Phys. Rev. B* **33** 2893(R).

58. W. Schülke 2007 Electron dynamics by inelastic x-Ray scattering. Oxford: Oxford University Press.

59. Koshino M and Ando T 2008 Magneto-optical properties of multilayer graphene *Phy. Rev. B* **77** 115313.

60. Koshino M and McCann E 2009 Trigonal warping and Berry's phase N pi in ABC-stacked multilayer graphene *Phy. Rev. B* **80** 165409.

61. McCann E and Koshino M 2013 The electronic properties of bilayer graphene *Rep. Prog. Phys.* **76** 056503.

62. Castro Neto A H, Guinea F, Peres N M R, Novoselov K S, and Geim A K 2009 The electronic properties of graphene *Rev. Mod. Phys.* **81** 109–162.

63. Chan K T, Neaton J B, and Cohen M L 2008 first-principles study of metal adatom adsorption on graphene *Rev. Mod. Phys.* **77** 235430.

64. Yao Y, Ye F, Qi X L, Zhang S C, and Fang Z 2007 Spin-orbit gap of graphene: first-principles calculations *Phys. Rev. B* **75** 041401(R).

65. Feng W, Yao Y, Zhu W, Zhou J, Yao W, and Xiao D 2012 Intrinsic spin Hall effect in monolayers of group-VI dichalcogenides: a first-principles study *Phys. Rev. B* **86** 165108.

66. Ezawa M 2012 Valley-polarized metals and quantum anomalous Hall effect in silicene *Phys. Rev. Lett.* **109** 055502.

67. Ostahie B and Aldea A 2016 Phosphorene confined systems in magnetic field, quantum transport, and superradiance in the quasiflat band *Phys. Rev. B* **93** 075408.

68. Zhou X Y, Zhang R, Sun J P, Zou Y L, Zhang D, Lou W K, et al. 2015 Landau levels and magneto-transport property of monolayer phosphorene *Sci. Rep.* **5** 12295.

69. Berman O L, Gumbs G, and Lozovik Y E 2008 Magnetoplasmons in layered graphene structures *Phys. Rev. B* **78** 085401.

70. Bychkov Yu A and Martinez G 2008 Magnetoplasmon excitations in graphene for filling factors $\nu \leq 6$ *Phys. Rev. B* **77** 125417.

71. Mao W L, Mao H k, Eng P J, Trainor T P, Newville M, Kao C, et al. 2003 Bonding changes in compressed superhard graphite *Science* **302** 425.

72. Guo W, Zhu C Z, Yu T X, Woo C H, Zhang B, and Dai Y T 2004 Formation of sp^3 bonding in nanoindented carbon nanotubes and graphite *Phys. Rev. Lett.* **93** 245502.

73. Lee D W and Seo J W 2011 sp^2/sp^3 Carbon ratio in graphite oxide with different preparation times *J. Phys. Chem. C* **115** 2705.

74. Guerin K, Pinheiro J P, Dubois M, Fawal Z, Masin F, Yazami R, et al. 2004 Synthesis and characterization of highly fluorinated graphite containing sp^2 and sp^3 carbon *Chem. Mater.* **16** 1786.

75. Hu A 2007 Direct synthesis of sp-bonded carbon chains on graphite surface by femtosecond laser irradiation *Appl. Phys. Lett.* **91** 131906.

76. Bernal J D 1924 The structure of graphite *Proc. R. Soc. London, Ser. A* **106** 749.

77. Novoselov K S, Geim A K, Morozov S V, Jiang D, Katsnelson M I, Grigorieva I V, et al. 2005 Two-dimensional gas of massless Dirac fermions in graphene *Nature* **438** 197.

78. Peng J, Gao W, Gupta B K, Liu Z, Romero-Aburto R, Ge L, et al. 2012 Graphene quantum dots derived from carbon fibers *Nano Lett.* **12** 844.

79. Iijima S 1991 Helical microtubules of graphitic carbon *Nature* **354** 56.

80. Taylor R and Walton D R M 1993 The chemistry of fullerenes *Nature* **363** 685.
81. Tomita S, Fujii M, Hayashi S, and Yamamoto K 1999 Electron energy-loss spectroscopy of carbon onions *Chem. Phys. Lett.* **305** 225.
82. Wilder J W G, Venema L C, Rinzler A G, Smalley R E, and Dekker C 1998 Electronic structure of atomically resolved carbon nanotubes *Nature* **391** 59.
83. Crespi V H, Cohen M L, and Rubio A 1997 In situ band gap engineering of carbon nanotubes *Phys. Rev. Lett.* **79** 2093.
84. Rubio A, Sanchez-Portal D, Artacho E, Ordejon P, and Soler J M 1999 Electronic states in a finite carbon nanotube: a one-dimensional quantum box *Phys. Rev. Lett.* **82** 3520.
85. Yan W, Liu M, Dou R F, Meng L, Feng L, Chu Z D, et al. 2012 Angle-dependent van Hove singularities in a slightly twisted graphene bilayer *Phys. Rev. Lett.* **109** 126801.
86. Havener R W, Zhuang H, Brown L, Hennig R G, and Park J 2012 Angle-resolved Raman imaging of interlayer rotations and interactions in twisted bilayer graphene graphene bilayer *Nano Lett.* **12** 3162.
87. Wu J B, Zhang X, Ijas M, Han W P, Qiao X F, Li X L, et al. 2014 Resonant Raman spectroscopy of twisted multilayer graphene graphene bilayer *Nat. Comm.* **5** 5309.
88. Ferrari A C, Meyer J C, Scardaci V, Casiraghi C, Lazzeri M, Mauri F, et al. 2006 Raman spectrum of graphene and graphene layers *Phys. Rev. Lett.* **97** 187401.
89. Brihuega I, Mallet P, Gonzalez-Herrero H, Trambly de Laissardiere G, Ugeda M M, Magaud L, et al. 2012 Unraveling the intrinsic and robust nature of van Hove singularities in twisted bilayer graphene by scanning tunneling microscopy and theoretical analysis *Phys. Rev. Lett.* **109** 196802.
90. Havener R W, Liang Y, Brown L, Yang L, and Park J 2014 van Hove singularities and excitonic effects in the optical conductivity of twisted bilayer graphene *Nano Lett.* **14** 3353.
91. Moon P and Koshino M 2012 Energy spectrum and quantum Hall effect in twisted bilayer graphene *Phys. Rev. B* **85** 195458.
92. Mele E J 2010 Commensuration and interlayer coherence in twisted bilayer graphene *Phys. Rev. B* **81** 161405.
93. Kim K S, Walter A L, Moreschini L, Seyller T, Horn K, Rotenberg E, and Bostwick A 2013 Coexisting massive and massless Dirac fermions in symmetry-broken bilayer graphene *Nat. Mater.* **12** 887.
94. Moon P and Koshino M 2013 Optical absorption in twisted bilayer graphene *Phys. Rev. B* **87** 205404.
95. Sanchez-Yamagishi J D, Taychatanapat T, Watanabe K, Taniguchi T, Yacoby A, and Jarillo-Herrero P 2012 Quantum Hall effect, screening, and layer-polarized insulating states in twisted bilayer graphene *Phys. Rev. Lett.* **108** 076601.
96. Cao Y, Fatemi V, Fang S, Watanabe K, and Taniguchi T 2018 Efthimios Kaxiras & Pablo Jarillo-Herrero Unconventional superconductivity in magic-angle graphene superlattices *Nature* **556** 43.
97. Jiang L, Wang S, Shi Z, Jin C, Utama M I B, Zhao S, et al. 2018 Manipulation of domain-wall solitons in bi- and trilayer graphene *Nat. Nanotech.* **13** 204.
98. Ju L, Shi Z, Nair N, Lv Y, Jin C, Velasco Jr J, et al. 2015 Topological valley transport at bilayer graphene domain walls *Nature* **520** 650.
99. Berger C, Song Z, Li X, Wu X, Brown N, Naud C, et al. 2006 Electronic confinement and coherence in patterned epitaxial graphene *Science* **312** 1191.
100. Pavesi L, Negro L D, Mazzoleni C, Franzo G, and Priolo F 2000 Optical gain in silicon nanocrystals *Nature* **408** 440.
101. McConnell H M, Owicki J C, Parce J W, Miller D L, Baxter G T, Wada H G, et al. 1992 The cytosensor microphysiometer: biological applications of silicon technology *Science* **257** 1906.

102. Madar R 2004 Silicon carbide in contention *Nature* **430** 974.
103. Cui Y and Lieber C M 2001 Functional nanoscale electronic devices assembled using silicon nanowire building blocks *Science* **291** 851.
104. Feng B, Ding Z, Meng S, Yao Y, He X, Cheng P, et al. 2012 Evidence of silicene in honeycomb structures of silicon on Ag(111) *Nano Lett.* **12** 3507.
105. Vogt P, Padova P D, Quaresima C, Avila J, Frantzeskakis E, Asensio M C, et al. 2012 Silicene: compelling experimental evidence for graphenelike two-dimensional silicon *Phys. Rev. Lett.* **108** 155501.
106. Meng L, Wang Y, Zhang L, Du S, Wu R, Li L, et al. 2013 Buckled silicene formation on Ir(111) *Nano Lett.* **13** 685.
107. Fleurence A, Friedlein R, Ozaki T, Kawai H, Wang Y, and Yamada-Takamura Y 2012 Experimental evidence for epitaxial silicene on diboride thin films *Phys. Rev. Lett.* **108** 245501.
108. Tao L, Cinquanta E, Chiappe D, Grazianetti C, Fanciulli M, Dubey M, et al. 2015 Silicene field-effect transistors operating at room temperature *Nat. Nanotechnol.* **10** 227.
109. Ni Z, Liu Q, Tang K, Zheng J, Zhou J, Qin R, et al. 2012 Tunable bandgap in silicene and germanene *Nano Lett.* **12** 113.
110. Padilha J E and Pontes R B 2015 Free-standing bilayer silicene: the effect of stacking order on the structural, electronic, and transport properties *J. Phys. Chem. C* **119** 3818.
111. Zhang C and Yan S 2012 First-principles study of ferromagnetism in two-dimensional silicene with hydrogenation *J. Phys. Chem. C* **116** 4163.
112. Osborn T H and Farajian A A 2012 Stability of lithiated silicene from first principles *J. Phys. Chem. C* **116** 22916.
113. Cai Y, Chuu C P, Wei C M, and Chou M Y 2013 Stability and electronic properties of two-dimensional silicene and germanene on graphene *Phys. Rev. B* **88** 245408.
114. Wu J Y, Chen S C, Gumbs G, and Lin M F 2016 Feature-rich electronic excitations of silicene in external fields *Phys. Rev. B* **94** 205427.
115. Kulish V V, Malyi O I, Persson C, and Wu P 2015 Adsorption of metal adatoms on single-layer phosphorene *Phys. Chem. Chem. Phys.* **17** 992.
116. Machado B F and Serp P 2012 Graphene-based materials for catalysis *Catal. Sci. Technol.* **2** 54.
117. Lin S Y and Lin M F 2018 Metal-adsorbed graphene nanoribbons *arXiv* **1806** 05290.
118. Leventis N, Sadekar A, Chandrasekaran N, and Sotiriou-Leventis C 2010 Synthesis of monolithic silicon carbide aerogels from polyacrylonitrile-coated 3D silica networks *Chem. Mater.* **22** 2790.
119. McCarthy M C, Apponi A J, and Thaddeus P 1999 Rhomboidal SiC^3 *J. Chem. Phys.* **110** 10645.
120. Fan S, Zhang L, Xu Y, Cheng L, Lou J, Zhang J, et al. 2007 Microstructure and properties of 3D needle-punched carbon/silicon carbide brake materials *Composites Science and Technology* **67** 2390.
121. Shi Y F, Meng Y, Chen D H, Cheng S J, Chen P, Yang H F, et al. 2006 Highly ordered mesoporous silicon carbide ceramics with large surface areas and high stability *Adv. Funct. Mater.* **16** 561.
122. Zhou W, Yan L, Wang Y, and Zhang Y 2006 SiC nanowires: a photocatalytic nanomaterial *Appl. Phys. Lett.* **89** 013105
123. Hua J Q, Bando Y, Zhan J H, and Golberg D 2004 Fabrication of ZnS/SiC nanocables, SiC-shelled ZnS nanoribbons (and sheets), and SiC nanotubes (and tubes) *Appl. Phys. Lett.* **85** 2932.
124. Romo-Herrera J M, Terrones M, Terrones H, Dag S, and Meunier V 2007 Covalent 2D and 3D networks from 1D nanostructures: designing new materials *Nano Lett.* **7** 570.

125. Togo A, Chaput L, Tanaka I, and Hug G 2010 First-principles phonon calculations of thermal expansion in Ti_3SiC_2, Ti_3AlC_2, and Ti_3GeC_2 *Phys. Rev. B* **81** 174301.

126. Fukumoto A 1996 First-principles calculations of p-type impurities in cubic SiC *Phys. Rev. B* **53** 4458.

127. Chisholm N, Mahfuz H, Rangari V K, Ashfaq Ad, and Jeelani S 2005 Abrication and mechanical characterization of carbon/SiC-epoxy nanocomposites *Composite Structures* **67** 115.

128. Mpourmpakis G, Froudakis G E, Lithoxoos G P, and Samios J 2006 SiC nanotubes: a novel material for hydrogen storage *Nano Lett.* **6** 1581.

129. Doán S, Teke A, Huang D, and Morkoc H 2003 4H-SiC photoconductive switching devices for use in high-power applications *Appl. Phys. Lett.* **82** 3107.

130. Melinon P, Masenelli B, Tournus F, and Perez A 2007 Playing with carbon and silicon at the nanoscale *Nat. Mater.* **6** 479.

131. Pochet P, Genovese L, Caliste D, Rousseau I, Goedecker S, and Deutsch T 2010 First-principles prediction of stable SiC cage structures and their synthesis pathways *Phys. Rev. B* **82** 035431.

132. Huckestein B 1995 Scaling theory of the integer quantum Hall effect *Rev. Mod. Phys.* **67** 357.

133. Xu Y, Yan B, Zhang H J, Wang J, Xu G, Tang P, et al. 2013 Large-gap quantum spin hall insulators in tin films *Phys. Rev. Lett.* **111** 136804.

134. Brune C, Liu C X, Novik E G, Hankiewicz E M, Buhmann H, Chen Y L, et al. 2011 Quantum Hall effect from the topological surface states of strained bulk HgTe *Phys. Rev. Lett.* **106** 126803.

135. Inoue J, Bauer G E W, and Molenkamp L W 2004 Suppression of the persistent spin Hall current by defect scattering *Phys. Rev. B* **70** 041303(R).

136. Lian K Y, Ji Y F, Li X F, Jin M X, Ding D J, and Luo Y 2013 Big bandgap in Highly reduced graphene oxides *J. Phys. Chem. C* **117** 6049.

137. Do T N, Lin C Y, Lin Y P, Shih P H, and Lin M F 2015 Configuration-enriched magnetoelectronic spectra of AAB-stacked trilayer graphene *Carbon* **94** 619.

138. Moler K A, Baar D J, Urbach J S, Liang R, Hardy W N, and Kapitulnik A 1994 Magnetic field dependence of the density of states of $YBa_2Cu_3O_{6.95}$ as determined from the specific heat *Phys. Rev. Lett.* **73** 2744.

139. Lee J H, Jang J, Choi J, Moon S H, Noh S, Kim J, et al. 2011 Exchange-coupled magnetic nanoparticles for efficient heat induction specific heat *Nat. Nanotechnol.* **6** 418.

140. Eisenstein J P, Gossard A C, and Narayanamurti V 1987 Quantum oscillations in the thermal conductance of GaAs/AlGaAs heterostructures Specific Heat *Phys. Rev. Lett.* **59** 1341.

141. Kashiwagia T, Grulkeb E, Hildingb J, Grotha K, Harrisa R, Butlera K, et al. 2001 Thermal and flammability properties of polypropylene/carbon nanotube nanocomposites *Polymer* **45** 4227.

142. Higginbotham A L, Lomeda J R, Morgan A B, and Tour J M 2009 Graphite oxide flame-retardant polymer nanocomposites *ACS Appl. Mater. Interfaces* **1** 2256.

143. Derivaz M, Dentel D, Stephan R, Hanf M-C, Mehdaoui A, Sonnet P, et al. 2015 Continuous germanene layer on Al(111) *Nano Lett.* **15** 2510.

144. Davila M E, Xian L, Cahangirov S, Rubio A, and Lay G L 2014 Germanene: a novel two-dimensional germanium allotrope akin to graphene and silicene *New J. Phys.* **16** 095002.

145. Zhuang J, Gao N, Li Z, Xu X, Wang J, Zhao J, et al. 2017 Cooperative electron phonon coupling and buckled structure in germanene on Au(111) *ACS Nano* **11** 3553.

146. Shin S Y, Hwang C G, Sung S J, Kim N D, Kim H S, and Chung J W 2011 Observation of intrinsic intraband π-plasmon excitation of a single-layer graphene *Phys. Rev. B* **83** 161403.

147. Generalov A V and Dedkov Yu S 2012 EELS study of the epitaxial graphene/Ni(1 1 1) and graphene/Au/Ni(1 1 1) systems *Carbon* **50** 183.

148. Gurtubay I G, Pitarke J M, Ku W, Eguiluz A G, Larson B C, Tischler J, et al. 2005 Electron-hole and plasmon excitations in 3d transition metals: ab initio calculations and inelastic x-ray scattering measurements *Phys. Rev. B* **72** 125117.

149. Shin S Y, Kim N D, Kim J G, Kim K S, Noh D Y, Kim K S, et al. 2016 Control of the π plasmon in a single layer graphene by charge doping *Appl. Phys. Lett.* **99** 082110.

150. Ju L, Geng B, Horng J, Girit C, Martin M, Hao Z, et al. 2011 Graphene plasmonics for tunable terahertz metamaterials *Nat. Nanotechnol.* **6** 630.

151. Volovik G E 2003 *Universe in a Helium Droplet*. New York: Oxford University Press.

152. Chiu C K, Teo J C Y, Schnyder A P, and Ryu S 2016 Classification of topological quantum matter with symmetries *Rev. Mod. Phys.* **88** 035005.

153. Wen X G 2017 Colloquium: zoo of quantum-topological phases of matter *Rev. Mod. Phys.* **89** 041004.

154. Landau L D and Lifschitz E M 1958 *Course of Theoretical Physics* vol. 5. London: Pergamon.

155. Klitzing K von, Dorda G, and Pepper M 1980 New method for high-accuracy determination of the fine-structure constant based on quantized Hall resistance *Phys. Rev. Lett.* **45** 494.

156. Thouless D J, Kohmoto M, Nightingale M P, and Nijs M den 1982 Quantized Hall conductance in a two-dimensional periodic potential *Phys. Rev. Lett.* **49** 405.

157. Avron J E, Seiler R and Simon B 1983 Holonomy, the quantum adiabatic theorem, and Berry's phase *Phys. Rev. Lett.* **51** 51.

158. Simon B 1983 Holonomy, the quantum adiabatic theorem, and Berry's phase *Phys. Rev. Lett.* **51** 2167.

159. Novoselov K S, Geim A K, Morozov S V, Jiang D, Katsnelson M I, Grigorieva I V, et al. 2005 Room-temperature quantum Hall effect in graphene *Nature* **438** 197.

160. Zhang Y, Tan Y W, Stormer H L, and Geim A K 2005 Room-temperature quantum Hall effect in graphene *Nature* **438** 201.

161. Zheng Y and Ando T 2002 Hall conductivity of a two-dimensional graphite system *Phys. Rev. B* **65** 245420.

162. Gusynin V P and Sharapov S G 2005 Unconventional integer quantum Hall effect in graphene *Phys. Rev. Lett.* **95** 146801.

163. Hatsugai, Y 2011 Topological aspect of graphene physics *J. Phys. Conf. Ser.* **334** 012004.

164. Kawarabayashi T, Morimoto T, Hatsugai Y, and Aoki H 2010 Anomalous criticality at the n=0 quantum Hall transition in graphene: the role of disorder preserving chiral symmetry *Phys. Rev. B* **82** 195426.

165. Kawarabayashi T, Hatsugai Y, and Aoki H 2010 Landau level broadening in graphene with long-range disorder—Robustness of the $n = 0$ level *Physica E* **42** 759.

166. Kopelevich Y, Torres J H S, da Silva R R, Mrowka F, Kempa H, and Esquinazi P 2003 Reentrant metallic behavior of graphite in the quantum limit *Phys. Rev. Lett.* **90** 156402.

167. Kempa H, Esquinazi P and Kopelevich Y 2006 Integer quantum Hall effect in graphite *Solid State Commun.* **138** 118.

168. Xu Y, Miotkowski I, Liu C, Tian J, Nam H, Alidoust N, et al. 2014 Observation of topological surface state quantum Hall effect in an intrinsic three-dimensional topological insulator *Nat. Phys.* **10** 956.

169. Yoshimi R, Tsukazaki A, Kozuka Y, Falson J, Takahashi K S, Checkelsky J G, et al. 2015 Quantum Hall effect on top and bottom surface states of topological insulator $(Bi_{1-x}Sb_x)_2Te_3$ films *Nat. Commun.* **6** 6627.

170. Ma J, Alfe D, Michaelides A, and Wang E 2009 Stone-Wales defects in graphene and other planar sp^2-bonded materials *Phys. Rev. B* **80** 033407.

171. Kim K, Sussman A, and Zettl A. 2010 Graphene nanoribbons obtained by electrically unwrapping carbon nanotubes *ACS Nano* **4** 1362.

172. Zhang W B, Song Z B, and Dou L M 2015 The tunable electronic structure and mechanical properties of halogenated silicene: a first-principles study *J. Mater. Chem. C* **3** 3087.

2 Experimental Measurements on Magnetic Quantization

Shih-Yang Lin,[e] *Thi-Nga Do,*[c,d] *Chiun-Yan Lin,*[a]
Jhao-Ying Wu,[b] *Po-Hsin Shih,*[a]
Ching-Hong Ho,[b] *Ming-Fa Lin*[a,f,g]

[a] Department of Physics, National Cheng Kung University,
Tainan 701, Taiwan
[b] Center of General Studies, National Kaohsiung University of
Science and Technology, Kaohsiung 811, Taiwan
[c] Laboratory of Magnetism and Magnetic Materials, Advanced
Institute of Materials Science, Ton Duc Thang University,
Ho Chi Minh City, Vietnam
[d] Faculty of Applied Sciences, Ton Duc Thang University,
Ho Chi Minh City, Vietnam
[e] Department of Physics, National Chung Cheng University,
Chiayi 621, Taiwan
[f] Quantum Topology Center, National Cheng Kung University,
Tainan 701, Taiwan
[g] Hierarchical Green-Energy Materials Research Center,
National Cheng Kung University, Tainan, Taiwan

CONTENTS

High-resolution experimental equipment has now been developed to fully explore the novel physical phenomena in the emergent layered materials. In general, they can measure the fundamental quantities very accurately and efficiently, covering the geometric, electronic, optical, transport, thermal, and Coulomb-excitation properties. Specifically, the observed phenomena will become complex and diverse in the presence of a uniform perpendicular magnetic field. The greatly diversified magnetic quantization is clearly revealed in the distinct physical properties, as thoroughly discussed in the following sections associated with the various experimental measurements.

2.1 SCANNING TUNNELING SPECTROSCOPY

Scanning tunneling microscopy (STM) has become one of the most important experimental techniques in resolving the diversified nanostructures since the first discovery by Binnig and Rohrer in 1982 [1]. STM is capable of revealing the surface topographies in the real space with both lateral and vertical atomic resolutions, e.g., the nanoscaled bond lengths, crystal symmetries, planar/nonplanar geometries, step edges, local vacancies, amorphous dislocations, adsorbed adatoms & molecules, and nanoclusters & nanoislands [1]. The STM instrument is composed of a conducting (semiconducting) solid surface and a sharp metal tip, where the distance of several angstroms is modulated by piezoelectric feedback devices. A very weak current will be generated from the quantum tunneling effect in the presence of a bias voltage between surface and tip. The tunneling current presents a strong dependence on their distance; furthermore, the relation roughly agrees with the exponential decay form. It flows from the occupied electronic states of tip into the unoccupied ones of the surface under the positive bias voltage $V > 0$, and vice versa. This current is required to serve as a feedback signal, so that the most commonly used mode, a constant tunneling one (current and voltage), will be operated to resolve the surface structure very delicately by using a piezoelectric device. By combining the tip of a metal probe and a precise scanning device, the spectroscopic information of surface morphology is thoroughly revealed at the preselected positions with the well-defined conditions. The structural response hardly depends on the background effects, being attributed to an ultrahigh vacuum environment. The up-to-date spatial resolution of STM measurements is ~ 0.1 Å. The rather accurate measurements on the graphene-related systems have clearly illustrated the complicated relations among the honeycomb lattice, the finite-size quantum confinement, the flexible feature, and the active chemical bondings of carbon atoms. For example, the experimental identifications on the rich and unique geometric structures cover the achiral (armchair and zigzag) and chiral graphene nanoribbons with the nanoscale-width planar honeycomb lattices [2,3], the curved [4], folded [5], scrolled [6] and stacked graphene nanoribbons [6], the achiral and chiral arrangements of the hexagons on cylindrical surfaces of carbon nanotubes [7], the AB [8], ABC [8–10] and AAB stackings [11, 12] in few-layer graphene systems, the twisted bilayer graphenes [13, 14], the corrugated substrate and buffer graphene layer [15], the rippled structures of graphene islands [16, 17], the adatom distributions on graphene surfaces [18], and the 2D networks of local defects on graphite surface [19, 20].

Scanning tunneling spectroscopy (STS), an extension of STM in Fig. 2.1, creates the tunneling current through the tip-surface junction under a constant height mode [details in Ref [1]]. The electronic properties are characterized by both I-V and dI/dV curves sweeping over the bias voltage V. In the STS measurements, the normalized differential conductance, being created by a small alternating current (AC) modulation of dV, could be measured by adding a lock-in amplifier. Furthermore, the noise

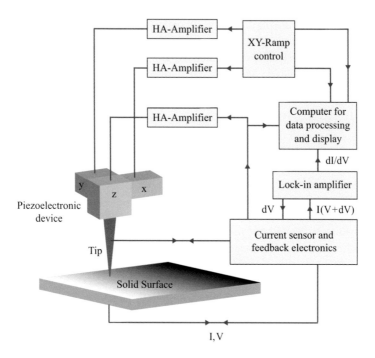

FIGURE 2.1 Equipment of scanning tunneling spectroscopy.

involved in the observed conductance is largely reduced. The up-to-now experiments present the highest resolution of $\sim$ 10 pA. STS is a very efficient method for examining the electronic energy spectra of condensed-matter systems. The tunneling differential conductance is approximately proportional to the density of states (DOS) and directly presents the main features, being useful in identifying the semiconducting, semimetallic, or metallic behaviors. For example, a zero-gap semiconducting monolayer graphene presents a V-shape DOS with a vanishing value at the Fermi level (the Dirac point) [21]. The semimetallic ABC-stacked graphene systems, a pronounced peak at the Fermi level is revealed in trilayer and pentalayer cases as a result of the surface-localized states [9, 10]. The high-resolution STS could serve as a powerful experimental method for investigating the magnetically quantized energy spectra of layered graphenes. The measured tunneling differential conductance directly reflects the structure, energy, number, and degeneracy of the Landau-level delta-function-like peaks. Part of the theoretical predictions on the Landau-level energy spectra have been confirmed by the STS measurements, such as the $\sqrt{B_z}$-dependent Landau-level energy in monolayer graphene [22, 23], the linear B_z-dependence in AB-stacked bilayer graphene [22], the coexistent square-root and linear B_z-dependences in trilayer ABA stacking [22], and the 2D and 3D characteristics of the Landau subbands in Bernal (AB-stacked) graphite [24, 25]. Specifically, STS is suitable for identifying the spin-split DOS, when the magnetic tips are utilized in the experimental measurements. The spin-polarized STS (SPSTS) is very powerful in fully comprehending the magneto-electronic properties, e.g., the spin-split Landau levels/Landau subbands due to the significant spin-orbital couplings and the gate voltage (perpendicular electric field, e.g., those in silicene and germanene systems [details in Chaps. 6 and 9]).

The high-resolution STS measurements under the specific energies can directly map the spatial probability distributions of wave functions in the absence and presence of a uniform perpendicular magnetic field. Up to now, they have confirmed the rich and unique electronic states in graphene-related systems. For example, the normal standing waves along the tubular axis, with the finite zero points, are clearly identified to survive in finite-length carbon nanotubes [26]. The topological edge states are created by the AB-BA domain wall of bilayer graphenes [27]. Moreover, the well-behaved Landau-level wave functions, being similar to those of an oscillator, come to exist in few-layer graphene systems with the distinct stacking configurations, such as, the $n^{c,v} = 0$, and ± 1 Landau levels for the monolayer and AB-stacked bilayer graphenes. The higher-n [28], undefined [29], and anti-crossing Landau levels [30] are worthy of a series of experimental examinations. The similar STS measurements are available in testing the theoretical predictions on the nonuniform magnetic quantization, establishing the direct relations between the band-edge state energies and the distribution range of the regular or irregular wave functions quantization, e.g., the main features of Landau subbands in Chap. 5.

2.2 MAGNETO-OPTICAL SPECTROSCOPIES

Optical spectroscopies are very powerful in characterizing the absorption [31], transmission [51], and reflectance [33] of any condensed-matter systems, especially for the diverse optical properties of emergent layered materials. They can provide the rather sufficient information on the single-particle and many-body vertical optical excitations, directly reflecting the main features of energy bands and strongly coupled excitons [electron-hole pairs attracted by the longitudinal Coulomb interactions; [31,51]], e.g., the experimental examinations on the predicted energy gap, the threshold excitation frequency, the band-edge state energies, the special absorption structures, the electric-field effects, and the magneto-optical selection rules. Absorption spectroscopy, which is based on the analytical method of measuring the fraction of incident radiation absorbed by a sample, is one of the most versatile and widely used techniques in the sciences of physics, chemistry, and materials. This well-developed tool is available within a wide frequency range. The measured spectral functions are frequency-dependent functions for characterizing the fundamental electronic properties of materials. The experimental setup related to the light source, sample arrangement, and detection technique is sensitive to the frequency range and the experimental purpose.

The most common optical setup, as clearly shown in Fig. 2.2, is to emit a radiation beam on the specific sample and then delicately measure the transmitted intensity that is responsible for the significant frequency-dependent absorption spectrum. In general, optical experiments are performed by a broadband light source, where the intensity and frequency could be adjusted within a broad range. According to the frequency range of optical operations, the most utilized light sources are classified into three kinds: (**I**) the xenon-mercury arc lamp under the high pressure in the far-infrared region [34, 35], (**II**) the black-body source of the heated SiC element in the mid- to near-infrared spectral range [36–38], and (**III**) the tungsten halogen lamp

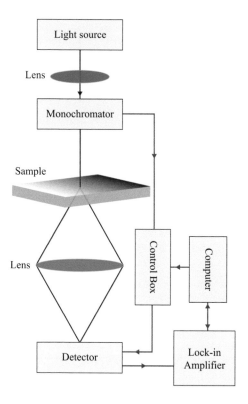

FIGURE 2.2 Optical spectroscopy.

in a continuous spectrum from the visible to near-ultraviolet [39, 40]. These optical sources are commonly adopted for the analytical characterizations of optical properties in emergent materials over a broad spectrum, since they are successfully operated even at a high temperature [> 3,000 K] under the inert-halogen mixture atmosphere.

Generally speaking, the optical absorption spectra are delicately measured using the Fourier transform spectrometer [41, 42], linear photodiode array spectrometer [43,44], or charge-coupled device (CCD) spectrometer [33,45]. Fourier transform spectrometer is based on the coherent observations of electromagnetic radiations, in either the time- or space-domain ones, e.g., Bruker IFS125 with the high-resolution linewidths narrower 0.001 cm [41, 42]. On the other side, the photodiode array spectrophotometer, which is built from hundreds of linear high-speed detectors integrated on a single chip, simultaneously measures the dispersive light over a wide frequency range. CCD spectrometer is also a multichannel detector; it can simultaneously detect as many frequency channels as the number of the individual resolution pixels. According to the linear dynamic Kubo formula [33,45], the optical absorption structures are closely related to electronic properties [energy spectra and wave functions], in which they have finite widths and specific forms arising from the initial and final band-edge states [the critical points in energy-wave-vector space].

Up to now, the high-resolution optical observations [47–49, 51, 59] on layered graphene systems are consistent with part of theoretical calculations [28] in the absence/presence of a uniform perpendicular electric field $[E_z \hat{z}]$. For example, the AB-stacked bilayer graphene, the ~ 0.3-eV shoulder structure under a zero field [46,47], the E_z-created semimetal-semiconductor transition and two low-frequency asymmetric peaks, and two very prominent π-electronic absorption peaks at middle frequency [46,47]. The similar examinations performed for the trilayer ABA-stacked graphene clearly show one shoulder at ~ 0.5 eV [48], the gapless behavior unchanged by the gate voltages, the E_z-induced low-frequency multi-peak structures [48], and several strong π-electronic absorption peaks at higher frequencies [48]. The identified features of absorption spectra in the trilayer ABC stacking cover two low-frequency characteristic peaks [48,49] and gap opening under the electric field [48]. The above-mentioned strong π peaks, which are due to all the valence states of carbon-$2p_z$ orbitals, have the most high frequency in Bernal graphite because of the infinite graphene layers [51, 59]. On the other hand, certain theoretical predictions deserve further experimental verifications, e.g., the rich and unique optical properties of the trilayer AAB and AAA stackings [28] and sliding bilayer graphene systems [28, 29].

Magneto-optical spectroscopies are available for a full exploration of the various magnetic quantization phenomena in the dimension-dependent condensed-matter systems, e.g., the 1D-3D graphene-related sp^2-bonding systems. A uniform magnetic field is generated by using the superconducting magnet, in which a 25-T cryogen-free superconducting magnet is successfully developed at the High Field Laboratories [52, 53]. Furthermore, the super high magnetic fields can be built under a semidestructive single-turn coil technique that provides a pulsed magnetic field larger than 100 T for pulse lengths of tens microseconds [54]. Up to now, the high-resolution magneto-optical measurements have confirmed the well-known Aharonov-Bohm effect due to the periodical boundary condition in cylindrical carbon nanotubes [55]. The infrared transmission spectra clearly identify the $\sqrt{B_z}$-dependent absorption frequencies of the interband Landau-level transitions in mono- and few-layer graphene systems [56,57]. Specifically, the magneto-Raman spectroscopy is utilized to observe the low-frequency Landau-level excitation spectra for the AB-stacked graphenes up to five layers [58]. Only the well-behaved selection rule of $\Delta n = 1$ is revealed in the above-mentioned inter-Landau-level excitation experiments. Apparently, the extra magneto-optical selection rules in few-layer graphene systems are worthy of a series of detailed examinations, e.g., those in trilayer ABC [28] and AAB [28] stackings, and the electric-field-applied AB-stacked systems [28]. As for the magneto-optical reflection/absorption/transmission/Raman experiments on the Bernal (AB-stacked) graphite [59–63], the vertical optical excitations, which mainly originate from a lot of 1D Landau subbands with the significant k_z-dispersions, are verified to present the monolayer- and bilayer-like graphene behaviors. Such phenomena respectively correspond to the magnetic quantizations initiated from K and H points ($k_z = 0$ and π).

2.3 QUANTUM TRANSPORT APPARATUS, AND DIFFERENTIAL THERMAL CALORIMETER/LASER FLASH ANALYSIS

In general, the Hall-bar method is available in fully exploring quantum transport properties since its first discovery by Edwin Herbert Hall in 1879 [64]. The measurements of electrical resistivities for the rectangular thin films are carried out by a five-probe

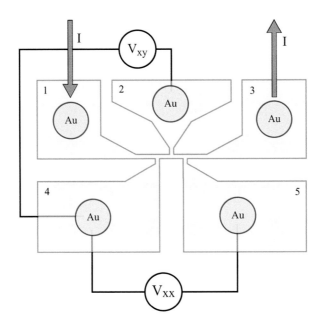

FIGURE 2.3 Transportant instrument for Hall conductivities.

technique, as clearly displayed in Fig. 2.3. There exist the typical dimensions of 2 mm in length, 1 mm in width, and 100 nm in thickness [65]. This equipment is very efficient and reliable in simultaneously measuring the longitudinal resistivity and the transverse Hall voltage (the positive or negative charges of free carriers and their density). Five gold-pads are evaporated onto the contact areas to ensure the excellent electrical contacts for the transverse Hall and longitudinal resistivity measurements using the standard direct current methods. The Hall voltages are delicately analyzed through reversing the field direction at a fixed temperature to eliminate the offset voltage associated with the unbalanced Hall terminals. The Hall coefficients are taken in a magnetic-field range of $\sim 1-10$ T, and a typical dc current density of $\sim 10^3-10^4$ A/cm^2 is applied along the longitudinal direction of the sample. In addition, another method, the four-probe measurement of van der Pauw, is frequently utilized for the irregular finite-width materials.

Recently, there are a lot of magneto-transport measurements on quantum Hall conductivities of few-layer graphene systems with the various layer numbers and stacking configurations under the low temperatures. Monolayer graphene has been clearly identified to present the unconventional half-integer transverse conductivity $\sigma_{xy} = (m + 1/2)4e^2/h$ [66], where m is an integer and the 4 factor represents the spin and the equivalence of A and B sublattices. This unique quantization mainly comes from the quantum anomaly of $n^{c,v} = 0$ Landau levels corresponding to the Dirac point, as discussed later in Sec. 3.3.1. As to bilayer AB stacking [67], the quantum Hall conductivity is verified to be $\sigma_{xy} = 4m'e^2/h$ (m' a non-zero integer). Furthermore, an unusual integer quantum Hall conductivity, a double step height of $\sigma_{xy} = 8e^2/h$, appears at zero energy under a sufficiently low magnetic field. This mainly comes from $n = 0$ and $n = 1$ Landau levels of the first group. The low-lying quantum Hall plateaus in trilayer ABA graphene are revealed as $2e^2/h$, $4e/h$, $6e^2/h$, and $8e^2/h$ [68], with a step height of $2e^2/h$, especially for the energy range of $\sim \pm 20$ meV. Such observation is consistent with the calculated Landau-level energy spectra [69]. The neighboring- and next-neighboring-layer hopping integrals, respectively, create the separated Dirac cone and parabolic bands, and the valley splitting of the latter. At low energy, these further lead to six quantized Landau levels with double spin degeneracy and thus the quantum Hall step of $2e^2/h$. Specifically, the higher-energy quantum transport properties in trilayer ABA stacking are roughly regarded as the superposition of monolayer and AB bilayer graphenes [70]. Moreover, the ABC-stacked trilayer graphene shows the significant differences in the main features of quantum Hall conductivities compared with the ABA trilayer system. Its quantum Hall conductivity [70] behaves as a sequence of $\sigma_{xy} = 4(\pm|m'| \pm 1/2)e^2/h$ in the absence of the $\sigma_{xy} = \pm 2e^2/h$, $\pm 4e^2/h$, and $\pm 8e^2/h$ plateaus. Specifically, a $\sigma_{xy} = 12e^2/h$ step comes to exist near zero energy, being created by the $n = 0, 1$, and 2 Landau levels of the first group due to the quantization of surface-localized at bands. [69]. There also exists the four-fold spin and valley degeneracy in ABC trilayer stacking, resulting in the quantum conductivity step height of $4e^2/h$. The above-mentioned interesting quantum transports open the door for fully exploring the configuration-modulated Hall conductivities in other layered graphenes, such as the trilayer AAB stacking, sliding bilayer systems, and twisted bilayer ones.

Up to now, two kinds of high-resolution instruments, differential scanning calorimeter [DSC; [71–73]] and laser flash analysis [LFA; [75,76]], are frequently utilized to measure the thermal properties, especially for the specific heats of condensed-matter systems. As to the former [Fig. 2.4(a)], the sample material is subjected to a linear temperature algorithm. The heat flow rate into the sample, which corresponds to the instantaneous specific heat, is under the continuous and delicate

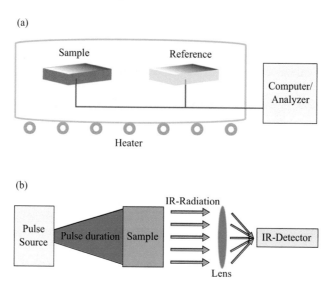

(a)

Sample   Reference

Computer/
Analyzer

Heater

(b)

IR-Radiation

Pulse
Source   Pulse duration   Sample   IR-Detector

Lens

FIGURE 2.4   (a) differential thermal calorimeter and (b) laser flash analysis.

measurements [71]. Specifically, two sample holders are symmetrically mounted inside an enclosure being normally held at room temperature. A primary temperature-control system dominates the average temperature of two sample holders, in which it covers platinum resistance thermometers and heating elements embedded in the sample holders. Furthermore, a secondary temperature-control system measures the temperature difference between the two sample holders, and generates the vanishing one by controlling a differential component of the total heating power. This differential power is accurately measured and recorded. Very interestingly, an extension of DSC, a modulated DSC (MDSC) [72, 73], is further developed to provide the more reliable results through measuring the heat capacity of the sample and the heat flow at the same time. MDSC utilizes a sinusoidal temperature oscillation instead of the traditional linear ramp. How to examine the magnetic quantization phenomena in thermal properties is a high-technique measurement. The close association of magnetic-field apparatus and DSC is the so-called PPMS calorimeter (a Quantum Design Physical Property Measurement System for magnetization); furthermore, this combined instrument can measure the T- and B_z-dependent and specific heat and magnetic susceptibility simultaneously. It should be noticed that the stationary and pulsed magnetic fields can achieve the strengths of 10 T and 50 T in standard laboratories, respectively.

LFA [73], as clearly shown in Fig. 2.4(b), represents a recent technical progress in identifying the crucial thermal phenomena. The sample is positioned on a robot being surrounded by a furnace. For the accurate measurements, the furnace remains at a predetermined temperature and a programmable energy pulse irradiates the backside of the specimen. These create a homogeneous temperature rise at the sample surface. The resulting temperature rise of the surface of the sample is measured by a very sensitive high-speed infrared detector. Both properties, the specific heat and thermal diffusivity, are simultaneously determined from the temperature versus time data. The thermal conductivity could also be evaluated from the estimated carrier density. In short, LFA covers the experimental operation processes: uniform heating by a homogenized laser beam through an optical fiber with a mode mixer, measuring the transient temperature of a sample with a calibrated radiation thermometer, analyzing a transient temperature curve with a curve fitting method, and achieving the differential laser flash calorimetry. Recently, LFA is commonly used for the measurements of thermal properties, mainly owing to the excellent performances in the high accuracy, good repeatability, and wide temperature range [71–73].

The high-resolution experimental measurements on specific heats of condensed-matter systems are very useful in examining the theoretical predictions about the electronic and phonon energy spectra, e.g., those measured for the various carbon-related materials. For example, The thermal examinations on the AAA- and ABC-stacked graphites [simple hexagonal and rhombohedral configurations] are lacking so far, while they have been conducted for the AB-stacked graphite (Bernal type), including electronic specific heat [77, 80] and lattice specific heat [78, 79]. The previous experiments clearly show that the former presents a linear T-dependence at very low temperature [$T < 1.2$ K; [77]], and the latter follows the T^3 law at low temperature and approximately T^2 law at higher temperature [78]. The applications of graphite systems promise an economical way to a new class of efficient thermal management materials, such as the relatively high thermal conductivities are very suitable for thermal interface applications [81, 82]. Specifically, the magneto-electronic specific heats of conventional quantum Hall systems have been successfully verified for many years. In the research on GaAs–GaAlAs [83], the thermal properties are accurately measured under the influences of temperature and magnetic field. For the temperature-dependent specific heat [84], the specific heat reveals a peak structure at a low critical temperature. Furthermore, the magnetic-field dependence heat obviously presents an oscillation behavior [83]. These two kinds of behaviors directly reflect the main characteristics of the Landau levels, such as the filling factors, the energy spacings, and the Zeeman effects. The similar examinations could be generalized to the emergent layered material in fully understanding the diverse thermal phenomena under the various magnetic quantizations.

The magneto-electronic specific heats could present another magnetic quantization phenomenon. The Landau levels in monolayer graphene are very suitable for a model investigation. It is well known that the hexagonal crystal structure accounts for the unique low-energy electronic properties, the isotropic and linear valence and conduction bands intersecting at the Dirac point and the square-dependent Landau-level energy spectrum of $E^{c,v} \propto \sqrt{n^{c,v}B_z}$. The Zeeman effect further splits the magneto-electronic states into the spin-up and spin-down ones. In general, the level spacings between two neighboring quantum numbers are much wider than the Zeeman splitting energy. This clearly indicates that the former is insensitive to temperature. Furthermore, the latter might be comparable to the thermal energy at low temperature. The Zeeman effect is expected to play an important role in the thermal properties. On the other hand, monolayer graphene could be rolled up to become a hollow cylinder (a single-walled carbon nanotube). Apparently, the periodic boundary condition is responsible for the creation of one-dimensional energy bands with linear and parabolic energy dispersions. However, it is almost impossible to induce the dispersionless Landau levels in nanotube surfaces under various high magnetic fields [85]. Dimensionality and magnetic quantization have a great influence on the van Hove singularities of the DOS, so the magneto-electronic specific heats would be very different between these two systems.

2.4   ELECTRON ENERGY LOSS SPECTROSCOPY AND LIGHT INELASTIC SCATTERINGS

Electron energy loss spectroscopy [EELS; [86–97] and inelastic x-ray scatterings [IXS; [107–124]] are the only two available tools in delicately examining excitation spectra of condensed-matter systems. They have successfully identified the reliable Coulomb excitations and phonon dispersion spectra in any dimensional materials, being supported by the theoretical predictions [88–90, 92, 94, 95]. In general, EELS is achieved by the scattered reflection [98–100] and transmission [101, 102, 104–106, 115] of a narrow electron beam, in which the former is very suitable for thoroughly exploring the low-energy excitation properties lower than 1 eV [98]. Reflection ELLS (RELLS) is chosen for a clear illustration. This outstanding instrument extracts the bulk or surface energy loss functions from the backward electrons scattered from a specific sample [115]. If the incident electron beam has a kinetic energy of several hundred eVs, the very efficient technique can provide the screened response functions with an energy resolution of a few meVs, being sufficient

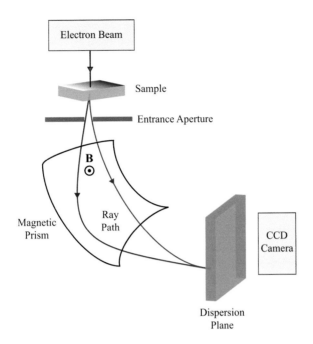

FIGURE 2.5   Reflection spectroscopy of electron energy loss.

in resolving atomic and electronic excitation modes [104, 115]. Up to date, EELS is widely utilized to explore the rich physical, chemical, and material properties of condensed-matter surfaces [102]. It frequently operates under a 25-meV energy resolution within the energy range between 15 and 70 eV [102], while controlling the momentum resolution down to 0.013 Å^{-1}, about 1% of a typical Brillouin zone [102]. Moreover, the energy resolution could be greatly enhanced to $\sim$ 1 meV under the specific condition, in which much weaker electron beams are adopted using a high-resolution monochromator at an ultrahigh vacuum base pressure ($\sim 2 \times 10^{-10}$ Torr). The momentum-dependent dispersion relations of collective excitation modes can be accurately detected by an angle-resolved EELS, which is performed at the low kinetic energy and utilizes an analyzer to measure the scattered electrons. The delicate analyzer consists of a magnetic-prism system, as clearly displayed in Fig. 2.5, where the commercially available Gatan spectrometer is installed beneath the camera and the basic interface and ray paths are shown as well. The surface of this prism is curved to largely reduce the spherical and chromatic aberrations. The scattered electrons in the drift tube are deflected by a uniform magnetic field into a variable entrance aperture (typical variation from 1 to 5 mm in diameter). The electrons that lose more energies deflect further away from the zero-energy-loss electrons according to the Lorenz force law. Furthermore, all the electrons in any direction are focused on the dispersion plane of the spectrometer. The magnetic prism projects the electron energy-loss spectrum onto a CCD camera, which is straightforward to capture the whole energy distribution simultaneously. It is possible to modulate the resolution of the transferred momentum by varying the half-angle of the incident beam in transmission electron microscope (TEM) and the scattering one in the spectrometer. For example, based on the angle variation of a few milliradians, the momentum resolution lies in the order of ~ 0.01 Å^{-1} [101, 102, 115].

The high-resolution EELS measurements are very powerful in thoroughly exploring electronic Coulomb excitation modes in various condensed-matter systems. They have clearly identified the significant single-particle and collective excitations in the sp^2-bonding graphene-related materials, such as 3D layered graphites [86–88], graphite intercalation compounds [89], 2D monolayer and few-layer graphene systems [90–93], 1D single- and multi-walled carbon nanotubes [94], graphene nanoribbons [95], 0D C_{60}-associated fullerenes [96], and carbon onions [97]. In general, the distinct dimensions, geometric symmetries, stacking configurations, interlayer atomic interactions, temperature, and chemical dopings might induce the high/some/few free conduction-electron [valence-hole] density, leading to the low-frequency acoustic or optical plasmon modes with frequencies lower than 1.5 eV. For room temperature, the AB-stacked graphite presents the low-frequency optical plasmons at ~ 45–50 meV and 128 meV under the long wavelength limit, corresponding to the electric polarizations parallel and perpendicular to the (x, y)-plane, respectively [86]. Furthermore, the detailed analyses on temperature-dependent energy loss spectra verify the strong temperature effects in the former [86]. The high-density free electrons and holes, which are, respectively, doped in the donor- and acceptor-type graphite intercalation compounds, lead to the ~ 1-eV optical plasmons [89]. This represents their coherent collective oscillations in the periodical graphitic layers, strongly depending on the transferred momenta [89]. Concerning the layered graphene systems, with the metal-adatom chemisorptions [90–93], the low-frequency plasmon modes become 2D acoustic ones, i.e., they have the $\sqrt{q}$ dependence at small transferred momenta [90–93]. The above-mentioned important results are consistent with the theoretical illustrations [88–90, 92, 94, 95].

IXS are capable of directly detecting the behaviors of dynamic charge screenings in crystal/non-crystal systems. They are successfully conducted on a wide range of physical phenomena, e.g., phonon energy dispersions in solids [107, 108], carrier dynamics of disordered materials [109–112] & biological structures [107], and electronic excitations in condensed-matter systems [113–119]. The transferred energy and momentum are available parameters and can cover the whole spectra of the dielectric responses. The uniform x-ray beam line is designed to provide a super high photon flux within the typical wave-vector range of the first Brillouin zone. Furthermore, photon energy is distributed from 4.9 to 15 keV, being accompanied by energy and momentum resolutions of ~ 70 meV and $0.02 - 0.03$ Å^{-1}, respectively. Specifically, the IXS instrument in the Swiss Light Source possesses an extremely good energy resolution of ~ 30 meV [125]. The super high resolution, ~ 10 meV, is expected to be achieved in the further development of new synchrotron sources. IXS is available in fully exploring all kinds of electronic excitations because the electronic charges strongly interact with the high-energy photon beam under the various transferred momenta and energies. Using hard x-ray synchrotron sources, this spectroscopy becomes a very powerful technique to observe the intrinsic properties of condensed-matter systems, and it is suitable for the external electric and magnetic fields [126]. The analyzer, which is built on the basis of Bragg optics, as depicted in

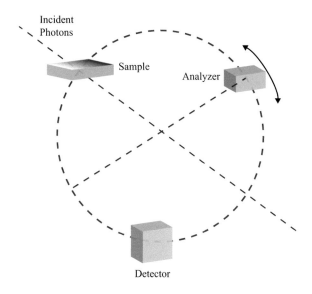

FIGURE 2.6   Inelastic X-ray scattering instrument.

Fig. 2.6, efficiently collects and analyzes the energy and momentum distributions of the scattered photons with a small solid angle $[d\Omega]$. This instrument can provide full information on the screened response function. To maximize the light intensity, a spherically bent analyzer ($\sim$ 10 cm in diameter) is frequently utilized to capture all the scattered radiation of the momentum-transferred photons in $d\Omega$. The transferred energy is projected onto a CCD detector, and then the full energy loss spectrum is scanned by varying the Bragg angle of the specific crystal. When IXS is operated in the Rowland circle geometry, the specific measurement of the double differential scattering cross-section corresponds to the screened electronic excitations in the characteristic energy-loss regime by using the dissipation-fluctuation theorem.

There exist certain important differences between EELS and IXS in terms of the physical environments. The incident electron and photon beams can be, respectively, focused on the spatial range of $\sim$ 10 Å [127] and 100 Å [128]. The former possesses more outstanding resolutions for the transferred momenta and energies during the many-particle interactions. Much more inelastic scattering events are conducted by using the EELS within a short time, i.e., the EELS measurements on the full excitation spectra are performed more quickly and accurately. EELS is very reliable for the low-dimensional materials and nano-scale structures, mainly owing to the simultaneous identifications on their positions. On the other hand, IXS, which arises from the continuous synchrotron radiation sources, presents a rather strong intensity with the tunable energies and momenta. Most importantly, the extreme surrounding environments, induced by the applications of magnetic and electric fields as well various temperatures and pressures, can be overcome under the inelastic light scattering. The external fields have strong effects on the incident charges and the sample chamber is too narrow; therefore, EELS cannot work under such environments. As a result, the high-resolution IXS measurements are very useful in thoroughly understanding the magneto-electronic single-particle and collective excitations in any condensed-matter systems, especially for those in graphene-related materials.

REFERENCES

1. Binnig G and Rohrer H 1986 Scanning tunneling microscopy *IBM J. Res. Dev.* **30** 355.
2. Miccoli I, Aprojanz J, Baringhaus J, Lichtenstein T, Galves L A , Lopes J M J, et al., 2017 Quasi-free-standing bilayer graphene nanoribbons probed by electronic transport *Appl. Phys. Lett.* **110** 051601.
3. Liu J Z, Li B W, Tan Y Z, Giannakopoulos A, Sanchez-Sanchez C, Beljonne D, et al., 2015 Toward cove-edged low band gap graphene nanoribbons *J. Am. Chem. Soc.* **137** 6097.
4. Kim K, Sussman A, and Zettl A. 2010 Graphene nanoribbons obtained by electrically unwrapping carbon nanotubes *ACS Nano* **4** 1362.
5. Vo T H, Shekhirev M, Kunkel D A, Morton M D, Berglund E, Kong L, et al., 2014 Large-scale solution synthesis of narrow graphene nanoribbons *Nat. Comm.* **5** 3189.
6. Patra N, Wang B, and Kral P 2009 Nanodroplet Activated and Guided Folding of Graphene Nanostructures *Nano Lett.* **9** 3766.
7. Song J J, Zhang H J, Cai Y L, Zhang Y X, Bao S N, and He P M 2016 Bottom-up fabrication of graphene nanostructures on Ru(10$\bar{1}$0) *Nanotechnology* **27** 055602.
8. Que Y, Xiao W, Chen H, Wang D, Du S, and Gao H-J 2015 Stacking-dependent electronic property of trilayer graphene epitaxially grown on Ru (0001) *Appl. Phys. Lett.* **107** 263101.
9. Xu R, Yin L J, Qiao J B, Bai K K, Nie J C, and He L 2015 Direct probing of the stacking order and electronic spectrum of rhombohedral trilayer graphene with scanning tunneling microscopy *Phys. Rev. B* **91** 035410.
10. Pierucci D, Sediri H, Hajlaoui M, Girard J C, Brumme T, Calandra M, et al., 2015 Evidence for flat bands near the Fermi level in epitaxial rhombohedral multilayer graphene *ACS Nano* **9** 5432.
11. Rong Z Y and Kuiper P 1993 Electronic effects in scanning tunneling microscopy: Moire pattern on a graphite surface *Phys. Rev. B* **48** 17427.
12. Campanera J M, Savini G, Suarez-Martinez I, and Heggie M I 2007 Density functional calculations on the intricacies of Moire patterns on graphite *Phys. Rev. B* **75** 235449.
13. Cherkez V, Trambly de Laissardiere G, Mallet P, and Veuillen J Y 2015 Van Hove singularities in doped twisted graphene bilayers studied by scanning tunneling spectroscopy *Phys. Rev. B* **91** 155428.
14. Yan W, Liu M, Dou R F, Meng L, Feng L, Chu Z D, et al., 2012 Angle-dependent van Hove singularities in a slightly twisted graphene bilayer *Phys. Rev. Lett.* **109** 126801.
15. Simonov K A, Vinogradov N A, Vinogradov A S, Generalov A V, Zagrebina E M, Svirskiy G I, et al., 2015 From graphene nanoribbons on Cu(111) to nanographene on Cu(110): critical role of substrate structure in the bottom-up fabrication strategy. *ACS Nano* **9** 8997.
16. de Parga A L V, Calleja F, Borca B, Passeggi M C G, Hinarejos J J, Guinea F, et al., 2008 Periodically rippled graphene: growth and spatially resolved electronic structure *Phys. Rev. Lett.* **100** 056807.
17. Varchon F, Mallet P, Veuillen J-Y, and Magaud L 2008 Ripples in epitaxial graphene on the Si-terminated SiC(0001) surface *Phys. Rev. B* **77** 235412.

18. Eelbo T, Wániowska M, Thakur P, Gyamfi M, Sachs B, Wehling T O, Forti S, Starke U, Tieg C, Lichtenstein A I, and Wiesendanger R 2016 Adatoms and clusters of 3d transition metals on graphene: electronic and magnetic configurations *Phys. Rev. Lett.* **110** 136804.
19. Xu P, Yang Y R, Barber S D, Schoelz J K, Qi D, Ackerman M L, et al., 2012 New scanning tunneling microscopy technique enables systematic study of the unique electronic transition from graphite to graphene *Carbon* **50** 4633.
20. Zeinalipour-Yazdi C D and Pullman D P 2008 A new interpretation of the scanning tunneling microscope image of graphite *Chem. Phys.* **348** 233.
21. Li G, Luican A, and Andrei E Y 2009 Scanning tunneling spectroscopy of graphene on graphite *Phys. Rev. Lett.* **102** 176804.
22. Yin L J, Li S Y, Qiao J B, Nie J C, and He L 2015 Landau quantization in graphene monolayer, Bernal bilayer, and Bernal trilayer on graphite surface *Phys. Rev. B* **91** 115405.
23. Miller D L, Kubista K D, Rutter G M, Ruan M, de Heer W A, First P N, et al., 2009 Observing the quantization of zero mass carriers in graphene *Science* **324** 924.
24. Matsui T, Kambara H, Niimi Y, Tagami K, Tsukada M, and Fukuyama H 2005 STS observations of Landau levels at graphite surfaces *Phys. Rev. Lett.* **94** 226403.
25. Li G and Andrei E Y 2007 Observation of Landau levels of Dirac fermions in graphite *Nat. Phys.* **3** 623.
26. Wilder J W G, Venema L C, Rinzler A G, Smalley R E, and Dekker C 1998 Electronic structure of atomically resolved carbon nanotubes *Nature* **391** 59.
27. Huang S, Kim K, Efimkin D K, Lovorn T, Taniguchi T, Watanabe K, et al., 2018 Topologically protected helical states in minimally twisted bilayer graphene *Phys. Rev. Lett.* **121** 037702.
28. Lin C Y, Chen R B, Ho Y H, and Lin M F 2018 Electronic and Optical Properties of Graphite-Related Systems. Boca Raton FL: CRC Press.
29. Huang Y K, Chen S C, Ho Y H, Lin C Y, and Lin M F 2014 Feature-rich magnetic quantization in sliding bilayer graphenes *Sci. Rep.* **4** 7509.
30. Lin C Y, Wu J Y, Chiu Y H, and Lin M F 2014 Stacking-dependent magneto-electronic properties in multilayer graphenes *Phys. Rev. B* **90** 205434.
31. Kravets V G, Grigorenko A N, Nair R R, Blake P, Anissimova S, Novoselov K S, et al., 2010 Spectroscopic ellipsometry of graphene and an exciton-shifted van Hove peak in absorption *Phys. Rev. B* **81** 155413.
32. Mak K F, Shan J, and Heinz T F 2011 Seeing many-body effects in single- and few-layer graphene: observation of two-dimensional saddle-point excitons *Phys. Rev. Lett.* **106** 046401.
33. Martinez L F L, Garcia R C, Navarro R E B, and Martinez A L 2009 Microreflectance difference spectrometer based on a charge coupled device camera: surface distribution of polishing-related linear defect density in GaAs (001) *Appl. Opt.* **48** 5713.
34. Epstein M S, Rains T C 1976 Evaluation of a xenon-mercury arc lamp for background corrextion in atomic-absorption spectrometry *Anal. Chem.* **48** 528.
35. Stamm G L, Denningh R L, and Rockman A G 1969 Some operating characteristics of a xenon and a xenon-mercury short-arc lamp immersed in water *Report of NRL Progress* **33**.
36. Ferguson L G and Dogan F 2001 Spectrally selective, matched emitters for thermophotovoltaic energy conversion processed by tape casting *J. Mater. Sci.* **36** 137.
37. Zhen H L, Li N, Xiong D Y, Zhou X C, Lu W, and Liu H C 2005 Fabrication and investigation of an upconversion quantum-well infrared photodetector integrated with a light-emitting diode *Chin. Phys. Lett.* **22** 1806.
38. Stewart J E and Richmond J C 1957 Infrared emission spectrum of silicon carbide heating elements *J. Res. Natl. Bur. Stand.* **59** 405.

39. Lu S H, Liu W C, and Liu J P 2015 High-axial-resolution, full-field optical coherence microscopy using tungsten halogen lamp and liquid-crystal-based achromatic phase shifter *Appl. Opt.* **54** 4447.

40. Wei J F, Hu X Y, Sun L Q, Zhang K, and Chang Y 2015 Technology for radiation efficiency measurement of high-power halogen tungsten lamp used in calibration of high-energy laser energy meter *Appl. Opt.* **54** 2289.

41. Keppler H, Dubrovinsky L S, Narygina O, and Kantor I 2008 Optical absorption and radiative thermal conductivity of silicate perovskite to gigapascals *Science* **322** 1529.

42. Albert S, Albert K K, Lerch P, and Quack M 2011 Synchrotron-based highest resolution Fourier transform infrared spectroscopy of naphthalene ($C_{10}H_8$) and indole (C_8H_7N) and its application to astrophysical problems *Farad. Discuss.* **150** 71.

43. Gasparian G A and Lucht H 2000 Indium gallium arsenide NIR photodiode array spectroscopy *Spectroscopy* **15** 16.

44. Dessy R E, Nunn W G, Titus C A, and Reynolds W R 1976 Linear photodiode array spectrometers as detector systems in automated liquid chromatographs *J. Chromatogr. Sci.* **14** 195.

45. Podobedov V B, Miller C C, and Nadal M E 2012 Performance of the NIST goniocolorimeter with a broad-band source and multichannel charged coupled device based spectrometer *Rev. Sci. Instrum.* **83** 093108.

46. Kuzmenko A B, van Heumen E, van der Marel D, Lerch P, Blake P, Novoselov K S, et al., 2009 Infrared spectroscopy of electronic bands in bilayer graphene *Phys. Rev. B* **79** 115441.

47. Zhang L M, Li Z Q, Basov D N, and Fogler M M 2008 Determination of the electronic structure of bilayer graphene from infrared spectroscopy *Phys. Rev. B* **78** 235408.

48. Lui C H, Li Z, Mak K F, Cappelluti E, and Heinz T F 2011 Observation of an electrically tunable band gap in trilayer graphene *Nat. Phys.* **7** 944.

49. Mak K F, Shan J, and Heinz T F 2010 Electronic structure of few-layer graphene: experimental demonstration of strong dependence on stacking sequence *Phys. Rev. Lett.* **104** 176404.

50. Taft E A and Philipp H R 1965 Optical properties of graphite *Phys. Rev.* **138** A197.

51. Mak K F, Shan J, and Heinz T F 2011 Seeing many-body effects in single- and few-layer graphene: observation of two-dimensional saddle-point excitons *Phys. Rev. Lett.* **106** 046401.

52. Awaji S, Watanabe K, Oguro H, Miyazaki H, Hanai S, Tosaka T, et al., 2017 First performance test of a 25 T cryogen-free superconducting magnet *Supercond. Sci. Technol.* **30** 065001.

53. Takahashi M, Iwai S, Miyazaki H, Tosaka T, Tasaki K, Hanai S, et al., 2017 Design and test results of a cryogenic cooling system for a 25-T cryogen-free superconducting magnet *IEEE Trans. Appl. Supercond.* **27** 4603805.

54. Sakakura R, Matsuda Y H, Tokunaga M, Kojima E, and Takeyama S 2010 Application of an electro-magnetic induction technique for the magnetization up to 100 T in a vertical single-turn coil system *J. Low Temp. Phys.* **159** 297.

55. Zaric S, Ostojic G N, Kono J, Shaver J, Moore V C, Strano M S, et al., 2004 Optical signatures of the Aharonov-Bohm phase in single-walled carbon nanotubes *Science* **304** 1129.

56. Orlita M, Faugeras C, Martinez G, Maude D K, Sadowski M L, and Potemski M 2008 Dirac fermions at the H Point of graphite: magnetotransmission studies *Phys. Rev. Lett.* **100** 136403.

57. Toyt W W and Dresselhaus M S 1977 Minority carriers in graphite and the H-point magnetoreflection spectra *Phys. Rev. B* **15** 4077.

58. Berciaud S, Potemski M, and Faugeras C 2014 Probing electronic excitations in mono- to pentalayer graphene by micro magneto-Raman spectroscopy *Nano Lett.* **14** 4548.

59. Taft E A and Philipp H R 1965 Optical properties of graphite *Phys. Rev.* **138** A197.

60. Chuang K-C, Baker A M R, and Nicholas R J 2009 Magnetoabsorption study of Landau levels in graphite *Phys. Rev. B* **80** 161410(R).

61. Orlita M, Faugeras C, Schneider J M, Martinez G, Maude D K, and Potemski M 2009 Graphite from the viewpoint of Landau level spectroscopy: an effective graphene bilayer and monolayer *Phys. Rev. Lett.* **102** 166401.

62. Goncharuk N A, Nádvornik L, Faugeras C, Orlita M, and Smrčka L 2012 Infrared magnetospectroscopy of graphite in tilted fields *Phys. Rev. B* **86** 155409.

63. Orlita M, Faugeras C, Barra A-L, Martinez G, Potemski M, Basko D M, et al., 2015 Infrared magneto-spectroscopy of two-dimensional and three-dimensional massless fermions: a comparison *J. App. Phys.* **117** 112803.

64. Hall E H 1879 On a new action of the magnet on electric currents *American Journal of Mathematics* **2** 287-292.

65. Nitta J, Akazaki T, Takayanagi H, and Enoki T 1997 Gate control of spin-orbit interaction in an inverted In0.53Ga0.47As/In0.52Al0.48 as heterostructure *Phys. Rev. Lett.* **78** 1335.

66. Novoselov K S, McCann E, Morozov S V, Fal'ko V I, Katsnelson M I, Zeitler U, et al., 2006 Unconventional quantum Hall effect and Berry's phase of 2π in bilayer graphene. *Nat. Phys.* **2** 177.

67. Sanchez-Yamagishi J D, Taychatanapat T, Watanabe K, Taniguchi T, Yacoby A, and Jarillo-Herrero P 2012 Quantum Hall effect, screening, and layer-polarized insulating states in twisted bilayer graphene *Phys. Rev. Lett.* **108** 076601.

68. Henriksen E A, Nandi D, Eisenstein J P 2012 Quantum Hall effect and semimetallic behavior of dual-gated ABA-stacked trilayer graphene *Phys. Rev. X* **2** 011004.

69. Yuan S, Roldan R, and Katsnelson M I 2011 Landau level spectrum of ABA- and ABC-stacked trilayer graphene *Phys. Rev. B* **84** 125455.

70. Kumar A, Escoffier W, Poumirol J M, Faugeras C, Arovas D P, Fogler M M, et al., 2011 Integer quantum hall effect in trilayer graphene *Phys. Rev. Lett.* **107** 126806.

71. Hohne G, Hemminger W F, and Flammersheim H J 2003 Differential scanning calorimetry *Springer Science & Business Media*, New York.

72. Gill P S, Sauerbrunn S R, and Reading M 1993 Modulated differential scanning calorimetry *J. Therm. Anal.* **40** 931.

73. Schuller M, Shao Q, and Lalk T 2015 Experimental investigation of the specific heat of a nitrate–alumina nanofluid for solar thermal energy storage systems *International J. Therm. Sci.* **91** 142.

74. Shinzato K and Baba T 2001 A laser flash apparatus for thermal diffusivity and specific heat capacity measurements *J. Therm. Anal. Calorim.* **64** 413.

75. Zhou K, Wang H P, Chang J, and Wei B 2015 Experimental study of surface tension, specific heat and thermal diffusivity of liquid and solid titanium *Chem. Phys. Lett.* **639** 105.

76. Kover M, Behulova M, Drienovsky M, and Motycka P 2015 Determination of the specific heat using laser flash apparatus *J. Therm. Anal. Calorim.* **122** 151.

77. Van der Hoeven Jr B J C and Keesom P H 1963 Specific heat of various graphites between 0.4 and 2.0 K *Phys. Rev.* **130** 1318.

78. Krumhansl J and Brooks H 1953 The lattice vibration specific heat of graphite *J. Chem. Phys.* **21** 1663.

79. DeSorbo W and Tyler W W 1953 The specific heat of graphite from 13 to 300 K *J. Chem. Phys.* **21** 1660.

80. Bowman J C and Krumhansl J A 1958 The low-temperature specific heat of graphite *J. Phys. Chem. Solids.* **6** 367.

81. Yu A, Ramesh P, Itkis M E, Bekyarova E, and Haddon R C 2007 Graphite nanoplatelet-epoxy composite thermal interface materials *J. Phys. Chem. C* **111** 7565.

82. Lin C and Chung D D L 2009 Graphite nanoplatelet pastes vs. carbon black pastes as thermal interface materials *Carbon* **47** 295.

83. Gornik E, Lassnig R, Strasser G, Stormer H L, Gossard A C, and Wiegmann W 1985 Specific heat of two-dimensional electrons in GaAs-GaAlAs multilayers *Phys. Rev. Lett.* **54** 1820.

84. Bayot V, Grivei E, Melinte S, Santos M B, and Shayegan M 1996 Giant low temperature heat capacity of GaAs quantum wells near Landau level filling $\nu = 1$ *Phys. Rev. Lett.* **76** 4584.

85. Lin M F and Shung K W K 1995 Magnetoconductance of carbon nanotubes *Phys. Rev. B* **51** 7592.

86. Zeppenfeld K 1971 Nonvertical interband transitions in graphite by intrinsic electron scattering *Z. Phys.* **243** 229.

87. Dovbeshko G I, Romanyuk V R, Pidgirnyi D V, Cherepanov V V, Andreev E O, Levin V M, et al., 2015 Optical properties of pyrolytic carbon films versus graphite and graphene *Nanoscale Res. Lett.* **10** 234.

88. Marinopoulos A G, Reining L, Olevano V, Rubio A, Pichler T, Liu X, et al., 2002 Anisotropy and interplane interactions in the dielectric response of graphite *Phys. Rev. Lett.* **89** 076402.

89. Fischer J E, Bloch J M, Shieh C C, Preil M E, and Jelley K 1985 Reflectivity spectra and dielectric function of stage-1 donor intercalation compounds of graphite *Phys. Rev. B* **31** 4773.

90. Wachsmuth P, Hambach R, Kinyanjui M K, Guzzo M, Benner G, and Kaiser U 2013 High-energy collective electronic excitations in free-standing single-layer graphene *Phys. Rev. B* **88** 075433.

91. Politanoa A, Radović I, Borkab D, Mišković Z L, Yu H K, Farías D, et al., 2017 Dispersion and damping of the interband π plasmon in graphene grown on Cu(111) foils *Carbon* **114** 70.

92. Liou S C, Shie C-S, Chen C H, Breitwieser R, Pai W W, Guo G Y, et al., 2015 π-plasmon dispersion in free-standing graphene by momentum-resolved electron energy-loss spectroscopy *Phys. Rev. B* **91** 045418.

93. Hage F S, Hardcastle T P, Gjerding M N, Kepaptsoglou D M, Seabourne C R, Winther K T, et al., 2018 Local plasmon engineering in doped graphene *ACS Nano* **12** 1837.

94. Knupfera M, Pichlera T, Goldena M S, Finka J, Rinzlerb A, and Smalley R E, Electron energy-loss spectroscopy studies of single wall carbon nanotubes *Carbon* **37** 733.

95. Suenaga K and Koshino M 2010 Atom-by-atom spectroscopy at graphene edge *Nature* **468** 1088.

96. Lucas A A, Henrad L, and Lambin Ph 1994 Computation of the ultraviolet absorption and electron inelastic scattering cross section of multishell fullerenes *Phys. Rev. B* **49** 2888.

97. Tomita S, Fujii M, Hayashi S, and Yamamoto K 1999 Electron energy-loss spectroscopy of carbon onions *Chem. Phys. Lett.* **305** 225.

98. Went M R, Vos M, and Werner W S M 2008 Extracting the Ag surface and volume loss functions from reflection electron energy loss spectra *Sur. Sci.* **602** 2069.

99. Werner W S M 2006 Dielectric function of Cu, Ag, and Au obtained from reflection electron energy loss spectra, optical measurements, and density functional theory *Appl. Phys. Lett.* **89** 213106.

100. Werner W S M, Went M R, and Vos M 2007 Surface plasmon excitation at a Au surface by 150–40000 eV electrons *Surf. Sci.* **601** L109.
101. Egerton R F 2007 Limits to the spatial, energy and momentum resolution of electron energy-loss spectroscopy *Ultramicroscopy* **107** 575.
102. Ibach H and Mills D L 1982 Electron energy-loss spectroscopy and surface vibrations spectroscopy, *Academic*, New York.
103. Brink H A, Barfels M M G, Burgner R P, and Edwards B N 2003 A sub-50 meV spectrometer and energy filter for use in combination with 200 kV monochromated (S)TEMs *Ultramicroscopy* **367** 367.
104. Su D S, Zandbergen H W, Tiemeijer P C, Kothleitner G, Havecker M, Hebert C, et al., 2003 High resolution EELS using monochromator and high performance spectrometer: comparison of V2O5 ELNES with NEXAFS and band structure calculations *Micron* **34** 235.
105. Terauchi M, Tanaka M, Tsuno K, and Ishida M 1999 Development of a high energy resolution electron energy-loss spectroscopy microscope *J. Microsc.* **194** 203.
106. Lazar S, Botton G A, and Zandbergen H W 2006 Enhancement of resolution in coreloss and low-loss spectroscopy in a monochromated microscope *Ultramicroscopy* **106** 1091.
107. W. Schülke 2007 Electron Dynamics by Inelastic X-ray Scattering. Oxford: Oxford University Press.
108. Mohr M, Maultzsch J, Dobardžić E, Reich S, Miloževič I, Damnjanovićq M, et al., 2007 Phonon dispersion of graphite by inelastic x-ray scattering, Phonon dispersion of graphite by inelastic x-ray scattering *Phys. Rev. B* **76** 035439.
109. Kimura K, Matsuda K, Hiraoka N, Fukumaru T, Kajihara Y, Inui M, et al., 2014 Inelastic x-ray scattering study of plasmon dispersions in solid and liquid Rb *Phys. Rev. B* **89** 014206.
110. Tirao G, Stutz G, Silkin V M, Chulkov E V, and Cusatis C 2007 Plasmon excitation in beryllium: inelastic x-ray scattering experiments and first-principles calculations *J. Phys.: Condens. Matter* **19** 046207.
111. Kimura K, Matsuda K, Hiraoka N, Kajihara Y, Miyatake T, Ishiguro Y, et al., 2015 Inelastic x-ray scattering study on plasmon dispersion in liquid Cs *J. Phys. Soc. Jpn.* **84** 084701.
112. Galambosi S, Soininen J A, Mattila A, Huotari S, Manninen S, Vanko Gy, et al., 2005 Inelastic x-ray scattering study of collective electron excitations in MgB_2 *Phys. Rev. B* **71** 060504.
113. Hambach R, Giorgetti C, Hiraoka N, Cai Y Q, Sottile F, Marinopoulos A G, et al., 2008 Anomalous angular dependence of the dynamic structure factor near Bragg reflections: graphite *Phys. Rev. Lett.* **101** 266406.
114. Charles Kittel 2004 Introduction to solid state physics 8th Edition Wiley, New York.
115. Brink H A, Barfels M M G, Burgner R P, and Edwards B N 2003 A sub-50 meV spectrometer and energy filter for use in combination with 200 kV monochromated (S)TEMs *Ultramicroscopy* **96** 367.
116. Lin M F, Huang C S, and Chuu D S 1997 Plasmons in graphite and stage-1 graphite intercalation compounds *Phys. Rev. B* **55** 13961.
117. Scholz A, Stauber T, and Schliemann J 2012 Dielectric function, screening, and plasmons of graphene in the presence of spin-orbit interactions *Phys. Rev. B* **86** 195424.
118. Schulke W, Bonse U, Nagasawa H, Kaprolat A, and Berthold A 1988 Interband transitions and core excitation in highly oriented pyrolytic graphite studied by inelastic synchrotron x-ray scattering: band-structure information *Phys. Rev. B* **38** 2112.

119. Zhang L, Schwertfager N, Cheiwchanchamnangij T, Lin X, Glans-Suzuki P-A, Piper L F J, et al., 2012 Electronic band structure of graphene from resonant soft x-ray spectroscopy: the role of core-hole effects *Phys. Rev. B* **86** 245430.

120. Gao X, Burns C, Casa D, Upton M, Gog T, Kim J, et al., 2011 Development of a graphite polarization analyzer for resonant inelastic x-ray scattering *Rev. Sci. Instrum.* **82** 113108.

121. Huotari S, Albergamo F, Vanko Gy, Verbeni R, and Monaco G 2006 Resonant inelastic hard x-ray scattering with diced analyzer crystals and positionsensitive detectors *Rev. Sci. Instrum.* **77** 053102.

122. Kotani A and Shin S 2001 Resonant inelastic x-ray scattering spectra for electrons in solids *Rev. Mod. Phys.* **73** 203.

123. Devereaux T P 2007 Inelastic light scattering from correlated electrons *Rev. Mod. Phys.* **79** 175.

124. Ament L J P, van Veenendaal M, Devereaux T P, Hill J P, and van den Brink J 2001 Resonant inelastic x-ray scattering studies of elementary excitations *Rev. Mod. Phys.* **83** 705.

125. Strocov V N, Schmitt T, Flechsig U, Schmidt T, Imhof A, Chen Q, et al., 2010 High-resolution soft x-ray beamline ADRESS at the Swiss Light Source for resonant inelastic x-ray scattering and angle-resolved photoelectron spectroscopies. *J. Synchrotron. Radiat.* **17** 631.

126. Qiao R, Li Q, Zhuo Z, Sallis S, Fuchs O, Blum M, et al., 2017 High-efficiency in situ resonant inelastic x-ray scattering (iRIXS) endstation at the Advanced Light Source *Rev. Sci. Instrum.* **88** 033106.

127. Egerton R F 2011 Electron energy-loss spectroscopy in the electron microscope *Springer Science & Business Media*, Plenum, New York and London.

128. Egerton R F 2007 Limits to the spatial, energy and momentum resolution of electron energy-loss spectroscopy *Ultramicroscopy* **107** 575.

3 Theoretical Models

Jhao-Ying Wu,[b] Chiun-Yan Lin,[a] Thi-Nga Do,[c,d]
Po-Hsin Shih,[a] Shih-Yang Lin,[e] Ching-Hong Ho,[b]
Ming-Fa Lin[a,f,g]

[a] Department of Physics, National Cheng Kung University,
Tainan 701, Taiwan
[b] Center of General Studies, National Kaohsiung University of
Science and Technology, Kaohsiung 811, Taiwan
[c] Laboratory of Magnetism and Magnetic Materials, Advanced
Institute of Materials Science, Ton Duc Thang University,
Ho Chi Minh City, Vietnam
[d] Faculty of Applied Sciences, Ton Duc Thang University,
Ho Chi Minh City, Vietnam
[e] Department of Physics, National Chung Cheng University,
Chiayi 621, Taiwan
[f] Quantum Topology Center, National Cheng Kung University,
Tainan 701, Taiwan
[g] Hierarchical Green-Energy Materials Research Center,
National Cheng Kung University, Tainan, Taiwan

CONTENTS

The further modifications on the theoretical models are necessary for the emergent layer materials. The modified & new theoretical framework is available in fully exploring the fundamental physical properties and thus creating the rich and unique phenomena, being clearly illustrated in the following sections.

3.1 THE GENERALIZED TIGHT-BINDING MODELS FOR TYPICAL LAYERED SYSTEMS

To fully comprehend the essential electronic properties of 2D condensed matter with the significant geometric symmetries, the intrinsic atomic interactions, and any external fields, we propose the generalized tight-binding model to diagonalize the various Hamiltonians efficiently. Furthermore, this model is accompanied with the concise physical pictures for the thorough understanding of the diverse physical phenomena. The main-stream typical systems, few-layer graphene, silicene, germanene, and phosphorene ones, are very suitable for a model study. The planar/buckled/puckered and layered structures, layer numbers, stacking configurations, and geometric modulations, are taken into account. The lattice- and atom-induced important interactions, the distinct site energies, the single- or multi-orbital hybridizations for the low-lying energy bands, the strong/weak spin-orbital couplings [SOCs], and the intralayer and interlayer hopping integrals are included in the various Hamiltonians. The field-induced independent Hamiltonian matrix elements will be derived in the analytic form, especially for those under a uniform perpendicular magnetic field. Finally, a giant magnetic Hermitian matrix can be accurately solved by the exact diagonalization method.

The bilayer AB-bt silicene system is very suitable for a full understanding of the magneto-electronic theoretical model, mainly owing to the observable buckling structure, the comparable intralayer and interlayer hopping integrals, and the significant layer-dependent spin-orbital interactions. The sp^2 chemical bondings still dominate the nonplanar honeycomb lattice; therefore, the π bondings due to the $3p_z$ orbitals of silicon atoms are responsible for the low-energy fundamental properties. The generalized tight-binding model will be developed to investigate the feature-rich electronic properties of AB-bt bilayer silicene system. The external magnetic and electric fields are included in the calculations simultaneously. This honeycomb structure, as shown in Figs. 3.1(a) and 3.1(b) (the top and side views), possesses four silicon atoms in a unit cell, in which the two primitive unit vectors, $\mathbf{a_1}$ and $\mathbf{a_2}$, have a lattice constant of $a = 3.86$ Å [1]. Apparently, the well-stacked bilayer silicene consists of four sublattices of (A^1, B^1) and (A^2, B^2). For each layer, the two sublattices lie on two different buckling planes with a separation of $l_z = 0.46$ Å. The B^1 and B^2 sublattices are situated at the higher and lower planes, respectively, where the interlayer distance is 2.54 Å. The buckled angle due to the intralayer Si-Si bond and the z axis is $\theta = 78.3°$.

The low-energy Hamiltonian, which is closely related to the silicon-$3p_z$ orbitals, built from the tight-binding model covering the intralayer and interlayer atomic interactions, and two kinds of layer-dependent spin-orbital interactions. Among all the intrinsic interactions, the intralayer hopping integral and two kinds of SOCs are similar to those in monolayer silicene [2, 3]. On the other side, such a bilayer system exhibits the layer-environment-induced SOCs, a vertical and two non-vertical interlayer atomic interactions, being absent in monolayer silicene. The very complex

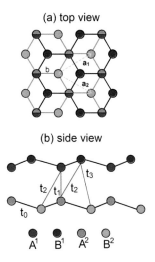

(a) top view

(b) side view

FIGURE 3.1 The geometric structure of bilayer AB-bt silicene systems: (a) top and (b) side views.

Hamiltonian is expressed as

$$H = \sum_{l,l} (\epsilon_l^l + U_l^l) c_{l\alpha}^{\dagger l} c_{l\alpha}^l + \sum_{I,J,\alpha,l,l'} \gamma_{IJ}^{ll'} c_{l\alpha}^{\dagger l} c_{J\alpha}^{l'}$$

$$+ \frac{i}{3\sqrt{3}} \sum_{\langle\langle I,J\rangle\rangle,\alpha,\beta,l} \lambda_l^{SOC} \gamma_l v_{IJ} c_{l\alpha}^{\dagger l} \sigma_{\alpha\beta}^z c_{J\beta}^l$$

$$- \frac{2i}{3} \sum_{\langle\langle I,J\rangle\rangle,\alpha,\beta,l} \lambda_l^R \gamma_l u_{IJ} c_{l\alpha}^{\dagger l} (\vec{\sigma} \times \hat{d}_{IJ})_{\alpha\beta}^z c_{J\beta}^l . \qquad (3.1)$$

$\epsilon_l^l(A^l, B^l)$ represents the sublattice-dependent site energy associated with the chemical environment difference (e.g., $\epsilon_l^l(A^l) = 0$; $\epsilon_l^l(B^l) = -0.12$ eV). $U_l^l(A^l, B^l)$ is the height-created Coulomb potential energy due to a uniform perpendicular electric field. The $c_{l\alpha}^l$ and $c_{l\alpha}^{\dagger l}$ operators, respectively, correspond to the annilation and creation of an electronic state with the spin polarization of α at the I-th site of the l-th layer. The atomic interactions in the second term include the nearest-neighbor intralayer hopping integral ($\gamma_0 = 1.13$ eV) and three interlayer hopping integrals due to (A^1, A^2), (B^1, A^2) or (A^1, B^2), and (B^1, B^2) [$\gamma_1 = -2.2$ eV, $\gamma_2 = 0.1$ eV, and $\gamma_3 = 0.54$ eV in Fig. 3.1(b)]. Quite importantly, the largest interlayer vertical hopping integral of t_1, which originates from the online hybridization of Si-$3p_z$ orbitals (the parallel interaction characterized by the orbital distribution and center-of-mass), induces very strong orbital hybridizations in bilayer silicene. Apparently, t_1 sharply contrasts with t_0 in the single-orbital bonding; that is. the latter belongs to the perpendicular atomic interactions. The traditional SOC (the third term) and the Bychkov-Rashba SOC (the fourth term) are only taken for the next-nearest-neighbor pairs $\langle\langle I, J\rangle\rangle$. $\vec{\sigma}$ is the Pauli spin matrix and $\hat{d}_{IJ} = \vec{d}_{IJ}/|d_{IJ}|$ denotes a unit vector linking the I- and J-th lattice sites. $v_{IJ} = 1/{-1}$ if the next-nearest-neighbor hopping is anticlockwise/clockwise with respect to the positive z axis. $u_{IJKL} = 1$ and -1, respectively, corresponds to the A and B

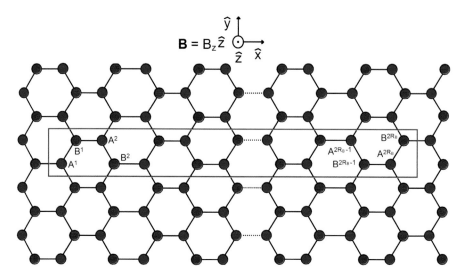

FIGURE 3.2   The enlarged rectangular unit cell of bilayer AB-bt silicene under a unifom perpendicular magnetic field.

sublattices. $\gamma_l = \pm 1$ presents the layer-dependent spin-orbital interactions due to the opposite buckled ordering of AB-bt bilayer silicene. Two kinds of SOCs come to exist in the diagonal elements of the Hermitian matrix. They are fitted as: $\lambda_1^{SOC} = 0.06$ eV, $\lambda_2^{SOC} = 0.046$ eV, $\lambda_1^R = -0.054$ eV, and $\lambda_2^R = -0.043$ eV so that the calculated low-lying energy bands are closed to those from the first-principles method [4,5].

A uniform perpendicular magnetic field creates an extra Peierls phase in the tight-binding function through the vector potential $\vec{A}$. The **B**-induced spatial period is chosen to be the integral times of honeycomb lattice, as clearly illustrated by a yellow-green rectangle in Fig. 3.2. The well-known phase is defined by $G_R = \frac{2\pi}{\phi_0}\int_R^r \vec{A} \cdot d\vec{l}$, where $\phi_0 = hc/e$ is the magnetic flux quantum and $\phi = B_z\sqrt{3}a^2/2$ is the magnetic flux through a hexagon. There are totally $8R_B$ ($R_B = \phi_0/\phi$) Si atoms in an enlarged unit cell. As a result, the magnetic Hamiltonian, being based on the tight-binding functions of the periodical $3p_z$ orbitals, is a $16R_B \times 16R_B$ Hermitian matrix. In general, its dimension is giant under the typical magnetic-field strength in laboratory. For example, the Hamiltonian is a $50,000 \times 50,000$ matrix at $B_z = 10$ T. It should be noticed that the Hamiltonian matrix elements might be real or imaginary numbers, being sensitive to the intrinsic geometric symmetries and orbital hybridizations. The eigenvalues and eigenfunctions are efficiently numerically solved by the exact diagonalization method: the band-like method [the rearrangement of the tight-binding functions in Refs.] and the spatial localizations of the magnetic wave functions [as discussed in Secs. 6.1 & 4.1].

Through the delicate calculations, the independent magnetic Hamiltonian matrix elements are expressed as

$$\langle A_m^{1\uparrow}|H|A_m^{1\uparrow}\rangle = -i\lambda_1^{SOC}(Q_1P_4 - Q_1^*P_4^*) + E_zl_1, \qquad (3.1.1)$$

$$\langle A_m^{1\uparrow}|H|A_m^{1\downarrow}\rangle = i\lambda_1^R(Q_1P_4 - Q_1^*P_4^*), \tag{3.1.2}$$

$$\langle A_m^{1\uparrow}|H|B_m^{1\uparrow}\rangle = \langle A_m^{1\downarrow}|H|B_m^{1\downarrow}\rangle = \gamma_0(Q_3P_2 + Q_3^*P_3), \tag{3.1.3}$$

$$\langle A_m^{1\downarrow}|H|A_m^{1\downarrow}\rangle = i\lambda_1^{SOC}(Q_1P_4 - Q_1^*P_4^*) + E_zl_1, \tag{3.1.4}$$

$$\langle B_m^{1\uparrow}|H|B_m^{1\uparrow}\rangle = i\lambda_2^{SOC}(Q_2P_4 + Q_2^*P_4^*) + E_zl_2 + \gamma_4, \tag{3.1.5}$$

$$\langle B_m^{1\uparrow}|H|B_m^{1\downarrow}\rangle = i\lambda_2^R(Q_2P_4 - Q_2^*P_4^*), \tag{3.1.6}$$

$$\langle B_m^{1\downarrow}|H|B_m^{1\downarrow}\rangle = -i\lambda_2^{SOC}(Q_2P_4 - Q_2^*P_4^*) + E_zl_2 + \gamma_4, \tag{3.1.7}$$

$$\langle A_m^{2\uparrow}|H|A_m^{2\uparrow}\rangle = i\lambda_1^{SOC}(Q_1P_4 - Q_1^*P_4^*) - E_zl_1, \tag{3.1.8}$$

$$\langle A_m^{2\uparrow}|H|A_m^{2\downarrow}\rangle = -i\lambda_1^R(Q_1P_4 - Q_1^*P_4^*), \tag{3.1.9}$$

$$\langle A_m^{2\uparrow}|H|B_m^{2\uparrow}\rangle = \langle A_m^{2\downarrow}|H|B_m^{2\downarrow}\rangle = \gamma_0P_1, \tag{3.1.10}$$

$$\langle A_m^{2\uparrow}|H|B_m^{2\uparrow}\rangle = \langle A_m^{2\downarrow}|H|B_m^{2\downarrow}\rangle = \gamma_0P_1, \tag{3.1.11}$$

$$\langle A_m^{2\downarrow}|H|A_m^{2\downarrow}\rangle = -i\lambda_1^{SOC}(Q_1P_4 - Q_1^*P_4^*) - E_zl_1, \tag{3.1.12}$$

$$\langle B_m^{2\uparrow}|H|B_m^{2\uparrow}\rangle = i\lambda_2^{SOC}(Q_5P_4 - Q_5^*P_4^*) - E_zl_2 + \gamma_4, \tag{3.1.13}$$

$$\langle B_m^{2\uparrow}|H|B_m^{2\downarrow}\rangle = i\lambda_2^R(Q_5P_4 + Q_5^*P_4^*), \tag{3.1.14}$$

$$\langle B_m^{2\downarrow}|H|B_m^{2\downarrow}\rangle = i\lambda_2^{SOC}(Q_5P_4 - Q_5^*P_4^*) - E_zl_2 + \gamma_4, \tag{3.1.15}$$

$$\langle A_m^{1\uparrow}|H|A_m^{2\uparrow}\rangle = \gamma_1, \tag{3.1.16}$$

$$\langle A_m^{1\uparrow}|H|B_m^{2\uparrow}\rangle = \gamma_2P_1, \tag{3.1.17}$$

$$\langle A_m^{1\downarrow}|H|A_m^{2\downarrow}\rangle = \gamma_1, \tag{3.1.18}$$

$$\langle A_m^{1\downarrow}|H|B_m^{2\downarrow}\rangle = \gamma_2P_1, \tag{3.1.19}$$

$$\langle B_m^{1\uparrow}|H|A_m^{2\uparrow}\rangle = \gamma_2(Q_3P_2^* - Q_3^*P_3^*), \tag{3.1.20}$$

$$\langle B_m^{1\uparrow}|H|B_m^{2\uparrow}\rangle = \gamma_3(Q_6P_2 + Q_6^*P_3), \tag{3.1.21}$$

$$\langle B_m^{1\downarrow}|H|A_m^{2\downarrow}\rangle = \gamma_2(Q_3P_2^* - Q_3^*P_3^*), \tag{3.1.22}$$

$$\langle B_m^{1\downarrow}|H|B_m^{2\downarrow}\rangle = \gamma_3(Q_6P_2 + Q_6^*P_3), \tag{3.1.23}$$

$$\langle A_m^{1\uparrow}|H|A_{m+1}^{1\uparrow}\rangle = i\lambda_1^{SOC}(P_6Q_4^* + P_5Q_4), \tag{3.1.24}$$

$$\langle A_m^{1\uparrow}|H|A_{m+1}^{1\downarrow}\rangle = i\lambda_1^R(P_5Q_4 - P_6Q_4^*), \tag{3.1.25}$$

$$\langle A_m^{1\downarrow}|H|A_{m+1}^{1\uparrow}\rangle = i\lambda_1^R\big(P_5Q_4e^{i\pi/6} + P_6Q_4^*e^{-i\pi/6}\big), \tag{3.1.26}$$

$$\langle A_m^{1\downarrow}|H|A_{m+1}^{1\downarrow}\rangle = -i\lambda_1^{SOC}(P_6Q_4^* + P_5Q_4), \tag{3.1.27}$$

$$\langle B_m^{1\uparrow}|H|A_{m+1}^{1\uparrow}\rangle = \gamma_2P_1, \tag{3.1.28}$$

$$\langle B_m^{1\uparrow}|H|B_{m+1}^{1\uparrow}\rangle = i\lambda_2^{SOC}(P_5Q_6 - P_6Q_6^*), \tag{3.1.29}$$

$$\langle B_m^{1\uparrow}|H|B_{m+1}^{1\downarrow}\rangle = i\lambda_2^R\big(P_5Q_6e^{-i\pi/6} + P_6Q_6^*e^{i\pi/6}\big), \tag{3.1.30}$$

$$\langle B_m^{1\uparrow}|H|A_{m+1}^{2\uparrow}\rangle = \gamma_2P_1, \tag{3.1.31}$$

$$\langle B_m^{1\downarrow}|H|A_{m+1}^{1\downarrow}\rangle = \gamma_2P_1, \tag{3.1.32}$$

$$\langle B_m^{1\downarrow}|H|B_{m+1}^{1\uparrow}\rangle = i\lambda_2^R\big(P_5Q_6e^{i\pi/6} - P_6Q_6^*e^{-i\pi/6}\big), \tag{3.1.33}$$

$$\langle B_m^{1\downarrow}|H|B_{m+1}^{1\downarrow}\rangle = i\lambda_2^{SOC}(P_5Q_6 + P_6Q_6^*), \tag{3.1.34}$$

$$\langle B_m^{1\downarrow}|H|A_{m+1}^{2\downarrow}\rangle = \gamma_2P_1, \tag{3.1.35}$$

$$\langle A_m^{2\uparrow}|H|A_{m+1}^{2\uparrow}\rangle = i\lambda_1^{SOC}(P_5Q_4 - P_6Q_4^*), \tag{3.1.36}$$

$$\langle A_m^{2\uparrow}|H|A_{m+1}^{2\downarrow}\rangle = i\lambda_1^R(P_6Q_4^* - P_5Q_4), \tag{3.1.37}$$

$$\langle A_m^{2\downarrow}|H|A_{m+1}^{2\uparrow}\rangle = i\lambda_1^R\big(P_6Q_4^*e^{-i\pi/6} - P_5Q_4e^{i\pi/6}\big), \tag{3.1.38}$$

$$\langle A_m^{2\downarrow}|H|A_{m+1}^{2\downarrow}\rangle = -i\lambda_1^{SOC}(P_5Q_4 + P_6Q_4^*), \tag{3.1.39}$$

$$\langle B_m^{2\uparrow}|H|A_{m+1}^{2\uparrow}\rangle = \gamma_0(P_2Q_6 + P_3Q_6^*), \tag{3.1.40}$$

$$\langle B_m^{2\uparrow}|H|B_{m+1}^{2\uparrow}\rangle = i\lambda_2^{SOC}(P_5Q_7 + P_6Q_7^*), \tag{3.1.41}$$

$$\langle B_m^{2\uparrow}|H|B_{m+1}^{2\downarrow}\rangle = i\lambda_2^R\big(P_6Q_7^*e^{i\pi/6} - P_5Q_7e^{-i\pi/6}\big), \tag{3.1.42}$$

$$\langle B_m^{2\downarrow}|H|A_{m+1}^{2\downarrow}\rangle = \gamma_0(P_2Q_6 + P_3Q_6^*), \tag{3.1.43}$$

$$\langle B_m^{2\downarrow}|H|B_{m+1}^{2\uparrow}\rangle = i\lambda_2^R\big(P_6Q_7^*e^{-i\pi/6} + P_5Q_7e^{i\pi/6}\big), \tag{3.1.44}$$

$$\langle B_m^{2\downarrow}|H|B_{m+1}^{2\downarrow}\rangle = -i\lambda_2^{SOC}(P_5Q_7 - P_6Q_7^*), \tag{3.1.45}$$

$$\langle B_m^{2\uparrow}|H|A_{m+1}^{1\uparrow}\rangle = \gamma_2(P_2Q_6 - P_3Q_6^*), \tag{3.1.46}$$

$$\langle B_m^{2\uparrow}|H|B_{m+1}^{1\uparrow}\rangle = \gamma_3P_1, \tag{3.1.47}$$

$$\langle B_m^{2\downarrow}|H|A_{m+1}^{1\downarrow}\rangle = \gamma_2(P_2 Q_6 - P_3 Q_6^*), \qquad (3.1.48)$$

$$\langle B_m^{2\downarrow}|H|B_{m+1}^{1\downarrow}\rangle = \gamma_3 P_1. \qquad (3.1.49)$$

Moreover, the wave-vector-dependent phase terms are given by

$$P_1 = exp[ik_x b],$$

$$P_2 = exp[i(k_x b/2 + k_y a/2)],$$

$$P_3 = exp[i(k_x b/2 - k_y a/2)],$$

$$P_4 = exp[ik_y a],$$

$$P_5 = exp[i(k_x 3b/2 + k_y a/2)],$$

$$P_6 = exp[i(k_x 3b/2 - k_y a/2)],$$

$$Q_1 = exp[i2\pi\psi(j-1)],$$

$$Q_2 = exp[i2\pi\psi(j-1+2/6)],$$

$$Q_3 = exp[i\pi\psi(j-1+1/6)],$$

$$Q_4 = exp[i\pi\psi(j-1+3/6)],$$

$$Q_5 = exp[i\pi\psi(j-1+4/6)],$$

$$Q_6 = exp[i\pi\psi(j-1+5/6)],$$

$$Q_7 = exp[i\pi\psi(j-1+7/6)],$$

in which j (an integer) indicates the lattice site of the atoms.

After the diagonalization of bilayer magnetic Hamiltonian, the Landau-level wave function, with a specific quantum number n, is given by

$$\Psi(n, \mathbf{k}) = \sum_{l=1,2} \sum_{l=1}^{R_B} \sum_{\alpha} [A_\alpha^{l,l}(n, \mathbf{k})|\psi_\alpha^{l,l}(A)\rangle + B_\alpha^{l,l}(n, \mathbf{k})|\psi_\alpha^{l,l}(B)\rangle]. \qquad (3.2)$$

$\psi_\alpha^{l,l}$ is the $3p_z$-orbital tight-binding function situated at the l-th site of the l-th layer with the spin α configuration; furthermore, $A_\alpha^{l,l}(n, \mathbf{k})$ [$B_\alpha^{l,l}(n, \mathbf{k})$] is the corresponding amplitude on the sublattice-dependent lattice site. Most importantly, all the amplitudes in an enlarged unit cell could be regarded as the continuous spatial distributions of the sub-envelope functions on the distinct sublattice, since the magnetic length ($l_B = \sqrt{hc/eB_z}$) is much smaller/longer than the length of the enlarged unit cell [Fig. 3.2]/the original lattice constant. Such subenvelope functions can provide much information for fully understanding the unusual Landau-level features, such as the sublattice-dominated Landau-level quantum number, the different localization centers, and the frequent crossing & anti-crossing phenomena. For bilayer silicene, the buckled honeycomb structure, the complicated intralayer and interlayer atomic interactions and the significant spin-orbital interactions need to be thoroughly taken into account in the theoretical model calculations. Such a system is expected to exhibit diverse physical properties under various external fields.

In general, there are two kinds of theoretical models in studying the magnetic quantization phenomena, namely, the low-energy elective-mass approximation and the tight-binding model. Concerning the low-energy perturbation method [6–8], the zero-field Hamiltonian matrix elements are expanded about the high-symmetry points (e.g., the K point in graphene). And then, the magnetic quantization is further done from an approximate Hamiltonian matrix, in which the magnetic Bloch wave function is assumed to be the superposition of the well-behaved and sublattice-dependent oscillator states [6]. The zero-field and magnetic Hamiltonian matrices present the same dimension. It should be noticed that certain significant interlayer hopping integrals in 2D layered materials will create much difficulty in the investigation of magnetic quantization; therefore, they are usually ignored under the effective-mass approximation. Consequently, the unique and diverse magnetic quantization phenomena are frequently lost within this method, e.g., the frequent Landau-level anti-crossing phenomenon in ABC-stacked trilayer graphene, and the extra magneto-optical selection rules [9,10]. In general, this perturbation method cannot solve the low-symmetry systems with the non-monotonous energy dispersions and the multi-pair band structures. For AB-bt bilayer silicene, it should be impossible to deal with the low-lying energy bands from the perturbation approximation, mainly owing to very complicated intrinsic interactions and energy bands. That is to say, the effective-mass model is not suitable for expanding the low-energy electronic states from the K and T points simultaneously [discussed later in Fig. 7.1]. This model becomes too cumbersome to generate the further magnetic quantization. Apparently, it is very difficult to comprehend the unusual Landau levels, being attributed to the unique Hamiltonian in bilayer silicene. It is in sharp contrast with the monolayer silicene case.

In the previous theoretical studies, the tight-binding model is developed using the $\vec{k}$-scheme, but not the $\vec{r}$-scheme. The magneto-electronic states are directly built from the original electronic states in the first Brillouin zone (the hexagonal Brillouin zone in silicene/graphene). However, it is not suitable to present the main features of Landau-level wave functions (oscillatory distributions in real space with the distinct localization centers; discussed later). Explicitly, the subenvelope functions could not be identified as the Landau-level wave functions since they are randomly distributed. This scheme is very difficult to explore the fundamental physical properties under the spatially modulated/nonuniform magnetic field, the modulated electric field, and the composite magnetic and electric fields, e.g., the magneto-optical properties and magneto-Coulomb excitations. For the generalized tight-binding model developed in this book, the calculations are based on the layer-dependent sublattices in an enlarged unit cell, in which the magnetic-electronic energy spectra and wave functions are closely linked to the well-known Kubo formula and the modified random-phase approximation [RPA].

3.2   MAGNETO-OPTICAL EXCITATION THEORY

The main features of magneto-electronic energy spectra and Landau-level wave functions will create the diverse optical excitation properties. When an electromagnetic wave exists in a condensed-matter systems, the occupied electronic states are excited to the unoccupied ones according to the conservation of momentum & energy and the Pauli principle. The optical excitations belong to the vertical channels, since the momenta of photon in the experimental frequency range are negligible, compared with the typical wave vectors of electrons. From the dynamic Kubo formula (the Fermi golden rule), the spectral functions in the presence and absence of external fields are

$$
A(\omega) \propto \sum_{h,h',m,m'} \int_{1stBZ} \frac{d\mathbf{k}}{(2\pi)^2} \left| \left\langle \Psi^{h'}(\mathbf{k},\mathbf{m}') \left| \frac{\hat{\mathbf{E}} \cdot \mathbf{P}}{m_e} \right| \Psi^{h}(\mathbf{k},\mathbf{m}) \right\rangle \right|^2
$$

$$
\times \operatorname{Im} \left[\frac{f(E^{h'}(\mathbf{k},\mathbf{m}')) - f(E^{h}(\mathbf{k},\mathbf{m}))}{E^{h'}(\mathbf{k},\mathbf{m}') - E^{h}(\mathbf{k},\mathbf{m}) - \omega - i\Gamma} \right]. \tag{3.3}
$$

h represents the valence or conduction band, $\mathbf{P}$ is the momentum operator, $f(E^{h}(\mathbf{k},\mathbf{m})$ the Fermi-Dirac distribution function; Γ the broadening parameter due to the various deexcitation mechanisms. The absorption spectrum is associated with the velocity matrix elements (the dipole perturbation; the first term) and the joint density of states (the second term). The former will determine whether the inter-Landau-level transitions are available during the optical excitations.

The velocity matrix elements, as successfully done for graphene-related materials [11] are evaluated under the gradient approximation in the form of

$$
\left\langle \Psi^{h'}(\mathbf{k},\mathbf{m}') \left| \frac{\hat{\mathbf{E}} \cdot \mathbf{P}}{m_e} \right| \Psi^{h}(\mathbf{k},\mathbf{m}) \right\rangle \cong \frac{\partial}{\partial k_y} \left\langle \Psi^{h'}(\mathbf{k},\mathbf{m}') \left| H \right| \Psi^{h}(\mathbf{k},\mathbf{m}) \right\rangle
$$

$$
= \sum_{\alpha,l,l'} \sum_{m,m'=1}^{2R_B} \left(c^*_{A^{l,m}_{\alpha,\mathbf{k}}} c_{A^{l',m'}_{\alpha,\mathbf{k}}} \frac{\partial}{\partial k_y} \left\langle A^{l,m}_{\alpha,\mathbf{k}} \left| H \right| A^{l',m'}_{\alpha,\mathbf{k}'} \right\rangle \right.
$$

$$
+ c^*_{A^{l,m}_{\alpha,\mathbf{k}}} c_{B^{l',m'}_{\alpha,\mathbf{k}'}} \frac{\partial}{\partial k_y} \left\langle A^{l,m}_{\alpha,\mathbf{k}} \left| H \right| B^{l',m'}_{\alpha,\mathbf{k}'} \right\rangle
$$

$$
+ c^*_{B^{l,m}_{\alpha,\mathbf{k}}} c_{A^{l',m'}_{\alpha,\mathbf{k}'}} \frac{\partial}{\partial k_y} \left\langle B^{l,m}_{\alpha,\mathbf{k}} \left| H \right| A^{l',m'}_{\alpha,\mathbf{k}'} \right\rangle
$$

$$
\left. + c^*_{B^{l,m}_{\alpha,\mathbf{k}}} c_{B^{l',m'}_{\alpha,\mathbf{k}'}} \frac{\partial}{\partial k_y} \left\langle B^{l,m}_{\alpha,\mathbf{k}} \left| H \right| B^{l',m'}_{\alpha,\mathbf{k}'} \right\rangle \right). \tag{3.4}
$$

Under this approximation, we do not need to really do the inner product of the left side in Eq. (3.4); that is, the wave functions of the $3p_z$ orbitals are not included in the calculations, but only amplitudes (subenvelope functions) are sufficient in the right-hand side of Eq. (3.4). It clearly indicates that the subenvelope functions can be used to investigate the magneto-absorption spectra. Additionally, the similar theoretical framework is very useful in understanding the quantum Hall conductivities. In general, one can fully explore the critical factors purely due to the characteristics of Landau levels, e.g., many symmetric delta-function-like absorption peaks with a uniform intensity in monolayer graphene [11].

The model calculations of magneto-absorption spectra are worthy of a closer examination, since they account for the existence of diverse selection rules. After solving the magneto-energy spectra and Landau-level wave functions, the matrices associated with them are directly utilized in evaluating the magneto-optical properties, e.g., the $1 \times 16R_B$ and $16R_B \times 16R_B$ matrices for bilayer AB-bt silicene system. Apparently, the accurate magneto-optical data consume very much computer time. In general, the band-like Hamiltonian matrix and the localization features of Landau-level wave functions are available in greatly enhancing the efficiency of numerical calculations. And then, how to classify a lot of magneto-absorption peaks and to propose the conceive physical pictures becomes critically important. Under the gradient approximation [Eq. (3.4)], the dipole matrix element is mainly determined by the variations of the Hamiltonian matrix elements with the specific wave vector, in which their magnitudes are related to the intralayer and interlayer hopping integrals. One can first examine the specific matrix elements (the specific two sublattices) which create the largest atomic interaction. Second, the initial Landau-level state associated with a certain one sublattice is excited to the final Landau-level state related to another sublattice. If the inner product is sufficiently strong, this channel is available during the vertical excitations. Third, the similar examinations are conducted on the second-largest terms, and the test results are compared with the above-mentioned ones. Finally, the magneto-optical selection rules could be obtained from the delicate analyses. In addition, any optical absorption rules might be absent, as a result of the thorough destructions in the spatial distribution symmetries of Landau levels.

The magneto-optical absorption spectrum of monolayer graphene is very suitable for illustrating a specific selection rule of $\Delta = n^c - n^v = \pm 1$. This gapless system, with the linearly isotropic Dirac-cone structure, exhibits the symmetric magneto-electronic energy spectrum about the Fermi level, as shown in Fig. 3.3(a) at $B_z = 40$ T. For the specific $(k_x = 0, k_y = 0)$ state, there are four-fold degenerate states being, respectively, localized at 1/6, 2/6, 4/6, and 5/6 [in the unit of a periodical length corresponding to an enlarged rectangle in Fig. 3.2]. Of course, each localization state makes the same contribution to any physical properties. For example, the 2/6-localization Landau levels in Fig. 3.3(b) possess the well-behaved spatial distributions, being similar to those an oscillator [9, 10, 11]. The quantum number is characterized from the dominating mode of the B sublattice, since its amplitude is finite at the Fermi level ($E_F = 0$). As a result, the low-lying Landau levels are initial from the Dirac point, i.e., the ordering of quantum is $n^{c,v} = 0, 1, 2, 3...$ The similar results are revealed for the 1/6-localization Landau levels, but they are dominated by the A sublattice. Most importantly, the number difference of zero points is just ± 1 for the B and A sublattices, directly reflecting their equivalence under a honeycomb lattice. The above-mentioned Landau-level features create a lot of delta-function-like prominent peaks due to the inter-Landau-level transitions from the valence to conduction states [Fig. 3.4]. The Landau-level energy spacing is proportional to the square root of the magnetic-field strength. Such symmetric peaks have a uniform intensity in the low-frequency range because of the identical group velocity/the Fermi velocity. Moreover, the available absorption peaks only arise from the n^v valence Landau level and the

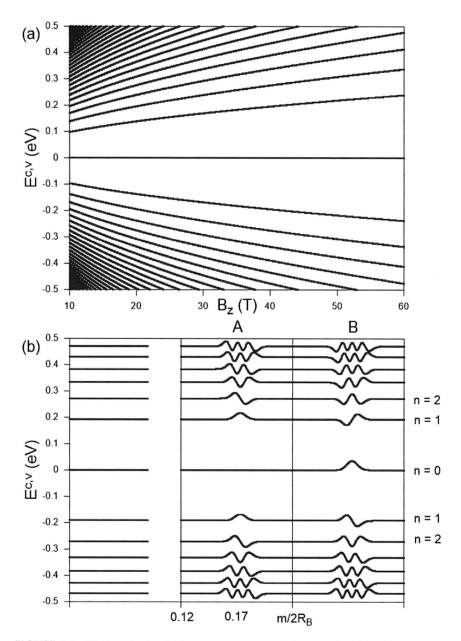

FIGURE 3.3 The Landau-level (a) energy spectrum and (b) wave functions for monolayer graphene.

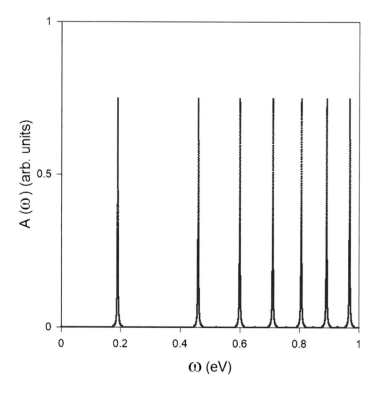

FIGURE 3.4 The magneto-optical absorption spectrum of monolayer graphene at $B_z = 40$ T.

$n^c = n^v + 1$ [$n^c = n^v - 1$] conduction Landau level for the 2/6 [1/6] localization center. These selection rules are associated with the B- and A-sublattice wave functions, respectively, corresponding to the initial and final Landau-level states. That only the nearest-neighbor hopping integral between the A and B sublattices is the main reason.

3.3 TRANSPORT THEORIES

3.3.1 QUANTUM HALL EFFECTS

Here the linearly static Kubo formula [11, 12], which directly combines with the generalized tight-binding model, is very suitable in fully exploring investigate the rich and unique quantum Hall conductivities of few-layer graphene systems with the high- and low-symmetry stacking configurations. The developed model will be very useful in the accurate identifications of the magneto-electronic selection rules under the static case; that is, the available scattering channels of the magneto-transport properties can be thoroughly examined and determined by the delicately numerical analysis [12]. The dependencies of quantum conductivity on the Fermi energy and magnetic-field strength are investigated in detail. The feature-rich Landau levels can create the extraordinary magneto-transport properties, such as those in the sliding bilayer graphenes, the trilayer ABC stacking, and AAB-stacked graphene. The unusual Hall conductivities, discussed later in Chap. 8, are predicted to cover the integer and non-integer conductivities, the zero and non-vanishing conductivities at the neutral point, the well-like, staircase, abnormal and composite/complex quantum structures. These results will be deduced to be dominated by the frequent crossing and anti-crossing energy spectra, and the spatial oscillation modes. Apparently, the low-energy perturbation approximation cannot generate the above-mentioned important phenomena [12].

Within the linear response, the transverse Hall conductivity is calculated from the static Kubo formula [11, 12].

$$\sigma_{xy} = \frac{ie^2\hbar}{S} \sum_{\alpha} \sum_{\alpha \neq \beta} (f_\alpha - f_\beta) \frac{\langle\alpha|\dot{\mathbf{u}}_x|\beta\rangle\langle\beta|\dot{\mathbf{u}}_y|\alpha\rangle}{(E_\alpha - E_\beta)^2}. \tag{3.5}$$

$|\alpha>/|\beta>]$ is the initial/final Landau-level state with energy E_α/E_β, S the area of the B_z-enlarged unit cell, f_α the Fermi-Dirac distribution functions, and $\dot{\mathbf{u}}_x$ the velocity operator along the x-direction. The matrix elements of the velocity operators, which can determine the available inter-Landau-level transitions, are solved under the gradient approximation, as discussed earlier in Eq. (3.4) [12]:

$$\langle\alpha|\dot{\mathbf{u}}_x|\beta\rangle = \frac{1}{\hbar}\langle\alpha|\frac{\partial\mathbf{H}}{\partial\mathbf{k}_x}|\beta\rangle$$

$$\langle\alpha|\dot{\mathbf{u}}_y|\beta\rangle = \frac{1}{\hbar}\langle\alpha|\frac{\partial\mathbf{H}}{\partial\mathbf{k}_y}|\beta\rangle. \tag{3.6}$$

As for few-layer graphene systems, such matrix elements are dominated by the intralayer nearest-neighbor intercation [12] so that the quantized mode of the initial state on the A^l sublattice must be identical to that of the final state on the B^l sublattice. Apparently, the well-behaved, perturbed and undefined Landau levels of layered graphenes [11], accompanied with various magnetic selection rules [10, 11], are expected to exhibit the rich and unique quantum conductivities.

Equation (3.5) is used to calculate the Hall conductivities of monolayer graphene, clearly indicating the typical quantum phenomenon. The Fermi energy-dependent σ_{xy}, as shown in Fig. 3.5(a), exhibits a lot of uniform-height plateau structures, but with the non-homogenous widths being inversely proportional to the square root of E_F. The quantum strictures directly reflect the fact that the Landau-level states become fully occupied or unoccupied at their initial and final positions. The same height of $4e^2/h$ should be closely related to all the available scattering events. By the detailed analysis on the enumerator in Eq. (3.5), the velocity matrix elements, which are determined by the initial/occupied and final/unoccupied Landau-level states, strongly depend on the well-behaved symmetries of spatial distributions on two equivalent A and B sublattices. As a result, the static selection rule of $\Delta n = \pm 1$ is identical to the dynamic ones [12]. Specifically, there exists a staircase of $4E^2/h$ height at the neutral point. This result is due to $n^v = 0 \rightarrow n^c = 1$, $n^v = 1 \rightarrow n^2 = 1$, $n^v = 2 \rightarrow n^c = 3 \ldots$ etc, in which the deeper-energy valence Landau levels would make the small contributions. It is consistent with the analytical calculations from the effective-mass approximation [6]. The similar results are revealed in the magnetic-field-dependent

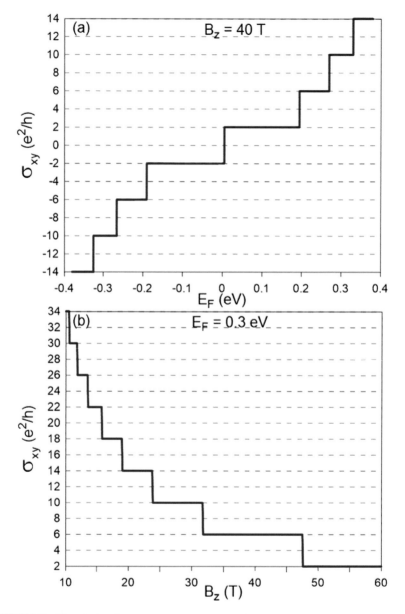

FIGURE 3.5   The (a) Fermi-energy- and (b) magnetic-field-dependent Hall conductivities in monolayer graphene under $B_z = 40$ T and $E_F = 0.3$ eV, respectively.

quantum Hall conductivities, e.g., σ_{xy} in Fig. 3.5(b). The widths of plateau structures are proportional to the square root of B_z, directly reflecting the key feature of the magneto-electronic energy spectra [11].

3.3.2 MAGNETO-HEAT CAPACITY

The electronic heat capacities in graphene-related systems are very sensitive to the changes in the magnetic-field strength and temperatures. The T-dependent total mean energy of an electronic spectrum under a uniform perpendicular magnetic field is characterized as

$$U(T, B_z) = \frac{3\sqrt{3}b^2}{4} N_A \sum_{\sigma,h,n} \int_{1stB.Z.} \frac{d^2k}{(2\pi)^2} [E_n^h(\vec{k}) - \mu] f((E_n^h(\vec{k}) - \mu)). \quad (3.7)$$

Furthermore, the electronic specific heat, being defined as the variation of the total energy with temperature, is

$$C(T, B_z) = \frac{\partial U(T, B_z)}{\partial T}$$

$$= \frac{3\sqrt{3}b^2}{4} \frac{N_A}{\beta T} \sum_{\sigma,h,n} \beta^2 (E_n^h - \mu)^2 \frac{exp[\beta(E_n^h - \mu)]}{\{1 + exp[\beta(E_n^h - \mu)]\}} \int_{1stB.Z.} \frac{d^2k}{(2\pi)^2}. \quad (3.8)$$

N_A is the Avogadro number. $h(= c \& v)$ represents the conduction and valence Landau levels, making the important contributions to the thermal property. $\beta = 1/k_B T$ and k_B is the Boltzmann constant.

The low-temperature magneto-electronic specific heat strongly depends on the Landau-level energy spectrum and the Zeeman spitting effect, even if the latter is very small (~ 1 meV) under $B_z \sim 10$ T. Each Landau-level state has a total energy, as expressed by

$$E_n^h(\sigma, B_z) = E_n^{c,v}(B_z) + E_z(\sigma, B_z), \quad (3.9)$$

and

$$E_z(\sigma, B_z) = (g\sigma\pi)/m^*(B_z/\phi_0). \quad (3.10)$$

The Zeeman splitting energy, $E_z(\sigma, B_z)$, plays a critical role in the low-energy free carrier distributions and thus the magneto specific heat. The g factor (~ 2) is identical to that of pure graphite [13], $\sigma (= \pm 1/2)$ stands for the electron spin, $\phi_0 (= hc/e)$ corresponds to the magnetic flux quantum, and m^* denotes the bare electron mass. For example, the specific heat in monolayer graphene is almost vanishing at $T \leq 10$ K in the absence of the Zeeman splitting [details in Sec. 8.3]; furthermore, there is no special structure at room temperature.

3.4   MODIFIED RANDOM-PHASE APPROXIMATION FOR MAGNETO-ELECTRONIC COULOMB EXCITATIONS

As to magneto-Coulomb excitations, monolayer germanene is chosen for a model study, since this system possesses an important SOC, a significant buckling structure, and the single-orbital-dominated low-lying energy bands. Furthermore, the external electric and magnetic fields could be applied to greatly diversify the single-particle excitations and the magnetoplasmon modes. Similar to graphene, germanene/silicene, as clearly shown in Fig. 3.2, consists of a honeycomb lattice with A and B sublattices. The latter presents the nonplanar structure, in which the two sublattice planes are separated by a distance of 2ℓ ($\ell = 0.33$ Å), as clearly illustrated in Fig. 3.6. The low-energy Hamiltonian, which is built from the spin-dependent $4p_z$-orbital tight-binding functions of Ge atoms, is given by [14].

$$H = -\gamma_0 \sum_{\langle I,J \rangle, \alpha} c_{I\alpha}^\dagger c_{J\alpha} + i\frac{\lambda_{SOC}}{3\sqrt{3}} \sum_{\langle\langle I,J \rangle\rangle, \alpha, \beta} v_{IJ} c_{I\alpha}^\dagger \sigma_{\alpha\beta}^z c_{J\beta}$$

$$- i\frac{2}{3}\lambda_{R2} \sum_{\langle\langle I,J \rangle\rangle, \alpha, \beta} u_{IJ} c_{I\alpha}^\dagger (\vec{\sigma} \times \hat{d}_{IJ})_{\alpha\beta}^z c_{J\beta}$$

$$+ \ell \sum_{I,\alpha} \mu_I E_z c_{I\alpha}^\dagger c_{I\alpha} . \tag{3.11}$$

Equation (3.11) could be obtained from Eq. (3.1) by ignoring the layer-dependent Coulomb potential energies, spin-orbital interactions and hopping integrals. The sums are carried out for nearest-neighbors $\langle I, J \rangle$ or next-nearest-neighbor lattice site pairs $\langle\langle I, J \rangle\rangle$. The first term (I) in Eq. (3.11) accounts for nearest-neighbor hopping with energy transfer $\gamma_0 = 0.86$ eV. The second term (II) describes the effective SOC for parameter $\lambda_{SOC} = 46.3$ meV. Additionally, we chose $v_{IJ} = \pm 1$ if the next-nearest-neighbor hopping is anticlockwise/clockwise with respect to the positive z-axis. In the third term (III), the intrinsic Bychkov-Rashba SOC is included in the next-nearest neighbor hopping through $\lambda_{R2} = 10.7$ meV, in which $u_{IJ} = \pm 1$ are for the A/B lattice sites, respectively. $\hat{d}_{IJ} = \vec{d}_{IJ}/|d_{IJ}|$ is a unit vector joining two sites I and J on the same sublattice as shown in Fig. 3.1(a). The staggered sublattice potential energy produced by the external electric field is characterized by the fourth term (IV), where $\mu_I = \pm 1$ for the A/B sublattice sites and $\ell = 0.33$ Å, referring to Fig. 3.1(b).

Monolayer germanene is assumed to be in a uniform perpendicular magnetic field. The magnetic flux, the product of the field strength and the hexagonal area, is $\Phi = (3\sqrt{3}b^2 B_z/2)/\phi_0$, where b ($=2.32$ Å) is the lattice constant. The vector potential, $\vec{A} = (B_z x)\hat{y}$, leads to a new period along the armchair direction, since it can create an extra magnetic Peierls phase, i.e., $\exp\{i[\frac{2\pi}{\phi_0} \int \vec{A} \cdot d\vec{r}]\}$. The unit cell is thus enlarged and its dimension is determined by $R_B = 1/\Phi$. The reduced first Brillouin zone has an area of $1/(3\sqrt{3}b^2 R_B)$. The enlarged unit cell contains $4R_B$ Ge atoms and the Hamiltonian matrix is a $8R_B \times 8R_B$ Hermitian matrix with the spin degree of freedom.

Corresponding to $B_z = 10$ T, where $R_B = 3,000$, the Hamiltonian has a dimension of $24,000 \times 24,000$.

Only the fourth interaction in Eq. 3.11 is a diagonal matrix regardless of the Peierls phase. By the detailed calculations, the independent Hamiltonian matrix elements, which are due to the extra position-dependent Peierls phases, are expressed in the Eqs. (3.12)–(3.20) (the sublattice-, site-, and SOC-dependent magnetic Hamiltonian matrix elements).

$$[I] \quad \langle B_J^\alpha | H | A_I^\beta \rangle = \gamma_0 \sum_{\langle I,J \rangle} \frac{1}{N} \exp\left[i\vec{k} \cdot (\vec{R}_{A_I^\beta} - \vec{R}_{B_J^\alpha}) \right]$$

$$\times \exp\left\{ i \left[\frac{2\pi}{\phi_0} \int_{\vec{R}_{A_I^\beta}}^{\vec{R}_{B_J^\alpha}} \vec{A} \cdot d\vec{r} \right] \right\}$$

$$= \gamma_0 t_{1,I} \delta_{I,J+1} \delta_{\alpha,\beta} + \gamma_0 s \delta_{I,J} \delta_{\alpha,\beta} , \qquad (3.12)$$

where $\vec{R}_{A_I^\beta}$ and $\vec{R}_{B_J^\alpha}$ denote the lattice sites for the A and B sublattices, with the spin polarizations β and α, respectively. Also, I [J] corresponds to the initial [final] sublattice site. Concerning the kinetic energy, the nearest-neighbor matrix element includes the phase-dependent $t_{1,I} = \exp\{i[-k_x\frac{b}{2} - k_y\frac{\sqrt{3}b}{2} + \pi\frac{\Phi}{\phi_0}(I - 1 + \frac{1}{6})]\} + \exp\{i[-k_x\frac{b}{2} + k_y\frac{\sqrt{3}b}{2} - \pi\frac{\Phi}{\phi_0}(I - 1 + \frac{1}{6})]\}$ and a specific term of $s = \exp[i(-k_x b)]$. The vector potential can create more complicated hopping phases in the SOC-related interactions, as presented in Eqs. (3.13)–(3.20). For example, $t_{2,1}$ ($t_{8,1}$) represents the phase term for the next-nearest-neighbor hopping from A_1^β to A_1^α (A_1^β to A_2^α), corresponding to the effective SOC (intrinsic Rashba SOC). An illustration of the SOC-induced hopping phases is clearly indicated in Fig. 1(d). Next,

$$[II] \quad \langle A_J^\alpha | H | A_I^\beta \rangle = \frac{\lambda_{SOC}}{3\sqrt{3}} \sum_{\langle\langle I,J \rangle\rangle} \frac{1}{N} \exp\left[i\vec{k} \cdot (\vec{R}_{A_I^\beta} - \vec{R}_{A_J^\alpha}) \right]$$

$$\times \exp\left\{ i \left[\frac{2\pi}{\phi_0} \int_{\vec{R}_{A_I^\beta}}^{\vec{R}_{A_J^\alpha}} \vec{A} \cdot d\vec{r} \right] \right\} + E_z \ell$$

$$= \frac{\lambda_{SOC}}{3\sqrt{3}} t_{2,I} \delta_{I,J} \delta_{\alpha,\beta} + E_z \ell , \qquad (3.13)$$

where $t_{2,I} = \exp i[k_y a + 2\pi\frac{\Phi}{\phi_0}(I-1)] - \exp i[-k_y a - 2\pi\frac{\Phi}{\phi_0}(I-1)]$. Furthermore,

$$\langle B_J^\alpha | H | B_I^\beta \rangle = \frac{\lambda_{SOC}}{3\sqrt{3}} \sum_{\langle\langle I,J \rangle\rangle} \frac{1}{N} \exp\left[i\vec{k}\cdot\left(\vec{R}_{B_I^\beta} - \vec{R}_{B_J^\alpha}\right)\right]$$

$$\times \exp\left\{ i\left[\frac{2\pi}{\phi_0} \int_{\vec{R}_{B_I^\beta}}^{\vec{R}_{B_J^\alpha}} \vec{A}\cdot d\vec{r} \right] \right\} - E_z\ell$$

$$= \frac{\lambda_{SOC}}{3\sqrt{3}} t_{3,I}\delta_{I,J}\delta_{\alpha,\beta} - E_z\ell \; , \tag{3.14}$$

where $t_{3,I} = \exp i\{-k_y a - 2\pi\frac{\Phi}{\phi_0}[(I-1)+\frac{1}{3}]\} - \exp i\{k_y a + 2\pi\frac{\Phi}{\phi_0}[(I-1)+\frac{1}{3}]\}$.

$$\langle A_J^\alpha | H | A_I^\beta \rangle = \frac{\lambda_{SOC}}{3\sqrt{3}} \sum_{\langle\langle I,J \rangle\rangle} \frac{1}{N} \exp\left[i\vec{k}\cdot\left(\vec{R}_{A_I^\beta} - \vec{R}_{A_J^\alpha}\right)\right]$$

$$\times \exp\left\{ i\left[\frac{2\pi}{\phi_0} \int_{\vec{R}_{A_I^\beta}}^{\vec{R}_{A_J^\alpha}} \vec{A}\cdot d\vec{r} \right] \right\}$$

$$= \frac{\lambda_{SOC}}{3\sqrt{3}} t_{4,I}\delta_{I,J-1}\delta_{\alpha,\beta} \; , \tag{3.15}$$

where $t_{4,I} = \exp i\{k_x\frac{3}{2}b - k_y\frac{a}{2} - \pi\frac{\Phi}{\phi_0}[(I-1)+\frac{1}{2}]\} - \exp i\{k_x\frac{3}{2}b + k_y\frac{a}{2} + \pi\frac{\Phi}{\phi_0}[(I-1)+\frac{1}{2}]\}$. In addition,

$$\langle B_J^\alpha | H | B_I^\beta \rangle = \frac{\lambda_{SOC}}{3\sqrt{3}} \sum_{\langle\langle I,J \rangle\rangle} \frac{1}{N} \exp\left[i\vec{k}\cdot(\vec{R}_{B_I^\beta} - \vec{R}_{B_J^\alpha})\right]$$

$$\times \exp\left\{ i\left[\frac{2\pi}{\phi_0} \int_{\vec{R}_{B_I^\beta}}^{\vec{R}_{B_J^\alpha}} \vec{A}\cdot d\vec{r} \right] \right\}$$

$$= \frac{\lambda_{SOC}}{3\sqrt{3}} t_{5,I}\delta_{I,J-1}\delta_{\alpha,\beta} \; , \tag{3.16}$$

where $t_{5,I} = \exp i\{k_x\frac{3}{2}b - k_y\frac{a}{2} - \pi\frac{\Phi}{\phi_0}[(I-1) + \frac{5}{6}]\} - \exp i\{k_x\frac{3}{2}b + k_y\frac{a}{2} + \pi\frac{\Phi}{\phi_0}[(I-1) + \frac{5}{6}]\}$.

$$[III] \quad \langle A_J^\alpha|H|A_I^\beta\rangle_{\alpha\neq\beta} = \frac{2}{3}\lambda_{R2}\sum_{\langle\langle I,J\rangle\rangle}\frac{1}{N}\exp\left[i\vec{k}\cdot(\vec{R}_{A_I^\beta} - \vec{R}_{A_J^\alpha})\right]$$

$$\times\exp\left\{i\left[\frac{2\pi}{\phi_0}\int_{\vec{R}_{A_I^\beta}}^{\vec{R}_{A_J^\alpha}}\vec{A}\cdot d\vec{r}\right]\right\}$$

$$= \frac{2}{3}\lambda_{R2}t_{6,I}\delta_{I,J}\,, \tag{3.17}$$

where $t_{6,I} = \exp i[k_ya + 2\pi\frac{\Phi}{\phi_0}(I-1) - \frac{\pi}{2}] + \exp i[-k_ya - 2\pi\frac{\Phi}{\phi_0}(I-1) + \frac{\pi}{2}]$. Similarly,

$$\langle B_J^\alpha|H|B_I^\beta\rangle_{\alpha\neq\beta} = \frac{2}{3}\lambda_{R2}\sum_{\langle\langle I,J\rangle\rangle}\frac{1}{N}\exp\left[i\vec{k}\cdot(\vec{R}_{B_I^\beta} - \vec{R}_{B_J^\alpha})\right]$$

$$\times\exp\left\{i\left[\frac{2\pi}{\phi_0}\int_{\vec{R}_{B_I^\beta}}^{\vec{R}_{B_J^\alpha}}\vec{A}\cdot d\vec{r}\right]\right\}$$

$$= \frac{2}{3}\lambda_{R2}t_{7,I}\delta_{I,J}\,, \tag{3.18}$$

where $t_{7,I} = \exp i\{k_ya + 2\pi\frac{\Phi}{\phi_0}[(I-1) + \frac{1}{3}] - \frac{\pi}{2}\} + \exp i\{-k_ya - 2\pi\frac{\Phi}{\phi_0}[(I-1) + \frac{1}{3}] + \frac{\pi}{2}\}$.

$$\langle A_J^\alpha|H|A_I^\beta\rangle_{\alpha\neq\beta} = \frac{2}{3}\lambda_{R2}\sum_{\langle\langle I,J\rangle\rangle}\frac{1}{N}\exp\left[i\vec{k}\cdot(\vec{R}_{A_I^\beta} - \vec{R}_{A_J^\alpha})\right]$$

$$\times\exp\left\{i\left[\frac{2\pi}{\phi_0}\int_{\vec{R}_{A_I^\beta}}^{\vec{R}_{A_J^\alpha}}\vec{A}\cdot d\vec{r}\right]\right\}$$

$$= \frac{2}{3}\lambda_{R2}t_{8,I}\delta_{I,J-1}\,, \tag{3.19}$$

where $t_{8,I} = \exp i\{k_x\frac{3}{2}b + k_y\frac{a}{2} + \pi\frac{\Phi}{\phi_0}[(I-1) + \frac{1}{2}] - \frac{\pi}{6}\} + \exp i\{k_x\frac{3}{2}b - k_y\frac{a}{2} - \pi\frac{\Phi}{\phi_0}[(I-1) + \frac{1}{2}] + \frac{\pi}{6}\}$. Finally, we have

$$\langle B_J^\alpha | H | B_I^\beta \rangle_{\alpha\neq\beta} = \frac{2}{3}\lambda_{R2} \sum_{\langle\langle I,J\rangle\rangle} \frac{1}{N} \exp\left[i\vec{k}\cdot(\vec{R}_{B_I^\beta} - \vec{R}_{B_J^\alpha})\right]$$

$$\times \exp\left\{i\left[\frac{2\pi}{\phi_0}\int_{\vec{R}_{B_I^\beta}}^{\vec{R}_{B_J^\alpha}} \vec{A}\cdot d\vec{r}\right]\right\}$$

$$= \frac{2}{3}\lambda_{R2}t_{9,I}\delta_{I,J-1} , \tag{3.20}$$

where $t_{9,I} = \exp i\{k_x\frac{3}{2}b + k_y\frac{a}{2} + \pi\frac{\Phi}{\phi_0}[(I-1) + \frac{5}{6}] - \frac{\pi}{6}\} + \exp i\{k_x\frac{3}{2}b - k_y\frac{a}{2} - \pi\frac{\Phi}{\phi_0}[(I-1) + \frac{5}{6}] + \frac{\pi}{6}\}$. To obtain the Landau-level energy spectra and wave functions efficiently, it needs to diagonalize a giant magnetic Hamiltonian matrix under the exact method.

The longitudinal Coulomb excitations of monolayer germanene is characterized by the magneto-dielectric function $[\epsilon(q, \omega, B_z)]$ and the energy loss function $[-1/\epsilon(q, \omega, B_z)]$, in which the former and the latter, respectively, represent the bare and screened response functions of valence and conduction Landau-level electrons. According to the definition, the ratio of the electric displacement [the external Coulomb potential] and the effective field [the effective Coulomb potential] is the longitudinal dielectric function of $\epsilon(q, \omega, B_z)$. When a 2D condensed-matter system experiences the external perturbation of the time-dependent Coulomb potential (e.g., the incident electron (light) beam), all the occupied electronic states simultaneously exhibit very complicated screening phenomena. Apparently, the dynamic electron–electron Coulomb interactions belong to the inelastic many-particle scattering. It is impossible to exactly solve the effective Coulomb potentials between two electrons under the various theoretical models, mainly owing to the infinite scattering processes. The previous studies show that a lot of approximate methods have been developed to investigate the induced Coulomb potential due to the screening charges, such as the random-phase [15–17], Hubbard [18], and Singwi-Sjolander approximations [19, 20]. And then, the total Coulomb potential is only the summation of the external and induced ones, or the dielectric function is easily obtained from their relations.

In general, the specific RPA, which only covers the bubble-like electron-hole pair excitations (a series of similar Feynman diagrams) in the absence of the vertex corrections [15–17], is frequently utilized to thoroughly explore the Coulomb excitations even for the low-dimensional systems with the strong correlation effects. Furthermore, it is very suitable for studying the magnet-electronic excitation spectra in the presence of a uniform perpendicular magnetic field, e.g., the various magnetoplasmon modes, the electric-field-enriched Coulomb excitations, and the diverse (momentum, frequency)-phase diagrams. Under this approximation, one can get

$$\epsilon(q, \omega, B_z) = \epsilon_1(q, \omega, B_z) + i\epsilon_2(q, \omega, B_z) = \epsilon_0 - v_q\chi^0(q, \omega, B_z). \tag{3.21}$$

$v_q = 2\pi e^2/q$ represents the in-plane Fourier transformation of the bare Coulomb potential energy, and $\epsilon_0 = 2.4$ (taken from that of graphite [15]) is the background dielectric constant due to the deep-energy electronic states. It should be noticed that the variation of ϵ_0 will lead to a vertical shift of the real part of the dielectric function and thus alter the peak intensity and position in the energy loss function. However, the main features of collective excitations keep the same. Such a form of dielectric function is frequently revealed in many research studies on two-dimensional systems, theoretically [16,17] and experimentally [21–23]. Furthermore, the induced Coulomb potential is proportional to the effective one within the linear response approximation, in which the coefficient is $-v_q\chi^0(q, \omega, B_z)$. The 2D bare response function [15] is expressed in the second term of Eq. (21). The bra and ket account for the initial and final Landau state wave functions with the respective quantum numbers of n and m, where the wave vectors are $\vec{k}$ and $\vec{k} + \vec{q}$. In the presence of $B_z\hat{z}$, the electronic states in a 2D system become fully quantized; therefore, the summation is done for all the available inter-Landau-level transitions at any temperature

$$\chi^0(q, \omega, B_z) = \frac{1}{3\sqrt{3}b^2R_B} \sum_{n,m} |\langle n; \vec{k} + \vec{q}|e^{i\vec{q}\cdot\vec{r}}|m; \vec{k}\rangle|^2$$
$$\times \frac{f(E_n) - f(E_m)}{E_n - E_m - (\omega + i\Gamma)}. \tag{3.22}$$

The pre-factor $1/(3\sqrt{3}b^2R_B)$ is a normalization constant, meaning the same contributions arising from the highly degenerate Landau-level states in the reduced first Brillouin zone. The equilibrium Fermi-Dirac distribution function is $f(E) = 1/[1 + exp(E - \mu/k_BT)]$. Γ is an energy broadening parameter induced by the various de-excitation mechanisms. μ is the temperature-dependent chemical potential, where the T-dependence might be negligible under the specific magnetic energy range in this work. The bare response function only relies on the magnitude of the momentum transfer under the isotropic Landau-level energy spectrum. The energy broadening of the Landau levels associated with the lattice structure is almost vanishing at low temperature. In addition, this work is focused on the low-energy magneto-electronic excitations. q is much smaller than the reciprocal lattice vector; that is, the local-field effects can be ignored [24].

The detailed calculations for the magneto-Coulomb matrix elements are clearly illustrated as

$$\langle n; \vec{k} + \vec{q}|e^{i\vec{q}\cdot\vec{r}}|m; \vec{k}\rangle$$
$$= \sum_{s=\alpha,\beta} \sum_{l=1-8R_{B_0}} \langle\phi_z(\vec{r} - \vec{R}_l)|e^{-i\vec{q}\cdot(\vec{r}-\vec{R}_l)}|\phi_z(\vec{r} - \vec{R}_l)\rangle[u_{nsl}(\vec{k} + \vec{q})u^*_{msl}(\vec{k})]. \tag{3.23}$$

Here, $\vec{R}_l$ defines the positions of atoms in a unit cell. $u^*_{msl}(\vec{k})$ and $(u_{nsl}(\vec{k} + \vec{q}))$ are the coefficients for the TB wave functions derived from Eqs. (3.11)–(3.20). $\langle\phi_z(\vec{r} - \vec{R}_l)|e^{-i\vec{q}\cdot(\vec{r}-\vec{R}_l)}|\phi_z(\vec{r} - \vec{R}_l)\rangle = C(q) = [1 + [\frac{qa_0}{Z}]^2]^{-3}$ was calculated by using hydrogenic wave function, where a_0 is the Bohr radius and Z is an effective core charge [15]. We note that the overlapping integrals between neighboring atoms are neglected, an approximation made originally in the 2D model by Blinowski et al [25].

For the small transferred momenta,, $C(q)$ is very close to 1. Since all the π-electronic states are included in the calculations, the strength and frequency of the resonances in $Im[-1/\epsilon(q, \omega)]$ can be correctly defined. Moreover, the calculations would be reliable in a wide range of the field strength and the chemical potential.

The screened response functions, the effective energy loss functions, will become very complicated when a 2D N-layer system is taken into account. All the perturbed Coulomb potentials on the different layers need to be included in the theoretical calculations, and so do the layer-dependent screened charge distributions (indicated Coulomb potentials). The linear response of RPA is utilized to establish a specific relation between the induced and effective Coulomb potentials [details in Ref. [16]]. The dielectric function is no longer a scaler function, while it is replaced by an $N \times N$ dielectric tensor under the Dyson equation. It should be noticed that this function is characterized by the layer indices, but not the energy subband ones being used in the previous studies for the multi-band systems [16]. And then, the effective energy loss spectra, being defined for the layered systems, are accurately derived under the inelastic scattering approximation. Moreover, the modified layer-dependent RPA is consistent with the layer- and sublattice-based generalized tight-binding model [discussed earlier in Sec. 3.1. That is to say, it is also suitable for the magneto-electronic single-particle and collective excitations. Such formula has been successfully utilized to fully explore the electronic excitations of few-layer graphene systems without/with the magnetic and electric fields, e.g., the diversified electron-hole excitation regions and plasmon/magnetoplasmon modes due to the number of layers and stacking configurations [17, 26].

REFERENCES

1. Zhao J, Liu H, Yu Z, Quhe R, Zhou S, Wang Y, et al. 2016 Rise of silicene: a competitive 2D material *Prog. Mater. Sci.* **83** 24.
2. Wu J Y, Chen S C, G Gumbs, and Lin M F 2017 Field-created diverse quantizations in monolayer and bilayer black phosphorus *Phys. Rev. B* **95** 115411.
3. Do T N, Shih P H, Gumbs G, Huang D, Chiu C W, and Lin M F 2018 Diverse magnetic quantization in bilayer silicene *Phys. Rev. B* **97** 125416.
4. Fu H X, Zhang J, Ding Z J, Li H, and Menga S 2014 Stacking-dependent electronic structure of bilayer silicene *Appl. Phys. Lett.* **104** 131904.
5. Padilha J E and Pontes R B 2015 Free-standing bilayer silicene: the effect of stacking order on the structural, electronic, and transport properties *J. Phys. Chem. C* **119** 3818.
6. Koshino M and McCann E 2011 Landau level spectra and the quantum Hall effect of multilayer graphene *Phys. Rev. B* **83** 165443.
7. Koshino M, Aoki H, Kuroki K, Kagoshima S, and Osada T 2001 Hofstadter butterfly and integer quantum Hall effect in three dimensions *Phys. Rev. Lett.* **86** 1062.
8. Taut M, Eschrig H, and Richter M 2005 Skyrmion in a real magnetic film *Phys. Rev. B* **72** 165304.
9. Lin C Y, Wu J Y, Chiu Y H, and Lin M F 2014 Stacking-dependent magneto-electronic properties in multilayer graphenes *Phys. Rev. B* **90** 205434.
10. Lin Y P, Lin C Y, Ho Y H, Do T N, and Lin M F 2015 Magneto-optical properties of ABC-stacked trilayer graphene *Phys. Chem. Chem. Phys.* **17** 15921.
11. Lin C Y, Do T N, Huang Y K, and Lin M F 2017 Electronic and optical properties of graphene in magnetic and electric fields *IOP Concise Physics*. San Raefel, CA, USA: Morgan & Claypool Publishers.

12. Do T N, Chang C P, Shih P H, and Lin M F 2017 Stacking-enriched magneto-transport properties of few-layer graphenes *Phys. Chem. Chem. Phys.* **19** 29525.
13. Shyu F L, Chang C P, Chen R B, Chiu C W, and Lin M F 2003 Magnetoelectronic and optical properties of carbon nanotubes *Phys. Rev. B* **67** 045405.
14. Shih P H, Chiu Y H, Wu J Y, Shyu F L, and Lin M F 2017 Coulomb excitations of monolayer germanene *Sci. Rep.* **7** 40600.
15. Shung K W K 1986 Lifetime effects in low-stage intercalated graphite systems *Phys. Rev. B* **34** 2.
16. Ho J H, Chang C P, and Lin M F 2006 Electronic excitations of the multilayered graphite *Phys. Lett. A* **352** 446.
17. Ho J H, Lu C L, Hwang C C, Chang C P, and Lin M F 2006 Coulomb excitations in AA- and AB-stacked bilayer graphites *Phys. Rev. B* **74** 085406.
18. Hubbard J 1963 The description of collective motions in terms of many-body perturbation theory. II. The correlation energy of a free-electron gas *Proc. R. Soc. Lond.* **243** 336.
19. Singwi K S, Tosi M P, Land R H, and Sjolander A 1968 Electron correlations at metallic densities *Phys. Rev.* **176** 589.
20. Vashishta P and Singwi K S 1972 Electron correlations at metallic densities. *Phys. Rev. B* **6** 875.
21. Generalov A V and Dedkov Yu S 2012 EELS study of the epitaxial graphene/Ni(1 1 1) and graphene/Au/Ni(1 1 1) systems *Carbon* **50** 183.
22. Shin S Y, Hwang C G, Sung S J, Kim N D, Kim H S, and Chung J W 2011 Observation of intrinsic intraband π-plasmon excitation of a single-layer graphene *Phys. Rev. B* **83** 161403.
23. Nelson F J, Idrobo J-C, Fite J D, Miskovic Z L, Pennycook S J, Pantelides S T, et al. 2014 Electronic excitations in graphene in the 1-50 eV Range: The π and $\pi + \sigma$ peaks are not plasmons *Nano Lett.* **14** 3827.
24. Wu J Y, Chen S C, Roslyak O, Gumbs G, and Lin M F 2011 Plasma excitations in graphene: their spectral intensity and temperature dependence in magnetic field *ACS Nano* **5** 1026.
25. Blinowski J, Hau N H, Rigaux C, Vieren J P, Toullee R le, Furdin C, et al. 1980 *J. Phys. (Paris)* **41** 47.
26. Lin C Y, Lee M H, and Lin M F 2018 Coulomb excitations in trilayer ABC-stacked graphene *Phys. Rev. B Rapid Communication* **98** 041408.

4 Twisted Bilayer Graphene Systems

Chiun-Yan Lin,[a] Thi-Nga Do,[c,d] Jhao-Ying Wu,[b]
Po-Hsin Shih,[a] Shih-Yang Lin,[e] Ching-Hong Ho,[b]
Ming-Fa Lin[a,f,g]

[a] Department of Physics, National Cheng Kung University,
Tainan 701, Taiwan
[b] Center of General Studies, National Kaohsiung University of
Science and Technology, Kaohsiung 811, Taiwan
[c] Laboratory of Magnetism and Magnetic Materials, Advanced
Institute of Materials Science, Ton Duc Thang University,
Ho Chi Minh City, Vietnam
[d] Faculty of Applied Sciences, Ton Duc Thang University,
Ho Chi Minh City, Vietnam
[e] Department of Physics, National Chung Cheng University,
Chiayi 621, Taiwan
[f] Quantum Topology Center, National Cheng Kung University,
Tainan 701, Taiwan
[g] Hierarchical Green-Energy Materials Research Center,
National Cheng Kung University, Tainan, Taiwan

CONTENTS

The achiral and chiral geometric structures play the critical roles in diversifying the essential properties, so they are one of the mainstream topics in basic researches [1,2], as well as highly potential applications [3]. A cylindrical carbon nanotube could be regarded as a rolled-up graphitic sheet; therefore, its structure is well characterized by two primitive vectors of monolayer graphene [1]. About twenty years ago, the hexagonal geometries and low-energy electronic properties of 1D single-walled carbon nanotubes are accurately identified from the simultaneous high-resolution measurements of scanning tunneling microscopy (STM) and scanning tunneling spectroscopy (STS) [4]. The close relations between the chiral angle (the arrangement of hexagons relative to the tubular axis) and the metallic/semiconducting behavior are thoroughly examined by the delicate experiments, directly verifying the theoretical predictions by the tight-binding model [5, 6] and first-principles calculations [7, 8]. That is, the low-lying π-electronic states in a carbon nanotube are sampled from those of monolayer graphene through the periodical boundary condition. However, the misorientation of $2p_z$ orbitals and their hybridizations with three σ orbitals, which appear in very small cylindrical surfaces, have very strong effects on the low-energy fundamental properties. The experimental examinations on the conducting or semiconducting behaviors have confirmed the critical mechanisms, the periodical boundary condition, and the curvature effects [9]. The similar geometric structures come to exist in the both ends of 1D graphene nanoribbons with the open boundary conditions [10]. The zigzag, armchair and chiral edge structures are clearly classified under the distinct STM experiments [4]. Also, all of them are confirmed to be semiconductors, in which their energy gaps are inversely proportional to the nanoribbon widths [4], as predicted by the theoretical calculations [5]. The semiconducting properties are principally determined by the open boundary, the nonuniform bond lengths/hopping integrals near two boundaries, and the zigzag-edge magnetism.

The twisted bilayer graphene systems have been successfully generated in experimental laboratories through the various methods, such as, the mechanical exfoliation [11], the chemical vapor deposition method [12, 13], and transfer-free method [14]. There exists a relative rotation angle (θ) between two honeycomb lattices, being clearly identified from the high-resolution STM measurements [15–17]. This significant parameter, being defined in Sec. 4.1, represents a specific Moire superlattice corresponding to an unit cell much larger than that (four carbon atoms) of the AA and AB stackings. Since two graphene layers are attracted by the van der Waals interactions, the low-energy essential physical properties are expected to be dominated by the very complicated interlayer hopping integrals of C-$2p_z$ orbitals in the absence of spin-orbit couplings. Furthermore, more pairs of valence and conduction bands will come to exist under the critical zone-folding effect. The rich and unique electronic structures are confirmed by the angle-resolved photoemission spectroscopy [ARPES] [18, 19], e.g., the coexistence of linear and parabolic energy dispersions in symmetry-broken bilayer graphene systems [18], and the semiconducting or semimetallic behaviors [19]. However, the further experimental examinations are required to identify more valence subbands in the reduced first Brillouin zone.

Both STS [16, 17] and transport experiments [20] on twisted bilayer graphenes could provide enough information on the low-energy physical properties. The former directly measure the band properties/magneto-electronic ones across the Fermi level and the structures of van Hove singularities (the local maxima & minima, and the saddle points in the energy-wave-vector space); that is, they are useful in identifying the complex effects due to the Moire superlattices and the external fields. Up to date, the experimental results clearly show three kinds of van Hove singularities, the V-shaped structure, shoulder, and prominent symmetric peaks at the negative and positive energies [16, 17], being very sensitive to the twisted angles. Obviously, such measurements indicate the different low-lying energy dispersions/critical points [thoroughly discussed in Sec. 4.1]. The unusual energy bands, being due to the zone-folding effects, will be magnetically quantized to the rich and unique Landau levels [one of the focuses of Chap. 4]. In addition, the STS and optical measurements on the magneto-electronic density of states are absent. Moreover, the quantum transport experiments display the Hall conductivities in analogy with those of bilayer AB stacking [details discussed in Sec. 9.1] or the superposition of two monolayer graphene systems. Very interestingly, certain twisted bilayer graphenes, with magic angles of $\sim 1.1°$ (relative to the bilayer AA stacking) [2], are identified to become the unusual superconductors under a low critical temperature ($T_c \sim 1.7$ K). It might need to develop the modified theoretical frameworks for the unique superconducting phenomena in emergent layered materials.

On the theoretical progress of twisted bilayer graphene systems, the first-principles method [17], the effective-mass approximation [21, 22], and the tight-binding model [23] are available in exploring their essential properties. An optimal geometric structure, with the lowest ground state energy, can only be determined by the numerical Vienna Ab initio simulation package (VASP) calculations [17]; that is, the complete relation between the total energy and the twisted angle is obtained under such evaluations. These three manners might be suitable/reliable in investigating band structures and density of states. In general, the calculated results are roughly consistent with the ARPES measurements, e.g., the modified linear dispersions and the emergence of parabolic dispersions [18, 19]. Furthermore, they clearly show that the band-edge states (the critical points in the energy-wave-vector space) create the V-shape form, shoulder and logarithmically divergent peaks, respectively, corresponding to the linear Dirac cone, the local maxima/minima and the saddle points. Such van Hove singularities have been identified from the high-resolution STS measurements (discussed earlier; Ref. [17]). The main features of electronic properties are directly reflected in optical spectra, such as, the number, frequency, and intensity of prominent absorption structures [24]. The optical spectroscopies are required to verify the theoretical predictions. However, it will be very difficult to investigate the twist-enriched magnetic quantization, mainly owing to the Moire superlattice with the complex interlayer hopping integrals. Only few studies by the effective-mass approximation and the tight-binding model are conducted on the electronic structure and Hall conductivities, e.g., the fractal energy spectrum [25], the magnetic-field-dependent butterfly diagram (the Aharonov-Bohm effect [23]), and quantum plateaus with the normal/irregular heights [20]. Apparently, one of the studying focuses, the characteristics of magnetic wave functions, are absent up to date, since their spatial distribution symmetries will dominate the magneto-selection rules in dynamic optical transitions and static scattering events [details in Secs. 3.2 and 3.3.1]. In short, the systematic investigations on the geometric symmetries, electronic properties, optical excitations, and magneto-electronic states are critical in providing a full understanding of the geometry-diversified physical phenomena for twisted bilayer graphene systems. Maybe, the further development of the generalized tight-binding model and its direct combination with the other modified theories are the most efficient ways, as discussed in the following paragraphs.

From the theoretical point of view, the generalized tight-binding model and the gradient approximation are available in fully understanding the essential electronic and optical properties. For twisted bilayer graphene systems, the low-energy electronic structures, van Hove singularities in density of states, optical absorption spectra, and magneto-electronic states, which are greatly diversified by the composite effects due to the zone folding, the twist-dependent interlayer hopping integrals, the stacking symmetries, the gate voltages, and the magnetic fields, will be investigated in detail. As to the low-energy physical properties, the studying focuses cover the semimetallic or semiconducting property across the Fermi level, the linear/parabolic/oscillatory energy dispersions, the gate-voltage-dominated state degeneracy, the creation of more band-edge states, the distinct special forms of 2D van Hove singularities, the twisting-enriched optical absorption spectra (the threshold and higher-frequency excitation structures), the pair/group number of valence and conduction bands/Landau levels, and the B_z- and V_z-dependent magneto-electronic energy spectra and wave functions with the non-crossing, crossing, and anti-crossing behaviors. Also, the concise physical pictures, being associated with to the significant cooperations/competitions among the zone-folding effect, the distinct interlayer hopping integrals, the gate voltages, and the magnetic fields, are proposed to account for the diverse physical phenomena. A detailed comparison with the sliding bilayer graphenes is made, clearly illustrating the significant differences in fundamental properties as a result of the distinct stacking symmetries. In addition, the magnetic quantization phenomena in magneto-optical properties, quantum Hall transports, and magnetoplasmon modes are deduced as open issues in Chap. 13.

4.1 ELECTRONIC PROPERTIES IN THE ABSENCE/PRESENCE OF GATE VOLTAGES

The specific geometry of a twisted bilayer graphene can be well defined by the primitive lattice vectors of honeycomb lattices, as done for any single-walled carbon nanotubes [1]. It is characterized by the relative rotation angle (θ) and translation vector between two graphitic sheets. The primitive unit cell of the commensurate graphene bilayer structure is clearly shown in Fig. 4.1, corresponds to the least

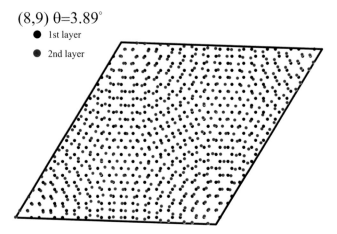

FIGURE 4.1   The geometric structure of a twisted bilayer graphene, such as for a (1,2) system with 14 carbon atoms in a unit cell for each layer.

common multiples of the two distinct graphene layers. The primitive unit vector G_1 expressed as

$$G_1 = m a_1^{(1)} + n a_2^{(1)} = m' a_1^{(2)} + n' a_2^{(2)}. \tag{4.1}$$

Furthermore, the G_2 has the same magnitude, but presents a relative rotation angle of $60°$. Both a_1 and a_2 are unit vectors of a honeycomb, in which the superscripts represent the first and second layers. Through the specific relation among them [22], the identical coefficients are revealed as $m = n'$ & $n = m'$. Apparently, a single pair of (m, n) is sufficient in describing a twisted bilayer graphene, leading to the rotation angle of $\cos(\theta), = (m^2 + n^2 + 4\,mn)/2(m^2 + n^2 + mn)$. In general, this angle is periodical within the range of $60° < \theta < 0°$, where the minimum and maximum ones are, respectively, the regular AA and AB stackings (the non-twisted systems). According to a hexagon with two carbons, the total number (N_M) in a Moire superlattice is $4(m^2 + n^2 + mn)$; that is, there are N_M $2p_z$-orbitals which dominate the low-energy essential properties, e.g., the band structures, optical excitations and magnetic quantizations. For example, an $N_M \times N_M$ Hamiltonian matrix, with the intrinsic atomic interactions in Eqs. (4.2) and (4.3), is sufficient in exploring electronic and optical phenomena under a vanishing magnetic field.

The low-energy tight-binding model, which mainly originates from the intralayer and interlayer atomic interactions of carbon-$2p_z$ orbitals, is expressed as

$$H = - \sum_{R_i, R_i} t(R_i, R_j)[c^+(R_i)c(R_j) + h.c.], \tag{4.2}$$

where $c^+(R_i)$ and $c(R_j)$ are creation and annihilation operators at positions R_i and R_j, respectively. The various hopping integrals are characterized under the empirical formula

$$-t(R_i, R_j) = \gamma_0 e^{-\frac{d-b}{\rho}}\left(1 - \left(\frac{d_0}{d}\right)^2\right) + \gamma_1 e^{-\frac{d-d_0}{\rho}}\left(\frac{d_0}{d}\right)^2, \tag{4.3}$$

where $d = |R_i - R_j|$ is the distance between two lattice points, $b = 1.42$ Å the in-plane C-C bond length, $d_0 = 3.35$ Å the interlayer distance, and $\rho = 0.184b$ the characteristic decay length. $\gamma_0 = -2.7$ eV is the intralayer nearest-neighbor hopping integral and $\gamma_1 = 0.48$ eV the interlayer vertical atomic interaction. This model is reliable and suitable for investigating the essential properties of the geometry-modulated bilayer graphene. In the presence of an external perpendicular electric field, the opposite onsite energies between two layers is used to simulate the layer-dependent Coulomb potentials.

The bilayer graphene systems, including the normal and twisted stacking configurations, exhibit the rich band structures, mainly owing to the regular/enlarged unit cells and distinct interlayer hopping integrals. For example, the AA and AB stackings ($\theta = 0°$ and $60°$), with only four carbon atoms in a primitive cell, possess two pairs of low-lying valence and conduction bands which are initiated from the K/K′ valleys, as

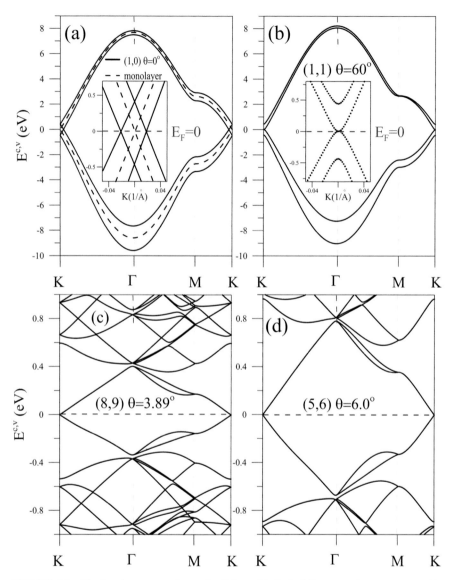

FIGURE 4.2   The band structures due to the C-$2p_z$ orbitals for bilayer graphene systems with the different twisted angles: (a) $\theta = 0°$ [AA], (b) $60°$ [AB], (c) $3.89°$ [(8,9)], and (d) $6.0°$ [(5,6)]. Also shown in (a) is that of monolayer graphene by the dashed curve.

clearly displayed in Figs. 4.2(a) and 4.2(b). The main features of the former cover two vertically separated Dirac-cone structures in the linear and isotropic form, and free valence holes and conduction electrons due to the AA stacking symmetry. The Dirac-point energies are dominated by the vertical interlayer hopping integral [details in Ref. [26]]. The energy spacing of two Dirac points gradually grows in the increase of gate voltage. This also leads to the significant enhancement of the Fermi-momentum states and thus the 2D free carrier densities. It should be noticed that such zero band-gap band structure will create an optical excitation gap [discussed later in Fig. 4.6]. As to the higher-/deeper-energy states ($\sim |E^{c,v}| \geq 2$ eV), the M valley is generated by the saddle point. On the other hand, electronic states of the latter, as shown in Fig. 4.2(b), have parabolic energy dispersions near the K point under a very weak valence and conduction overlap (inset). Two regular stackings belong to semimetals because of a finite density of states at the Fermi level [Fig. 4.5]. Very interestingly, a direct band gap is opened by the gate voltage; that is, the semimetal-semiconductor transition occurs during the variation of V_z. Furthermore, the monotonous energy dispersions of the first pair are dramatically changed into the oscillatory ones, where the extreme band-edge states consist of the unusual 1D constant-energy loops in the energy-wave-vector space.

Generally speaking, the valence and conduction bands are drastically changed by the twisted angles. At low energies near the Fermi level, the $\theta \neq 0°$ & 60° bi-layer graphene systems, as clearly indicated in Figs. 4.2(c)–4.2(d), 4.3(a), and 4.4(a), exhibit the degenerate Dirac-cone structures at the K/K' valleys with the almost isotropic energy spectra. The Fermi velocities are different for the conduction and valence states; furthermore, their magnitudes quickly decline as θ decrease as a re-sult of more complex interlayer hopping integrals. Such results could be roughly re-garded as the direct superposition of those in separate monolayer graphene within the reduced hexagonal first Brillouin zone; that is, the Moire superlattice of bilayer graphenes creates the nonuniform interlayer hopping integrals, while it hardly affects the low-energy fundamental properties, being quite different from the AA and AB stackings [Figs. 4.2(a) and 4.2(b)]. All the twisted materials only belong to zero-gap semiconductors, in analogy with a single-layer graphene. Each θ-dependent system presents $N_M/2$ pairs of valence and conduction subbands within the whole energy range, being directly reflected in the density of states, optical absorption spectra and magneto-electronic properties (discussed later).

Apparently, the significant energy ranges, which correspond to the different val-leys, are very sensitive to the variation of twisted angles. In addition to the lowest K-point valence/conduction valleys including the Fermi level, the first M-point valleys come to exist in the lower-/shallower-energy region for the conduction/valence bands, compared to the AA and AB stackings [Figs. 4.2(a) and 4.2(b)]. Furthermore, their state energies quickly grow in the increment of θ, such as, (0.15 eV, −0.18 eV), (0.42 eV, 0.45 eV), (0.52 eV, 0.60 eV) and (1.60 eV, −1.65 eV), respectively, originating from $\theta = 3.89°$, $\theta = 6.00°$ $\theta = 9.40°$, and.$\theta = 21.8°$ [Figs. 4.2(c), 4.2(d), 4.3(a), and 4.4(a)]. Very interestingly, there exist four conduction/valence subbands as-sociated with the M-point valleys along MΓ (marked by the green circles), and they are doubly degenerate under the direction of MK. Such energy dispersions might have the rich saddle points and local extreme ones, being situated at the M point, or close to it along MΓ. However, few of them do not present any band-edge states. With the increase of state energy, several Γ-related valleys, which in-cludes those along ΓK, are stably formed, in which their observations are rel-atively easy for the smaller-θ bilayer graphene systems, e.g., $\theta = 3.89°$ in Fig. 4.2(c). In addition, the extra few valleys are built form the K point. It should be noticed that both θ- and $60°-\theta$-dependent systems present the almost identical

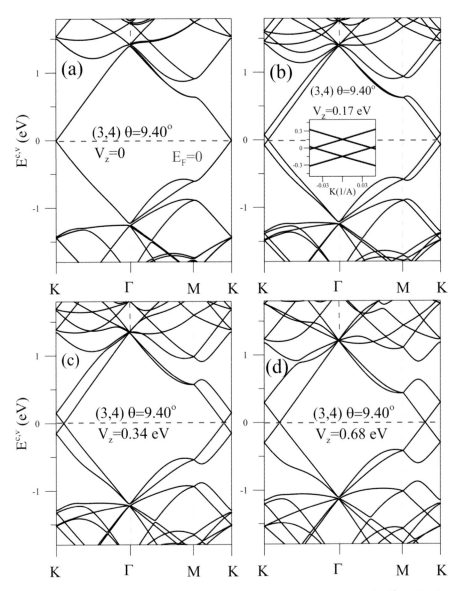

FIGURE 4.3 The band structures of the twisted (3,4) bilayer graphene [$\theta = 9.40°$] under the various gate voltages: (a) $V_z = 0$, (b) 0.17 eV, (c) 0.34 eV, and (d) 0.68 eV.

band structures, such as, electronic energy spectra of the (1,2) and (1.4) bilayer structures [the solid and dashed curves in Fig. 4.4(a)]. Apparently, the band-edge states in the different valleys are expected to exhibit the unusual van Hove singularities and initiate the highly degenerate Landau levels, and their diversified features are due to the combined effects of the zone folding and the complex stacking symmetry.

is principally responsible for the low-energy state splitting. The Fermi-momentum states gradually deviate from the K point in the increase of gate voltage. At zero V_z, the free carriers fully disappear as a result of the absence of valence and conduction band overlap. However, a uniform perpendicular electric field has successfully generated the vertical separation of two isotropic Dirac-cone structures, clearly illustrating their significant overlap across the Fermi level. This result indicates that both valence holes and conduction electrons, respectively, corresponding to the upper and lower Dirac cones, come to exist simultaneously. Furthermore, the 2D free carrier densities grow with the increasing/decreasing of gate voltages/twisted angles. Very interestingly, the cooperation of complex interlayer interactions and layer-dependent Coulomb potentials lead to the split K and K' valleys, such as the different Fermi momenta for free electrons and holes in the insets of Figs 4.4(c)–4.4(d). This result clearly illustrates the critical role of the latter in breaking the equivalence of (A^1, B^1) sublattices. It is further expected to create the valley-related Landau levels [discussed later in Sec. 4.3]. In addition, the splitting of electronic states might appear in other valleys. The above-mentioned features of energy bands will be directly reflected in more complicated van Hove singularities and optical absorption structures.

The ARPES, which possesses the very high resolutions in the measurements of energies and momenta, is the only tool in identifying the wave-vector-dependent quasi-particle energy spectra and widths for the occupied electronic states [details in Refs. [18, 19]]. The measured energy dispersion relations can directly examine the theoretical calculations from the tight-binding model and the first-principles method. In general, the ARPES chamber is accompanied with the instrument of sample synthesis to measure the *in-situ* band structures. Up to now. the experimental measurements have confirmed the feature-rich occupied band structures in the graphene-related systems with sp^2 bondings, as observed under the various dimensions [27, 28], layer numbers [29], stacking configurations [29], substrates [31], and adatom/molecule chemisorptions [32]. The rich and unique electronic energy spectra cover the 1D parabolic energy subbands with the specific energy spacings and band gap in graphene nanoribbons [27], the downward valence Dirac cone in monolayer graphene [31], two pairs of 2D parabolic bands in bilayer AB stacking [18] the coexisting linear and parabolic dispersions in symmetry-broken bilayer graphenes [18], the linear and parabolic bands in tri-layer ABA stacking [29], the linear, partially flat and sombrero-shaped bands in tri-layer ABC stacking [29], the substrate-induced large energy spacing between the π and π^* bands in bilayer AB stacking [30], the substrate-induced oscillatory bands in few-layer ABC stacking [31], the semimetal-semiconductor transitions and the tunable low-lying energy bands after the molecule/adatom adsorptions on graphene surface [32], the 3D band structure, with the bilayer- and monolayer-like energy dispersions, respectively, at $k_z = 0 = 0$ and 1 (K and H points in the 3D first Brillouin zone) and the strong wrapping effect along the KH axis, for the AB-stacked (Bernal) graphite [28]. Specifically, the theoretical predictions on the low-lying valence bands of the twisted bilayer graphene systems are verified by the delicate ARPES analysis and examinations, such as the valence Dirac-cone structure of the K-point valley and the parabolic dispersions near the saddle M-point [23]. The other stable valleys [Figs. 4.2–4.3], which are associated with the band-edges states at the deeper energies, are worthy of the further verifications.

Furthermore, the gate-voltage-enriched energy dispersions. the V_z-induced splitting of K and K$'$ valleys, and semiconductor-semimetal transitions require the experimental checks. They can provide the full information on the complicated interlayer hopping integrals.

The main features of energy bands in bilayer graphene systems result in the rich van Hove singularities. Their densities of states, which are the C-$2p_z$ orbitals, as shown in Figs. 4.5(a)–4.5(f), sharply contrast between the normal and twisted stackings. The AA bilayer stacking [the black curve in Fig. 4.5(a)] has one plateau and a pair of cusp structures across the Fermi level, directly reflecting the semimetallic behavior of two separated vertical Dirac cones near the K/K$'$ point [Fig. 4.2(a)]. Furthermore, another pair of prominent peaks, which are divergent in the logarithmic form, originate from the saddle M-point valleys at $\sim$ 2.00 eV & 2.80 eV/$\sim$ -2.50 eV & -3.50 eV for conduction-/valence-band states. But for the AB-stacked system [Fig. 4.5(b)], only a pair of neighboring shoulders, with a narrow energy spacing ($\sim$20 meV in the inset), appear at the left- and right-hand sides of $E_F = 0$. It also belongs to a semimetal, but presents the lower free carrier density compared with the AA case. At higher/deeper energies, the energy spacings of strong symmetric peaks become smaller, and even vanishing under the merged conduction valley [Fig. 4.2(b)]. Apparently, the special structures of the AA stacking are somewhat widened and lowered during the increment of gate voltage (the red and blue curves). On the other hand, the AB stacking exhibits the dramatic transformation in the low-lying structures and the obvious splitting of prominent conduction peaks. That is to say, a conduction/valence shoulder structure is replaced by an asymmetric peak in the square root divergent form and a shoulder with an observable band gap. Specifically, such peaks are created by the extreme constant-energy-loops, since their (k_x, k_y)-states consist of the effective 1D parabolic energy dispersions.

Because of the $\theta \neq 0°$ twisting effects, these bilayer graphenes exhibit the unique characteristics in the density of states, as clearly indicated by Figs. 4.5(c)–4.5(f). All the twisted systems have a V-shape structure with a zero DOS at the Fermi level and the different slopes in the left- and right-hand sides. This result directly reflects the degenerate Dirac cones in the distinct Fermi velocities [inversely proportional to DOS; Figs. 4.2(c), 4.2(d), 4.3(a), and 4.4(a)]; therefore, the vanishing valence and conduction band overlaps dominate the zero-gap semiconducting behaviors. When state energies become higher/deeper, a prominent symmetric peak in the logarithmic form comes to exist in the positive and negative regions, being followed by the obvious shoulders. Such van Hove singularities mainly originate from the saddle and extreme band-edge states near the M point. Furthermore, their energies are very sensitive to the twisted angles. For example, the separated/composite special structures are more close to the Fermi level for a smaller θ. Apparently, they also appear at the other range ranges as a result of the saddle M-point valleys and the extreme M-, Γ-, and K-point ones. Specifically, the gate voltages create the splitting of band-edge states and thus might induce some split van Hove singularities. The specific unusual structure, which crosses the Fermi level, is revealed as a plateau and two-side cusp structures. It represents a semimetallic system in the presence of a significant overlap behind two vertical Dirac-cone energy spectra/two pairs of valence and conduction bands.

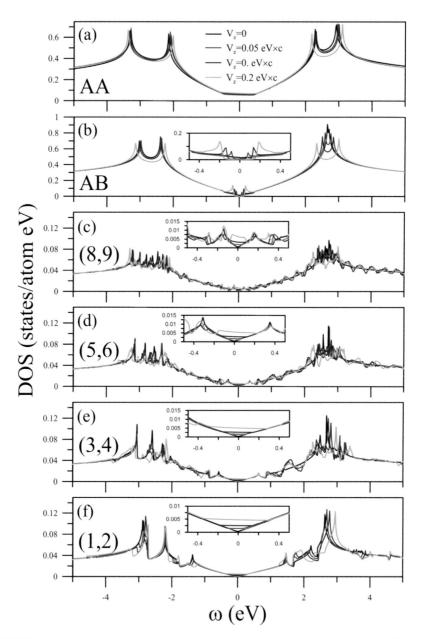

FIGURE 4.5   The twisted-angle- and gate-voltage-dependent densities of states for the bilayer graphene systems under $V_z = 0$. 0.05 eV, 0.10 eV, and 0.20 eV: (a) AA, (b) AB, (c) (8,9), (d), (5,6), (e) (3,4), and (f) (1,2).

The high-resolution STS measurements [details in Sec. 2.1], which can clearly identify the van Hove singularities due to the low-lying critical points of valence and conduction states, are very useful in examining the relation between stacking config- urations and energy bands in the twisted bilayer graphene systems. Up to now, the experimental examinations are consistent with the theoretical predictions [Fig. 4.5 ; Refs. [16, 17]]. As to the various θ-dependent bilayer materials, they have verified a pair symmetric peaks in the logarithmic form below and above the Fermi level, obviously arising from the lower-energy saddle M-point [Figs. 4.2 and 4.4]. For ex- ample, the energy spacings between the valence and conduction peak structures are observed to be 12 meV, 82 meV, and 430 meV for $\theta = 1.16°$, $1.79°$, and $3.40°$, re- spectively [16]. Apparently, the saddle-point energies are very sensitive to the twisted angles, as further confirmed by mre cases [17]. From the theoretical and experimental results, the very large Moire superlattices, with the significant zone-folding effects, are capable of greatly reduce the saddle-point energies, such as the obvious difference between the twisted bilayer and monolayer graphene ($\sim \pm 2$ eV) systems.

4.2   OPTICAL ABSORPTION SPECTRA

The rich electronic energy spectra and wave functions are responsible for the unusual optical properties in the twisted bilayer graphene systems. The vertical photon exci- tations are the transition channels from the occupied states to the unoccupied ones with the same wave vectors; therefore, they strongly depends on the joint density of states (JDOS) related to the initial and final states and the dipole matrix elements. The former, JDOS, might be dominated by the van Hove singularities of the band-edge states; that is, the frequency, form, and intensity of absorption structures would be determined by them. Furthermore, the wave functions of such critical points could create the finite dipole moments and thus the available optical excitations due to the almost symmetric valence→conduction channels. The typical stacking configurations are chosen for a model study, and the significant differences among the θ-dependent bilayer materials are explored in detail. As discussed earlier in the V_z-diversified band structures [Figs. 4.3 and 4.4], the gate voltage is expected to induce the dramatic trans- formation in absorption spectra, especial for the threshold structure associated with the semiconductor-semimetal transition.

The optical excitation spectra of graphene systems exhibit the diverse absorp- tion structures through the twisting-induced stacking configurations. A single-layer graphene, as shown in Fig. 4.6(a) by the dashed black curve, exhibits the linear ω-dependence at the low absorption frequency. An optical gap, the threshold fre- quency, is vanishing; furthermore, absorption structures are absent below ($2\gamma \sim 5$ eV). These results directly reflect energy spectrum and wave functions of the valence and conduction Dirac cones. It should be noticed that the low-frequency dipole matrix el- ement is proportional to the Fermi velocity ($v_F = 3\gamma_0 b/2$) [26]. In addition, there ex- ists a strong π-electronic absorption peak at $\omega = 2\gamma_0$, being initiated from the saddle M-point. Such prominent symmetric structure is frequently observed in the carbon- sp^2 bonding systems [26].

The AA-stacked bilayer graphene with $\theta = 0°$, presents the unusual low- frequency absorption spectra, as indicated in Fig. 4.6(a) by the solid black curve.

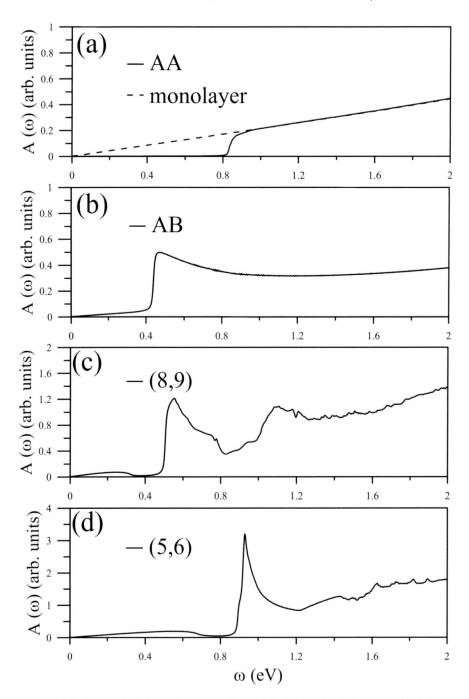

FIGURE 4.6 The optical absorption spectra for (a) AA, (b) AB, (c) (8,9), and (d) (5,6). Also shown in (a) is that of monolayer graphene by the dashed curve.

The critical picture is the well-behaved two pairs of valence and conduction Dirac cones almost symmetric about the Fermi level [Fig. 4.2(a)]. Their wave functions are the symmetric or antisymmetric linear superposition of the layer- and sublattice-dependent four tight-binding functions [26]. As a result, the available excitation channels only originate from the same Dirac-cone structure, in which their diploe moments are roughly same with that of monolayer graphene [26]. That is to say, the inter-Dirac-cone vertical transitions are forbidden during the vertical optical excitations, mainly owing to vanishing dipole matrix elements. The first and second Dirac-cone structures, respectively, possess the band-overlap-induced free holes and electrons; therefore, the Fermi-Dirac distribution functions will dominate the threshold absorption frequency and structure. From the analytic energy spectrum of the bilayer AA stacking [33], its optical gap, being accompanied with the initial shoulder structure, is examined to be approximately double of that the vertical interlayer hopping integral ($\sim 2\gamma_1$; [33]). The absorption gap is gradually enhanced by the external gate voltages. The comparable absorption frequencies are predicted for the AA-stacked graphenes with the even layer numbers. However, the N-odd few-layer systems have the zero threshold frequency, since the Dirac point of the middle cone structure touches with the Fermi level. For example, the trilayer material presents the composite absorption spectra of monolayer and bilayer ones. Apparently, the low- and middle-frequency optical citations are, respectively, dominated by the electronic states close to the K and M points.

The $\theta = 60°$ bilayer AB stacking, as illustrated in Fig. 4.6(b), clearly displays the unique absorption spectra. According to band structure in Fig. 4.2(b), the vertical threshold excitations are closely related to the first pair of parabolic valence and conduction bands across the Fermi level. Their dipole matrix elements near the band-edge states are finite, so that both optical and band gaps are zero. Specifically, those, which are due to the second pair of parabolic energy dispersions, do not create the finite contribution, i.e., the absence of absorption shoulder structure at $\omega \sim 2\gamma_1$. However, there is a broadened discontinuity, being revealed at $\omega \gamma_1$. This specific shoulder structure arises from the band-edge-state excitations of the first/second valence band and the second/first conduction one. In addition, an optical gap is presented when the external gate voltage is sufficiently large [26].

The $\theta \neq 0°$ and $6°$ twisted bilayer graphene systems display a lot of absorption structures below the middle-frequency π-electronic absorption peak ($\omega < 5$ eV), as indicated in Figs. 4.6(c), 4.6(d), and 4.7(a) by the solid black curves. The lower-frequency absorption spectra, which are due to the degenerate valence and conduction Dirac cones of the K/K′ valley [Figs. 4.2(c), 4.2(d), and 4.4], reveals the linear ω-dependence in the spectral intensity. The similar result is presented in monolayer graphene [the dashed curve in Fig. 4.6(a)] because of the almost identical dipole moments. With the increasing absorption frequency, three observable/four prominent symmetric peaks come to exist as a result of the electronic states near the saddle M-point [Fig. 4.7(b)]. For example, the (1,2) bilayer graphene exhibit the specific absorption peaks at $\omega_a \sim 3.11$ eV, 3.35 eV, and 3.45 eV [Fig. 4.7(a)]. By the delicate analysis using the absorption spectrum and joint density of states (the red curve), the first, second, and third peaks originate from the vertical transitions of the band-edge states near saddle M-point, respectively, corresponding to (the shallower

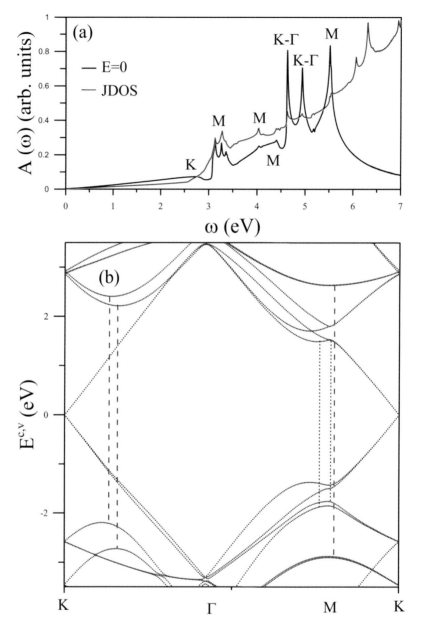

FIGURE 4.7 (a) The joint densities of state [the red curve] and optical excitation [the black curve] for the (1,2) bilayer graphene, and (b) the significant vertical transition channels corresponding to the strong absorption structures through the dashed blue and purple lines for the initial two and others, respectively.

valence band, the lower conduction one). (the shallower/deeper valence band, the higher/lower conduction one), and (the deeper valence band, the higher conduction one), as clearly indicated by the dashed blue lines in Fig. 4.7(b). That is, all the vertical excitations are available in the first and second pairs of parabolic valence and conduction bands. The absorption frequency of the initial prominent peaks is obviously reduced through the decrease of the twisted angle, e.g., ω_a 0.62 eV and 0.28 eV for the bilayer (5,6) and (8,9) bilayer graphene systems, respectively [Figs. 4.6(c) and 4.6(d)]. Apparently, the frequency, intensity, and number of the π-electronic absorption peaks in bilayer graphene materials are very sensitive to the change of twisted angle, especially for the wide-range variation of absorption frequency ($\omega_a \sim 0.1 - 6.0$ eV). Generally speaking, it is relatively easy to observe the initial strong peaks in the small-θ bilayer systems from the high-resolution optical measurements [Sec. 2.2]. Also, there exist other obvious absorption peaks and shoulders at higher excitation frequencies, such as those in the (1,2) bilayer system indicated by the dashed purple lines in Fig. 4.7(b). Such absorption structures are clearly identified from the critical points near the K, M, and γ valleys. They become more complicated for the twisted bilayer graphenes with smaller θs [Figs. 4.6(c) and 4.6(d)].

The vertical optical excitations are greatly enriched by the external gate voltages through the strong cooperations/competitions between the layer-dependent hopping integrals and Coulomb potential energies, e.g., the V_z-modified absorption spectra of the (1,2) bilayer system. The optical gap of the AA bilayer stacking is getting large in the increment of V_z [the dashed red curve in Fig. 4.9], mainly owing to the enlarged energy spacing of two vertical Dirac points. As for the bilayer AB stacking, the threshold absorption frequency is required to match with the opening of energy gap [the dashed blue curve in Fig. 4.9]; that is, it is associated with the specific semimetal-semiconductor transition, as identified in the experimental transport measurements [34]. Very interestingly, the unusual semiconductor-semimetal transition, in any twisted bilayer systems [e.g., the (1,2) system in Fig. 4.4], creates an optical gap even in the presence of gate voltages, as clearly illustrated in Figs. 4.8(a)–4.8(d). The destruction of the double degeneracy in valence and conduction Dirac cones, which is two vertical Dirac ones and is responsible for this dramatic transformation. The similar result is revealed in the pristine AA case [Fig. 4.6(a)]. This threshold absorption frequency grows with the increasing gate voltage [Fig. 4.9]. Moreover, the other/extra absorption structures are modified/created during the variation of V_z. For example, the (1,2) bilayer graphene exhibits an extra absorption peak, being due to the band-edge states in the first pair of valence and conduction bands near the M point [Fig. 4.8(d)], might appear and become an initial one, such as, the $\omega_a = 2.95$-eV absorption peak at $V_z = 0.1$ eV [Fig. 4.8(c)]. Also, the gate-voltage dependence of its excitation frequencies due to the initial four absorption structures is very important since the calculated could provide the critical V_zs in observing the emergent transition channels.

The optical absorption [35], transmission [36], and reflection [37] spectroscopies are available in examining the theoretical predictions on the vertical transitions [details in Sec. 2.2]. Up to date, the high-resolution measurements on bilayer graphene systems have confirmed the rich absorption structures only for the AB stacking, covering the ~ 0.3-eV shoulder structure under a vanishing field [36], the

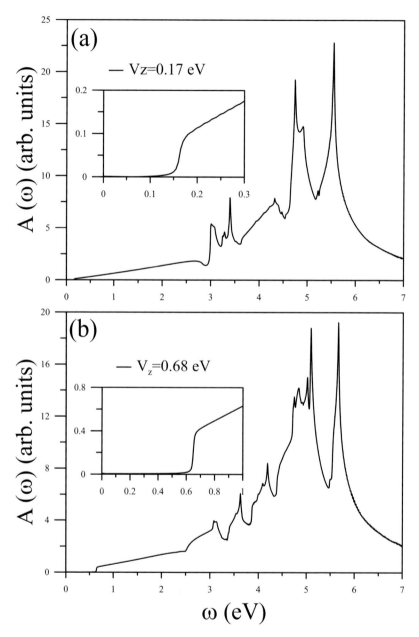

FIGURE 4.8 (a)(b) The gate-voltage-enriched absorption spectra of the (1,2) bilayer graphene at (a) $V_z = 0.17$ eV, (b) $V_z = 0.68$ eV, and (c) $V_z = 0.34$ eV; furthermore, (d) certain vertical transitions under the (c) case.

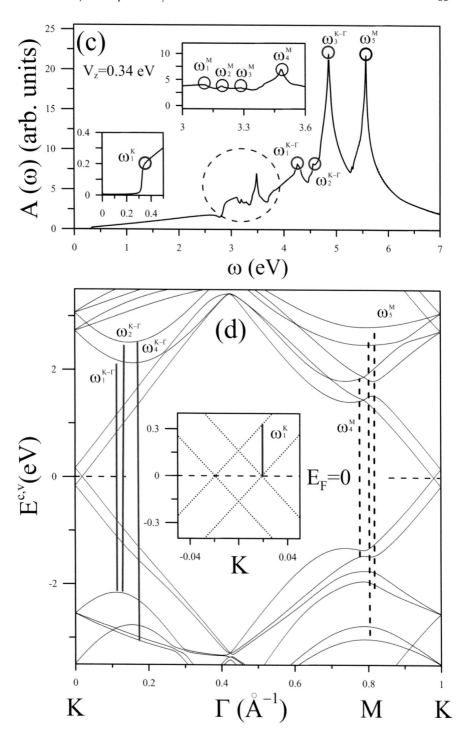

FIGURE 4.8 (c)(d) The gate-voltage-enriched absorption spectra of the (1,2) bilayer graphene at (a) $V_z = 0.17$ eV, (b) $V_z = 0.68$ eV and (c) $V_z = 0.34$ eV; furthermore, (d) certain vertical transitions under the (c) case.

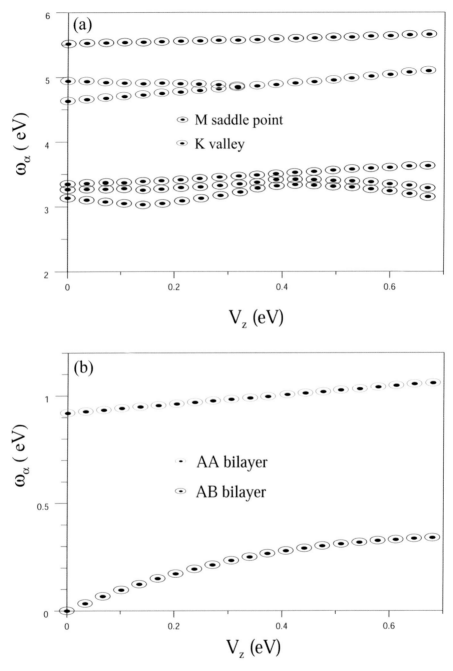

FIGURE 4.9 The gate-voltage-dependent excitation frequencies corresponding to (a) the low-lying K valley and M saddle structure in the bilayer (1,2) graphene system and (b) the optical gaps of the AA and AB stackings.

semimetal-semiconductor transition and two low-frequency asymmetric peaks for the sufficiently large gate voltages, and two rather strong π-electronic absorption peaks at middle frequency [36]. The delicate optical experiments are required to verify the theoretical predictions for the $\theta \neq 0°$ and $60°$ twisted bilayer graphenes, e.g., the zero threshold frequency for any systems, the initial prominent absorption peaks strongly depending on θ, the other higher-frequency excitation structures, the gate-voltage-created optical gaps similar to the AA case, and the V_z-enriched/modified transition channels. Such optical ions, as done by the ARPES and STS measurements, clearly identify the composite effects due to the zone folding and complex interlayer hopping integrals in the Moire superlattices.

4.3 MAGNETICALLY QUANTIZED LANDAU LEVELS FROM THE MOIRE SUPERLATTICE

The generalized tight-binding model is reliable in characterizing the magnetic wave functions, while the opposite is true for the effective-mass approximation. By the delicate calculations, one can derive the analytic formulas for the independent matrix elements of the magnetic Hamiltonian. The moire superlattices are very different for the twisted bilayer graphene systems; therefore, it is impossible to get the θ-dependent equations. Only the specific cases, which are full of the important features in magnetic quantizations, are suitable for the model studied. For example, the (1,2)-twisted bilayer graphene has the smallest number of the tight-binding functions, with the fourteen (A_i^l, B_i^l) sublattices in each layer [Fig. 4.1]. The new spatial period, which is due to the magnetic Peierls phase [the path integration of the vector potential in Sec. 3.1], is commensurate to that of the Moire superlattice, e.g., the dimension of $\sim 3,000 \times 3,000$ for the Hamiltonian Hermitian matrix at $B_z = 100$ T. Furthermore, the nonvanishing matrix elements are complex numbers even for the $(k_x = 0, k_y = 0)$ state, so the numerical calculations are more complicated in the twisted systems, but not under the normal stackings (AA, AB, ABC, and AAB configurations [26]). It should be noticed that the Coulomb gauge $\mathbf{A} = \frac{\sqrt{3}\mathbf{B_z}}{6n_0} \times [(2m + n)\widehat{x} - \sqrt{3}n\widehat{y}] \times [(n - m)\widehat{x} + \sqrt{3}(m + n)\widehat{y}$ is along G_1 and utilized for the twisted bilayer graphene.

First, the main features of magnetic quantizations are presented for monolayer graphene and bilayer AA and AB stackings. With the chosen gauge of $\mathbf{A} = (-\mathbf{B_z}y, 0, 0)$, the magnetic-field-modified unit cell is an enlarged unit cell along the armchair direction, as revealed in Fig. 3.2. For monolayer systems (AA and AB configurations), there are $4R_B$ carbon atoms/C-$2p_z$ tight-bindig functions ($8R_B$ carbon atoms), being classified into both A and B sublattices [(A^1, B^1) & (A^2, B^2)]. Their low-energy magneto-electronic states are directly quantized from the upward conduction K/K' valleys and the downward valence ones [Figs. 4.2(a) and 4.2(b)]. Each (k_x, k_y) state has the eight-fold degeneracy in terms of four localization centers (related to $\pm B_z\widehat{z}$ and K and K' valleys) and spin degree of freedom. Under $k_x = 0$ and $k_y - 0$, the oscillatory Landau-level wave functions are localized at the 1/6, 2.6, 4/6, and 5/6 of the enlarged unit cell, e.g., the first case in Figs. 4.10(a)–4.10(c). Apparently, the dominating sublattice in monolayer graphene belongs to the A/B sublattice for the 1/6/2/6 localization center; that is, the quantum number of Landau level ($n^{c,v}$) is characterized by its zero-point number [Fig. 4.10(a)]. The significant difference of zero points in A and B sublattices is equal to one, indicting the full equivalence between them. This property remains unchanged the normal AA and AB stackings because of the identical carbon-atom honeycomb lattices. When the interlayer chemical & physical environments are taken into account, the oscillation modes of four sublattices are quite different in these two systems. For the former, both (A^1, A^2) and (B^1, B^2) exhibit the totally same oscillations [Fig. 4.10(b)]. However, the latter [Fig. 4.10(c)] presents only the similar oscillations on (A^1, A^2) sublattices and the mode difference of 2 on (B^1, B^2) sublattices with the distinct C-atom projections. Concerning the magneto-electronic energy spectra (monolayer, bilayer AA) and AB, respectively, display the massless and massive behaviors (the $\sqrt{n^{c,v}B_z}$ and $n^{c,v}B_z$ dependences). In addition, there are two groups of valence and conduction Landau levels associated with the stable K/K′ valleys.

The (1,2) bilayer graphene is very suitable for a model study on the rich magnetic magnetization due to the Moire superlattice. Apparently, this system will present fourteen subgroups of conduction and valence Landau level, as magnetically quantized from the zero-field band structure [Fig. 4.4(a)]. Here, only the two subgroups, which are closely related to the low-energy Dirac cones arising from the K and K′ valleys, are fully explored for their main features [Figs. 4.11 and 4.14]. After diagonalizing the exact Hamiltonian matrix with a lot of complex elements, the Landau-level wave functions and the B_z-dependent energy spectra are obtained through the reliable and efficient way. In addition, an approximate Hamiltonian is not suitable for investigating the magneto-electronic properties. Very interestingly, there exists the eight-fold degeneracy in each Landau, covering the same first and the second subgroups, the double-degenerate composite localization centers (due to $\pm B_z\hat{z}$), and the independent spin-up and spin-down configurations (ignored later). For example, at $B_z = 100$ T [the second red point in Fig. 4.14], the second conduction Landau levels exhibit the unusual probability distributions in the magnetic-field-enlarged unit cell, as clearly indicated by (α, β, γ, δ) in Fig. 4.11 for the four-fold degenerate Landau states without spin arrangements. The oscillatory magnetic wave functions, which consist of fourteen subenvelope functions in each graphene layer, are localized about (2/6, 4/6) simultaneously on the upper/lower layer (the black/red curves), and near 2/6 or 4/6 singly under the opposite cases. That is to say, the α (β) state, respectively, exhibits (2/6, 4/6) and 2/6 (4/6 and (2/6 4/6)) localization modes on the upper and lower layers. Furthermore, the interchange of 2/6 and 4/6 localization centers in the α and β degenerate states just corresponds to the γ and δ ones. The critical factors in creating the four-fold degenerate Landau levels come from the equivalence of two layers and localization centers (the linear superposition of the tight-binding functions related to them).

The main features of magnetic oscillation modes in the twisted (1,2) bilayer graphene deserve a closer examination, especially for the unusual relations with those due to monolayer, and bilayer AA & AB stackings. From the viewpoint of 7 A^l_i/7 B^l_i sublattices on two single-layer graphenes [Fig. 4.12(a)], the nonuniform physical/chemical environment, which creates the complex interlayer hopping integrals, is response for the modified/non-identical probability distributions, as shown in

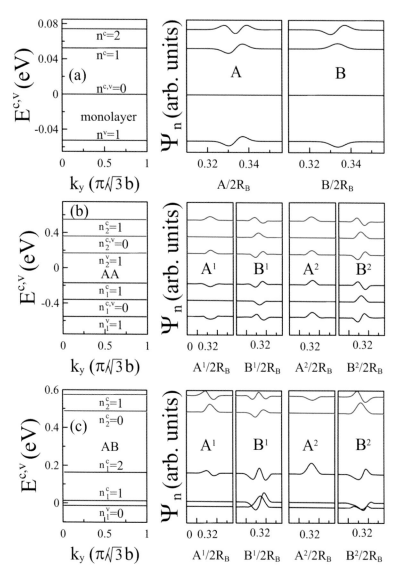

FIGURE 4.10 The Landau-level energies and subenvelope functions for the (a) monolayer and bilayer (b) AA and (c) AB staclings.

Fig. 4.11. For example, this result is clearly illustrated by the simultaneous localization of (2/6, 4/6) on the upper/lower layer of the α-/β-state Landau level. Generally speaking, all the magnetic wave functions do not exhibit to the symmetric or antisymmetric well-behaved spatial distributions; that is, they should belong to the perturbed type with the major (n) and minor ($n \pm 1$) modes. The dominating oscillation modes of the α-state Landau level, being associated with two localization centers, are summarized on the fourteen lattice sites of the upper/lower graphene layer [Fig. 4.12(b)].

$E^c(n^c=1)[\alpha,\beta,\gamma,\delta]=0.25$ eV

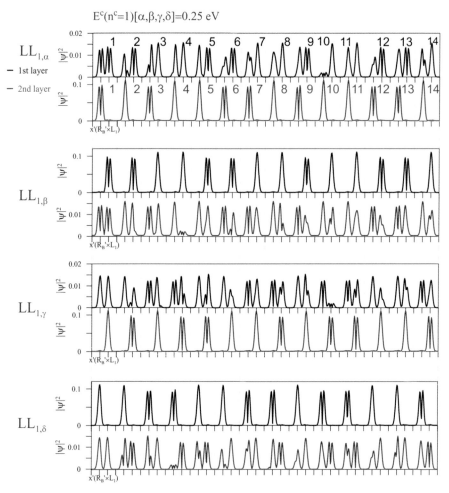

FIGURE 4.11 The magneto-electronic wave functions of the twisted (1,2) bilayer graphene, being built from the amplitudes of the 14-sublattices tight-binding functions on the upper and lower layers [the black and red curves], for the second conduction Landau levels of the first and second subgroups with the four degenerate states α, β, γ, and δ.

They present three specific relations. First, the lower layer, as shown by the red dots, only has the 2/6 localization probability distributions with the n and $n-1$ modes for the A_i^2 and B_i^2 sublattices, respectively. The equivalent results are revealed in the dominating modes of the 4/6 localization on the upper layer (the black dots). This property behaves like that of monolayer graphene [Fig. 4.10(a)]. While the planar projection displays a short distance <b/3, the two lattice sites on two distinct layers possess the same oscillation mode about the 2/6 localization center, as revealed in bilayer AA stacking [Fig. 4.10(b)]. Specifically, the 10th carbon atom on the upper layer [the balack dot in Fig. 4.12(a)] is just projected into the center of the lower hexagon; therefore, its $n+1$ mode about the 2/6; localization center is accompanied

(a)

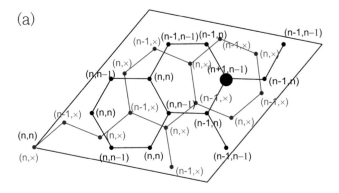

(b)

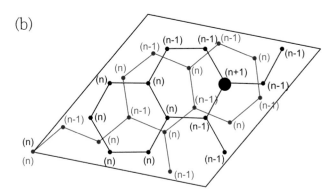

FIGURE 4.12 (a): For the (1,2) bilayer graphene, the oscillation modes of the 14 sublattices in the upper and lower layers, being illustrated for the α state at $B_z = 100$T and (a) $V_z = 0$ and (b) $V_z = 0.34$ eV. (b): For the (1,2) bilayer graphene, the oscillation modes of the fourteen sublattices in the upper and lower layers, being illustrated for the α state at $B_z = 100$T and (a) $V_z = 0$ and (b) $V_z = 0.34$ eV.

with the n and $n - 1$ modes on another layer [Fig. 4.1]. Such behavior corresponding to the bilayer AB stacking [Fig. 4.10(c)]. The similar phenomena are obtained for the β-, γ-, and δ-state Landau levels. In short, the rich magnetic quantizations in the θ-dependent twisted bilayer graphene systems originate from the highly hybridized characteristics of monolayer system, and AA & AB stackings.

A perpendicular Coulomb potential leads to the layer-dependent site energies and thus the obvious splitting of Dirac-cone structures/the separated Landau-level energy spectra, as clearly illustrated in Figs. 4.4 and 4.13. As for the K-/K'-induced magneto-electronic states, the four-fold degeneracy dramatically changes into the doubly one. That is to say, the $(\alpha, \beta)/(\gamma, \delta)$ states have the same quantized energy under the specific symmetry of $\pm B_z \hat{z}$, where the former/the latter display the higher/lower one. The magnetic wave functions are only localized about 2/6 or 4/6 of the B_z-enlarged unit

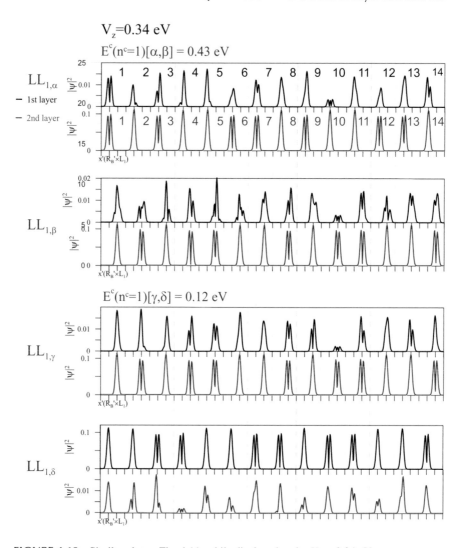

FIGURE 4.13 Similar plot as Fig. 4.11, while displayed under $V_z = 0.34$ eV.

cell [Fig. 4.13]. The coexistence of two localization centers on the upper or lower layer [Fig. 4.11] is absent through the full breaking of the mirror symmetry about the $z = 0$ plane. This phenomenon is examined in detail and should be independent of the magnitude of gate voltage (not shown). Roughly speaking, all the subenvelope functions on the twisting-induced multi-sublattices present the major $n^{c,v}$, $n^{c,v} + 1$, or $n^{c,v}1$, or they belong to the hybridized states of three oscillation modes. Furthermore, the robust relations on the intralayer & interlayer neighboring A_i^l and $B_j^{l'}$ subalttices could also be understood from those of monolayer; bilayer AA and AB stackings, as done for the $V_z = 0$ case [Fig. 4.12(b)]. It is also noticed that few lattice positions, with the AB stacking configuration, will lead to the destruction of the ± 1 mode difference

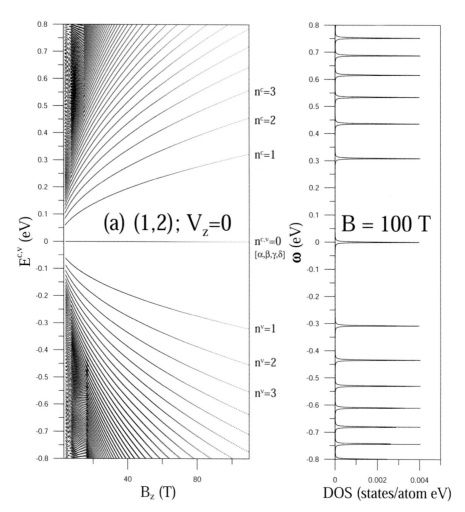

FIGURE 4.14 (a) The magnetic-field-dependent Landau-level energy spectrum of the (1,2) bilayer graphene system, being accompanied with (b) the density of states.

for the nearest sublattices on the same layer, e.g., those of the 10 & 9 sites [the first row in Fig. 4.13; the black dots in Figs. 4.12(a) and 4.12(b)]. Apparently, the equivalence of A^l and B^l sublattices is strongly modified by the gate voltage, or the competition among three kinds of geometric symmetries is enhanced by V_z.

The magnetic-field-dependent electronic energy spectra, as clearly illustrated in Figs. 4.14 and 4.15, are closely related to the main features of band structure [Figs. 4.2–4.4]. The low-lying Landau levels, being induced by the stable K/K' valleys, exhibit the mono-layer-like behavior at a zero gate voltage. For example, the non-crossing Landau level energies of the (1,2) bilayer graphene are roughly proportional to ($\sqrt{n^{c,v}B_z}$) in the wide energy range $|E^{c,v}| < 1.20$ eV [Fig. 4.14(a) at $V_z = 0$]. Such range is very sensitive to twisted angles and will quickly decline in the decrease

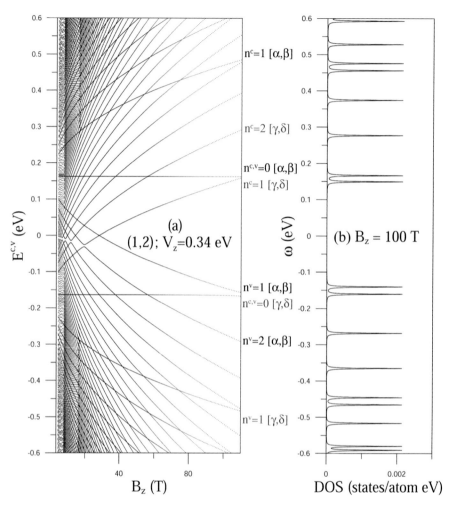

FIGURE 4.15   Similar plot as Fig. 4.14, but shown at (a) $V_z = 0.10$ eV with (b) the density of states.

of θ. The gate voltage creates two vertical Dirac-cone structures and thus the composite energy spectra of two monolayer-like ones with the frequent crossing phenomena, e.g., Fig. 4.15(a) at $V_z = 0.34$ eV, as observed in the bilayer AA case [discussed later in Sec. 9.1]. Very interestingly, the highly degenerate Landau levels shows the delta-function-like peaks in density of states, in which their energies/energy spacings, intensities and numbers strongly depend on the magnetic fields [Fig. 4.14(b)], and gate voltages [Fig. 4.15(b)]. For the deeper/higher valence/conduction states, the magnetic quantizations are expected to become very complicated as a result of the non-monotonous dispersion relations and the multi-constant energy loops, e.g., the magneto-electronic states associated with the saddle M-points. Whether the anti-crossing behaviors come to exist is under the current investigations. The theoretical predictions could be verified from the high-resolution spin-polarized STS [details in Sec. 2.1].

The main features of magneto-electronic energy spectra and wave functions are reflected in density of states, absorption spectra, Hall transport conductivities, and magnetoplasmon modes, where the essential properties are worthy of the further theoretical calculations. For example, the frequent Landau-level crossings and/or anti-crossings might be created through the unusual dispersion relations near the saddle M-points [e.g., Figs. 4.2(c) and 4.2(d)]. Obviously, this will lead to the abnormal delta-function-like van Hove singularities in terms of the magnetic-field-dependent energies and numbers [26], the coexistence of the regular and extra magneto-optical selection rules [26], the integer and non-integer quantum plateaus [38]; the inter-Landau-level single-particle & collective excitations and the 2D electron-gas-like magnetoplasmons. Except for the first term, how to solve the technical barriers due to the evaluations of the other essential properties are under the current investigations. On the experimental side, only the quantum transport measurements are conducted on the twisted bilayer graphenes [20], and the Hall conductivities are the superposition of these in two monolayer systems. This result is consistent with the quantized Landau levels from two completely degenerate Dirac cones crossing the Fermi level [Figs. 4.4, 4.11, and 4.14].

4.4 SIGNIFICANT DIFFERENCES BETWEEN TWISTED AND SLIDING SYSTEMS

There exist significant differences between the twisted and sliding bilayer graphenes in the fundamental physical properties. The former and the latter are, respectively, characterized by the specific twistings with two commensurate honeycomb lattices, and the relatively displacement of two graphene layers along the armchair direction and then the zigzag one [Figs. 4.16(a)–4.16(f); details in Sec. 9.1 [39]]: AA→AB→AA′ & AA′→AA. As a result, their primitive unit cells are mainly determined by the greatly enlarged Moire superlattice and the AA-like ones, respectively, including many and four carbon atoms. The numerical calculations are relatively easy for the sliding-induced electronic, optical, and transport properties, as well as their magnetic quantizations, mainly owing to the smaller-dimension Hamiltonian matrices [39]. Through modulating the various stacking configurations/symmetries, the physical phenomena become very rich and unique.

Apparently, the zone-folding effects thoroughly disappear in the sliding bilayer systems, clearly illustrating that their two pairs of $2p_z$-orbital-induced valence and conduction bands come to exist in the almost identical energy ranges. The low-energy essential properties are dominated by the low-lying electronic structures closely related to the K-/K′-point valleys, regardless of the higher/deeper M-point valleys. The weak band overlaps clearly indicate the semimetallic behaviors in all the sliding bilayer graphene systems. On the other hand, the θ-dependent bilayer materials are zero-gap semiconductors with the monolayer-like [doubly degenerate] Dirac-cone structures, and the obvious saddle points might appear at low energies for the smaller-θ ones [e.g., (8,9) system in Fig. 4.2(b)]. The relative shifts along the armchair direction can also create the dramatic transformations among the regular energy spectra [details in Sec. 9.1]: two vertical Dirac cones in AA stacking, two pairs of parabolic

armchair direction

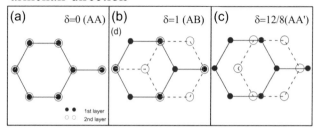

zigzag direction

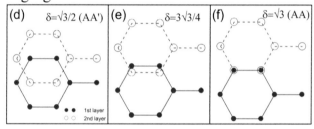

FIGURE 4.16 The sliding bilayer graphene systems with the relative shifts along the armchair and zigzag directions: AA→AB→AA′ & AA′→AA.

energy dispersions in AB configuration, and the non-vertical & tilted Dirac cones of AA′. That is to say, the linear & isotropic Dirac cones are thoroughly destroyed, and then the other irregular/normal energy bands are formed as the intermediate electronic structures. The main features of zero-field band structures cover the serious distortions/hybridizations in energy dispersions/bands, the creation of arc-shaped stateless regions, more band-edge states near the K/K′ points (saddle and extreme points), the strong dependence of band overlap on the stacking configuration, the finite densities of states at the Fermi level (semimetals), and the gate-voltage-enriched characteristics.

Concerning the magnetic quantization phenomena, the sliding bilayer graphene systems exhibit two groups of $2p_z$-orbital-crated and conduction Landau levels and the shift-induced three kinds of Landau levels. The magneto-electronic states are directly magnetically quantized from two pairs of energy bands based on the tight-binding functions of four sublattices [details in Ref. [39]]. The significant oscillation behaviors are characterized by the zero-point number of the localized subenvelope functions on the dominant sublattice. Obviously, the rich Landau levels can be classified into the well-e, perturbed, and undefined modes, in which they, respectively, possess a specific zero point, a major and some minor zero points, and the magnetic-field-dependent ones. The second kind, closely related to the non-monotonous energy dispersions/the non-single constant-energy loops, is accompanied by certain anti-crossings. Furthermore, the third kind, which mainly arises from the thorough destruction of Dirac-cone structures, leads to any anti-crossings between the first and second groups of conduction/valence Landau levels. Such a phenomenon could not survive under the twisted cases. As for the magneto-optical absorption spectra, the first, second, and third kinds of Landau levels generate the $\Delta n = \pm 1$ magneto-optical selection rules, the specific & extra ones, and the random vertical transitions with a lot of weaker absorption peaks. However, there are more Landau-level subgroups due to the specific Moire zone folding. As a result, the state degeneracy might be higher under the normal sliding, compared with the case of angle twisting. The magneto-optical properties and quantum Hall conductivities are worthy of the further systematic studies using the combined generalized tight-binding model and Kubo formulas. In short, each stacking configuration in sliding bilayer materials presents the independent and unusual phenomenon, while that of twisted systems shows the hybridized characteristics associated with those in monolayer, and AA & AB stackings (the complex relations among the A_i^l and B_i^l sublattices on the same and different graphene layers).

REFERENCES

1. Iijima S 1991 Helical microtubules of graphitic carbon *Nature* **354** 56.
2. Cao Y, Fatemi V, Fang S, Watanabe K, and Taniguchi T 2018 Efthimios Kaxiras & Pablo Jarillo-Herrero Unconventional superconductivity in magic-angle graphene superlattices *Nature* **556** 43.
3. Fan Z, Yan J, Zhi L, Zhang Q, Wei T, Feng J, et al. 2010 A three-dimensional carbon nanotube/graphene sandwich and its application as electrode in supercapacitors *Adv. Mater.* **22** 3723.
4. Wilder J W G, Venema L C, Rinzler A G, Smalley R E, and Dekker C 1998 Electronic structure of atomically resolved carbon nanotubes *Nature* **391** 59.
5. Shyu F L and Lin M F 2002 Electronic and optical properties of narrow-gap carbon nanotubes *J. Phys. Soc. Jpn.* **71** 1820.
6. Saito R, Fujita M, Dresselhaus G, and Dresselhaus M S 1998 Electronic structure of chiral graphene tubules *Appl. Phys. Lett.* **60** 2204.
7. Liu H J and Chan C T 2002 Properties of 4A carbon nanotubes from first-principles calculations *Phys. Rev. B* **66** 115416.
8. Choi H J, Ihm J, Louie S G, and Cohen M L 2000 Defects, quasibound states, and quantum conductance in metallic carbon nanotubes *Phys. Rev. Lett.* **84** 2917.
9. Novoselov K S, Geim A K, Morozov S V, Jiang D, Zhang Y, Dubonos S V, et al. 2004 Electric field effect in atomically thin carbon films *Science* **306** 666.
10. Chung H C, Chang C P, Lin C Y, and Lin M F 2016 Electronic and optical properties of graphene nanoribbons in external fields *Phys. Chem. Chem. Phys.* **18** 7573.
11. Schmidt H, Ludtke T, Barthold P, McCann E, Fal'ko V I, and Haug R J 2008 Tunable graphene system with two decoupled monolayers *Appl. Phys. Lett.* **93** 172108.
12. Berger C, Song Z, Li X, Wu X, Brown N, Naud C, et al. 2006 Electronic confinement and coherence in patterned epitaxial graphene *Science* **312** 1191.
13. Hass J, Feng R, Millan-Otoya J E, Li X, Sprinkle M, First P N, et al. 2007 Structural properties of the multilayer graphene/4H-SiC(000$\bar{1}$)system as determined by surface x-ray diffraction *Phys. Rev. B* **75** 214109.
14. Yan Z, Peng Z, Sun Z, Yao J, Zhu Y, Liu Z, et al. 2011 Growth of bilayer graphene on insulating substrates *ACS Nano.* **5** 8187.
15. Hass J, Varchon F, Millán-Otoya J E, Sprinkle M, Sharma N, de Heer W A, et al. 2008 Why multilayer graphene on 4H-SiC(000$\bar{1}$) behaves like a single sheet of graphene *Phys. Rev. Lett.* **100** 125504.

16. Li G, Luican A, Lopes dos Santos J M B, Castro Neto A H, Reina A, Kong J, et al. 2010 Observation of Van Hove singularities in twisted graphene layers *Nature Physics* **6** 109.

17. Brihuega I, Mallet P, Gonzalez-Herrero H, Trambly de Laissardiere G, Ugeda M M, Magaud L, et al. 2012 Unraveling the intrinsic and robust nature of van Hove singularities in twisted bilayer graphene by scanning tunneling microscopy and theoretical analysis *Phys. Rev. Lett.* **109** 196802.

18. Kim K S, Walter A L, Moreschini L, Seyller T, Horn K, Rotenberg E, et al. 2013 Coexisting massive and massless Dirac fermions in symmetry-broken bilayer graphene *Nat. Mater.* **12** 887.

19. Ohta T, Robinson J T, Feibelman P J, Bostwick A, Rotenberg E, and Beechem T E 2012 Evidence for interlayer coupling and moire periodic potentials in twisted bilayer graphene *Phys. Rev. Lett.* **109** 186807.

20. Fallahazad B, Hao Y, Lee K, Kim S, Ruoff R S, and Tutuc E 2012 Quantum Hall effect in Bernal stacked and twisted bilayer graphene grown on Cu by chemical vapor deposition *Phys. Rev. B* **85** 201408.

21. Morell E S, Pacheco M, Chico L, and Brey L 2013 Electronic properties of twisted trilayer graphene *Phys. Rev. B* **87** 125414.

22. Mele E J 2010 Commensuration and interlayer coherence in twisted bilayer graphene *Phys. Rev. B* **81** 161405.

23. Moon P and Koshino M 2012 Energy spectrum and quantum Hall effect in twisted bilayer graphene *Phys. Rev. B* **85** 195458.

24. Moon P and Koshino M 2013 Optical absorption in twisted bilayer graphene *Phys. Rev. B* **87** 205404.

25. Wang Z F, Liu F, and Chou M Y 2012 Fractal Landau-level spectra in twisted bilayer graphene *Nano Lett.* **12** 3833.

26. Lin C Y, Do T N, Huang Y K, and Lin M F 2017 Electronic and optical properties of graphene in magnetic and electric fields *IOP Concise Physics*. San Raefel, CA, USA: Morgan & Claypool Publishers.

27. Ruffieux P, Cai J, Plumb N C, Patthey L, Prezzi D, Ferretti A, et al. 2012 Electronic structure of atomically precise graphene nanoribbons *ACS Nano* **6** 6930.

28. Sugawara K, Sato T, Souma S, Takahashi T, and Suematsu H 2006 Fermi surface and edge-localized states in graphite studied by high-resolution angle-resolved photoemission spectroscopy *Phys. Rev. B* **73** 045124.

29. Bao C, Yao W, Wang E, Chen C, Avila J, Asensio M C, et al. 2017 Stacking-dependent electronic structure of trilayer graphene resolved by nanospot angle-resolved photoemission spectroscopy *Nano Lett.* **17** 1564.

30. Zhou S Y, Gweon G H, Fedorov A V, First P D, De Heer W A, Lee D H, et al. 2007 Substrate-induced bandgap opening in epitaxial graphene *Nature Materials* **6** 770.

31. Knox K R, Wang S, Morgante A, Cvetko D, Locatelli A, Mentes T O, et al. 2008 Spectromicroscopy of single and multilayer graphene supported by a weakly interacting substrate *Phys. Rev. B* **78** 201408.

32. Starodub E, Bostwick A, Moreschini L, Nie S, Gabaly F E, McCarty K F, et al. 2011 In-plane orientation effects on the electronic structure, stability, and Raman scattering of monolayer graphene on Ir(111) *Phys. Rev. B* **83** 125428.

33. Chang C P 2011 Exact solution of the spectrum and magneto-optics of multilayer hexagonal graphene *J. Appl. Phys.* **110** 013725.

34. Li M Y, Chen C H, Shi Y, and Li L J 2016 Heterostructures based on two-dimensional layered materials and their potential applications *Materials Today* **19** 322.

35. Nair R R, Blake P, Grigorenko A N, Novoselov K S, Booth T J, Stauber T, et al. 2008 Fine structure constant defines visual transparency of graphene *Science* **320** 1308.

36. Orlita M, Faugeras C, Plochocka P, Neugebauer P, Martinez G, Maude DK, et al. 2008 Approaching the Dirac point in high-mobility multilayer epitaxial graphene. *Phys. Rev. Lett.* **101** 267601.
37. Wang F, Zhang Y, Tian C, Girit C, Zettl A, Crommie M, et al. 2008 Gate-variable optical transitions in graphene *Science* **320** 206.
38. Do T N, Chang C P, Shih P H, and Lin M 2017 Stacking-enriched magneto-transport properties of few-layer graphenes *Phys. Chem. Chem. Phys.* **19** 29525.
39. Huang Y K, Chen S C, Ho Y H, Lin C Y, and Lin M F 2014 Feature-rich magnetic quantization in sliding bilayer graphenes *Sci. Rep.* **4** 7509.

5 Stacking-Configuration-Modulated Bilayer Graphene

Chiun-Yan Lin,[a] *Thi-Nga Do,*[c,d] *Jhao-Ying Wu,*[b]
Po-Hsin Shih,[a] *Shih-Yang Lin,*[e] *Ching-Hong Ho,*[b]
Ming-Fa Lin[a,f,g]

[a] Department of Physics, National Cheng Kung University,
Tainan 701, Taiwan
[b] Center of General Studies, National Kaohsiung University of
Science and Technology, Kaohsiung 811, Taiwan
[c] Laboratory of Magnetism and Magnetic Materials, Advanced
Institute of Materials Science, Ton Duc Thang University,
Ho Chi Minh City, Vietnam
[d] Faculty of Applied Sciences, Ton Duc Thang University,
Ho Chi Minh City, Vietnam
[e] Department of Physics, National Chung Cheng University,
Chiayi 621, Taiwan
[f] Quantum Topology Center, National Cheng Kung University,
Tainan 701, Taiwan
[g] Hierarchical Green-Energy Materials Research Center,
National Cheng Kung University, Tainan, Taiwan

CONTENTS

Stacking configurations and external electric & magnetic fields can greatly diversify the essential physical properties. Three typical categories of bilayer graphenes, the geometry-modulated, sliding and twisted systems, clearly display the artificial manipulations of stacking symmetries. The up-to-date experimental synthesis methods on the stacking modulations of bilayer graphene systems cover the mechanical exfoliation [1], and chemical vapor deposition [2], and tips of scanning tunneling microscopy [STM] [3]. Very importantly, the former is capable of tuning the width and position of domain wall (DW) between the normal stackings by manipulating atomic force microscopy (AFM) tip [4]. On the theoretical side, the first-principles method [5] and the tight-binding model [6] have been used to explore energy bands of the AB/DW/BA/DW bilayer graphene, with a very narrow DW width (only comparable to one C-C bond length) and a middle period. Whether the stacking-modulated band structures present quasi-2D or quasi-1D will be clarified in this study. It is well known that a uniform perpendicular electric field leads to an opening of band gap in bilayer AB stacking, confirmed by the theoretical [7] and experimental results [8]. Periodical gate voltages (V_z's), with the opposite Coulomb potentials in the neighboring ones, could be achieved under the delicately experimental designs [8]. This will be very efficient in modulating the low-energy electronic properties. Furthermore, the theoretical calculations predict the existence of 1D topological states [5].

As to bilayer graphene systems, there exist three types of well-behaved stacking configurations in bilayer graphenes [7]. According to the low-lying electronic structures, AA, AA′, and AB stackings, respectively, exhibit the vertical & non-vertical two Dirac-cone structures, and two pairs of parabolic bands. Such band structures are clearly illustrated along the high-symmetry points of the hexagonal first Brillouin zone, so that 2D behaviors are responsible for the other fundamental physical properties. For example, the density of states and optical spectra obviously reveal the 2D van Hove singularities, where the special structures are very sensitive to the dimension. Similar 2D phenomena appear in the sliding and twisted systems [details in Chaps. 4 and 9], while the important differences appear between them [6,9]. Random stacking configurations in the former show the highly distorted energy dispersions with an eye-shaped stateless region accompanied by saddle points. Furthermore, the Morié superlattice of the latter possesses a lot of carbon atoms in a primitive the unit cell and thus creates many 2D energy subbands. Most importantly, the physical properties of these two systems are expected to be in sharp contrast to the geometry-modulated bilayer graphenes.

In this chapter, we first investigate electronic properties and optical excitations in the geometry- and electric-field-modulated bilayer graphene systems, respectively, by using the tight-binding model and gradient approximation [details in Sec. 3.2]. The modulation of stacking configuration will lead to a great enlargement of the unit cell. The significant effects of zone folding on energy subbands, subenvelope functions on distinct sublattices, the density of states, and absorption structures are fully explored under the accurate calculations and delicate analysis. The concise physical pictures are proposed to account for the complex relations among the geometric modulation, the gate voltage, and the 1D behaviors. Very importantly, the current study clearly illustrates the diverse 1D phenomena, a lot of energy subbands with various band-edge states, the metallic behavior even in an electric field, the unusual van Hove singularities, the forbidden vertical excitation channels under the specific linear relations of the layer-dependent sublattices, the prominent asymmetric absorption peaks in absence of selection rule, and the DW- and V_z-created dramatic variations in optical absorption structures. Obviously, the nonuniform magnetization due to the stacking modulation can only be solved by the generalized tight-binding model, but not the effective-mass approximation [details in Sec. 3.1]. Most importantly, the strong competition between the geometric symmetry and magnetic field is studied in detail. This method is very reliable even or the complicated energy bands with the oscillatory dispersions and the multi-constant loops. The non-homogeneous interlayer hopping integrals and Peierls phases are simultaneously included in the diagonalization of the giant magnetic Hamiltonian matrix, where they have induced the high barriers in the numerical calculations. The magneto-electronic properties, energy spectrum, density of states, and wave functions, directly link one another by the competitive/cooperative pictures. The oscillatory Landau subbands are predicted to come to exist under the nonuniform environment, being thoroughly different from the dispersionless Landau levels in the well-stacked graphene systems [7]. The above-mentioned significant 1D features are expected to sharply contrast with the 2D ones in the twisted [10] and sliding [9] bilayer graphenes. Furthermore, they are thoroughly compared with the dimensionality-induced phenomena, e.g., the important differences of Landau subbands among the stacking-modulated bilayer graphenes [6], graphene nanoribbons [11], and AB- and ABC-stacked graphites [12]. Such theoretical predictions require a series of experimental examinations from the STS, optical and magneto-optical spectroscopies.

5.1 ELECTRONIC PROPERTIES AND ABSORPTION SPECTRA

Electronic and optical properties of the AB/DW/BA/DW bilayer graphene are greatly diversified by the manipulation of geometry and gate voltage. The periodical modulation along $\hat{x}$ induces the dramatic 2D to 1D changes, covering a lot of energy subbands and well-behaved/irregular standing waves, various van Hove singularities in density of states (the double- and single-peak structures, a pair of rather strong peaks, and a plateau across E_F), and the metallic behavior. The optical gaps vanish in any systems. A pristine bilayer AB stacking exhibits the featureless optical spectrum at lower frequency. As to the geometry-modulated systems, the observable absorption peaks could survive only under the destruction of the symmetric/antisymmetric linear superposition due to the interlayer (A^1, A^2)/(B^1, B^2) sublattices. The frequency, number, and intensity of absorption structures are very sensitive to the modulation DW and Coulomb potential.

A bilayer graphene, as clearly shown in Fig. 5.1(a), is periodically modulated along the $\hat{x}$ direction, in which the translation symmetry in the y-axis remains unchanged. This special geometric structure covers the wide AB and BA stacking regions, respectively, at the left- and right-hand sides, and the DWs in between them. The slowly stacking transformation is assumed to appear in DWs along the armchair direction by a uniform variation of the C-C bond lengths. Obviously, random stacking configurations appear within the ranges of DWs, leading to the nonuniform environment and thus the destruction in most rotation symmetries. A typical period within (60,10,60,10) (in the unit of $3b$; with b denoting the C-C bond length), with 1,120 carbon atoms, is very suitable for a full exploration of the modulation-diversified essential properties. That is to say, the $1,120 \times 1,120$ Hamiltonian matrix, which covers the various intralayer and interlayer hopping integrals in Eq. (4.2) [6], is available in fully exploring the gate-voltage-dependent electronic properties and optical absorption spectra. Such a bilayer system possesses a periodical boundary condition with a long period, so that the dimension-enriched electronic and optical properties will frequently come to exist. The total carbon atoms in a primitive unit cell can be classified into the four (A^1, B^1, A^2, B^2) sublattices. In the numerical calculations, the

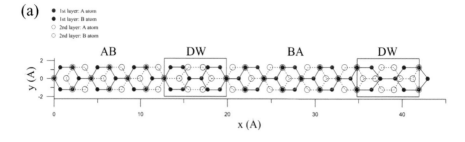

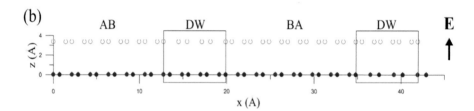

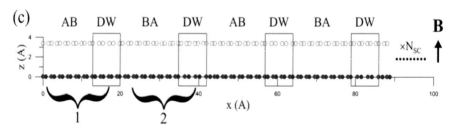

FIGURE 5.1 Geometric structures of the stacking-modulated bilayer graphenes associated with the AB configuration under the (a) top, (b) side views with a uniform perpendicular electric field, and (c) an enlarged unit cell in a commensurate magnetic field.

wave-vector-dependent wave functions strongly depend on the odd or even indices of each sublattice. The fundamental physical properties are very sensitive to the application of an external electric field/gate voltage [Fig. 5.1(b)]. V_z can generate the layer-dependent Coulomb potentials and thus destroy the mirror symmetry about the $z = 0$ plane.

The low-lying electronic structures of symmetric and asymmetric bilayer graphenes exhibit the unusual features. Obviously, a pristine bilayer AB stacking possesses two pairs of valence and conduction bands, with parabolic energy dispersions initiated from the K & K' valleys [details in Refs. [7,12]]. The first pair presents a slight overlap around $(2\pi/\sqrt{3}a, 2\pi/3a)$ and $(2\pi/\sqrt{3}a, -2\pi/3a)$, where $a = \sqrt{3}b$ is the lattice constant of monolayer graphenes. Furthermore, the second pair appears at $\sim \pm 0.42$ eV, mainly owing to the dominance of the interlayer vertical hopping integral. The lower-energy bands of the former are chosen as the studying focus. In order to compare the main features of electronic structures between the pure and geometry-modulated bilayer systems, energy bands of the former, corresponding to an enlarged unit cell, are folded into a specific valley around $k_y a = 2\pi/3$ [the K point in Fig. 5.2(a)]. That is, there are a lot of 1D parabolic dispersions due to the zone-folding effects. Since this system presents a translational invariance along the $\hat{x}$ direction, all valence and conduction bands are doubly degenerate except for those two bands nearest the Fermi level [Fig. 5.2(a)]. The particle-hole symmetry is obviously broken under the interlayer atomic interactions. According to state energies measured from E_F, the first, second and third pairs are denoted as (v_1, c_1), $(v_2, c_2)/(v_2', c_2')$; $(v_3, c_3)/(v_3', c_3')$, and so on. All band-edge states in distinct energy bands are very close to the K point.

Specifically, the wave functions at the K point, as indicated in Fig. 5.3(a), clearly illustrate the unique characteristics of the spatial distributions for various energy sub-bands. The subenvelope functions on four sublattices $((A^1, B^1, A^2, B^2)$ by the red, green, black, and blue colors, respectively) are responsible for four components of Bloch wave functions. The v_1 and c_1 states in Figs. 5.3(a) and 5.3(b), which have constant values for each component, are the antisymmetric and symmetric superposition of only the A^1- and A^2-dependent tight-binding functions, respectively. Although v_1 and c_1 present the different dominating components, these two states are degenerate at the K point [Fig. 5.2(a)] and might have other ratios between the weights of A^1- and A^2-sublattices through the linear combination. Notice that, due to arm-chair shape in the x-direction, each sublattice is further split into the odd and even indices. The eigenfunctions between two distinct indices within the same sublattice only exhibits a π phase difference. The odd-index components will account for a clear presentation. With the increasing 1D subband indices, the wavefunctions become the well-behaved standing waves instead of the uniform spatial distributions. For $(v_2, c_2)/(v_2', c_2')$ in Figs. 5.3(c)–5.3(d)/Figs. 5.3(e)–5.3(f), the K states have the dominant components in the A^1- and A^2-sublattices and minor weights in B^1- and B^2-sublattices. Apparently, the four subenvelope functions show the standing-wave behaviors with two nodes in each component of the spatial distributions, arising from the linear combination of the $\pm k_x$ states, as mentioned earlier. The $\pi/2$-phase difference for each sublattice between v_2 and v_2' (c_2 and c_2') is consistent with a uniform distribution in a 2D pristine system. The calculated results display that the subenvelope functions in the A^1- and A^2-sublattices are almost in-phase for conduction bands [Figs. 5.3(d) and 5.3(f)] and out-of-phase for valence ones [Figs. 5.3(c) and 5.3(e)], and the opposite is true for those between the B^1- and B^2-sublattices. The slight phase shift between A^1 and A^2 or between B^1 and B^2 is intrinsic, which depends on calculation parameters. The phase shift becomes less obvious for modes with higher energy. As to the next energy subbands of v_3 and c_3, the eigenfunctions reveal the four-node standing waves in the well-behaved forms [Figs. 5.3(g) and 5.3(h)]. The number of nodes, which indicates the quantization behavior along $\hat{x}$, is expected to continuously grow in the increment of subband index.

Apparently, the drastic changes of band structures come to exist for the geometry-asymmetry bilayer graphene, as clearly indicated in Figs. 5.2(b) and 5.2(c). They cover band asymmetry, band overlap, state splitting, energy dispersions, band-edge states, significant subband hybridizations, and distorted standing-wave behaviors. The asymmetry of energy spectrum about $E_F = 0$ becomes more obvious in the

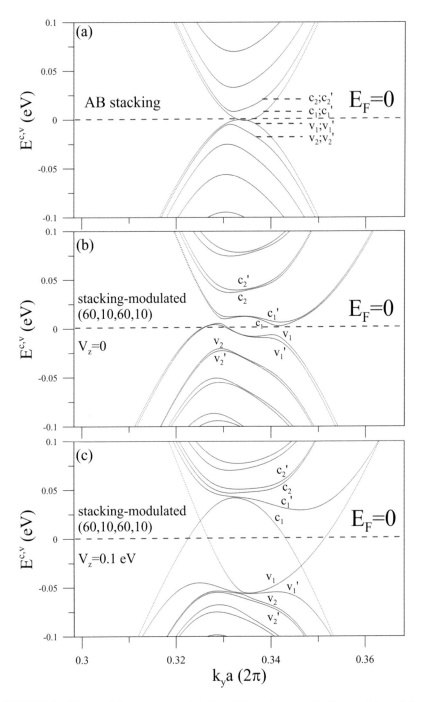

FIGURE 5.2 The low-lying energy bands initiated from the $k_y a = 2\pi/3$ state: (a) a pristine bilayer AB stacking, (b) the stacking-modulated AB/DW/BA/DW (60,10,60,10), and (c) the geometry- and electric-field-manipulated system under gate voltage of $V_z = 0.1$ eV.

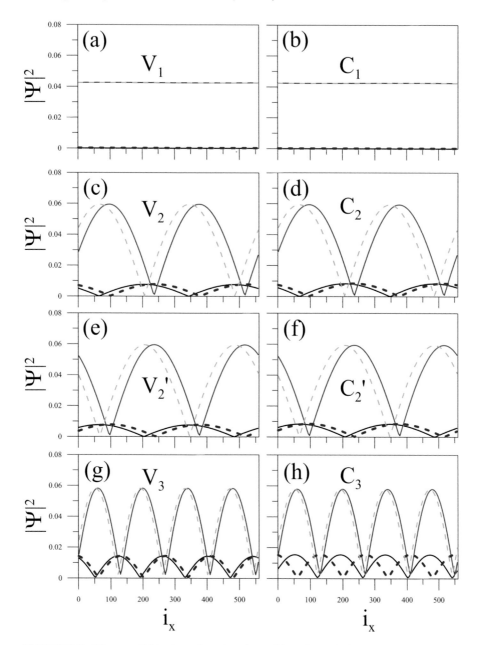

FIGURE 5.3 The zero-field subenvelope functions of the AB-stacked bilayer graphene in the (A^1, B^1, A^2, B^2) sublattices for the K point under an enlarged unit cell: the (a) v_1, (b) c_1, (c) v_2, (d) c_2, (e) v_2', (f) c_2', (g) v_3, and (h) c_3 energy subbands.

presence of the geometric modulation. This result is purely due to more nonuniform interlayer hopping integrals [details in Eq. (5.2)]. The overlap of valence and conduction energy bands is getting larger, and so do the free electron and hole densities. The destruction of the (x, y)-plane inversion symmetry leads to the splitting of doubly degenerate states, and therefore, there exist more pairs of neighboring energy subbands. The splitting electronic states are expected to induce more excitation channels and absorption spectrum structures. Most of the energy bands present parabolic dispersions, while the (v_1, c_1) and (v'_1, c'_1) energy subbands around the half filling exhibit the oscillating and crossing behaviors. In general, the band-edge states in various energy subbands seriously deviate from the K point. They will be responsible for the main features of van Hove singularities in the density of states and optical absorption spectrum. Most importantly, the strong hybridizations exist between the neighboring 1D energy subbands, as clearly identified from the wave functions in Fig. 5.4 (discussed later). As a result, it needs to redefine the energy subband indices in the ordering of (v_1, c_1), (v'_1, c'_1), (v_1, c_2), (v'_2, c'_2), ... etc.

The spatial distributions of Bloch wave functions belong to the unusual standing waves, especially for the DW regions. The symmetric and antisymmetric standing waves in a unit cell thoroughly disappear under the modulation of stacking configuration, as clearly indicated in Figs. 5.4(a)–5.4(h). Apparently, the oscillation modes are totally changed under the stacking modulation, as identified from a detailed comparison between Figs. 5.4 and 5.3. The number of nodes is fixed for all the four sublattices. Furthermore, it grows with the increasing state energies, two and four zero points of the (v_1, c_1), (v'_1, c'_1), (v_2, c_2) (v'_2, c'_2) energy subbands, respectively shown in Figs. 5.4(a)–5.4(b), 5.4(c)–5.4(d), 5.4(e)–5.4(f), and 5.4(g)–5.4(h). The subenvelope functions might be vanishing at the specific wave vectors in the AB, DW, BA, or DW regions. Usually, the weights within DWs are relatively small, compared with those in the normal stacking regions. When the weight is large for one region, it becomes small for the others. This clearly illustrates the drastic changes for the bilayer graphene with DWs. Moreover, the tight-binding functions on four sublattices roughly have a linear superposition relationship in the AB and BA stacking regions, being sensitive to the modulated DW regions. For conduction band at the K point, the subenvelope functions on the A^1 and A^2 (B^1 and B^2) sublattices are roughly in-phase for the normal stacking region with the dominant weight and out-of-phase for another one. These simple relations might be seriously destroyed in the DWs. Furthermore, those on the (A^1, B^1) and (A^2, B^2) sublattices present the approximately symmetric and antisymmetric distributions, respectively. On the other hand, the K states of valence energy subbands exhibit the opposite behaviors. The above-mentioned irregular standing waves will strongly affect the optical vertical excitations, e.g., the absence of specific optical selection rules.

Electronic energy spectra are greatly diversified by a uniform perpendicular electric field, as clearly illustrated in Fig. 5.2(c). That is, the nonuniform environment is largely enhanced by V_z. A gate voltage across bilayer graphene can create the layer-dependent Coulomb potentials or on-site energies. It leads to the breaking of mirror symmetry about the $z = 0$ plane, and so does the inversion symmetry on the (x, y) plane. For a normal bilayer AB stacking, a gate voltage opens a band gap [13]. However, the geometry-modulation system might present the semimetallic behavior as a

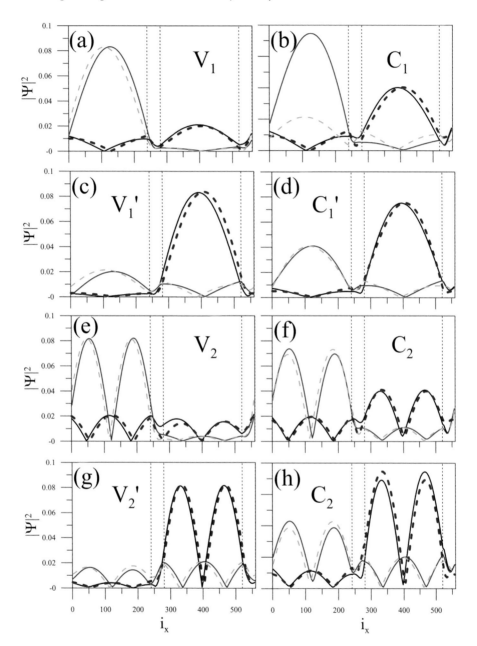

FIGURE 5.4 Similar plots as Fig. 5.3, but shown for the AB/DW/BA/DW bilayer graphene at the specific K point of the energy subbands: (a) v_1, (b) c_1, (c) v_1', (d) c_1', (e) v_2, (f) c_2, (g) v_2', and (h)) c_2', according to the energy ordering.

result of the finite, but the low density of states at the Fermi level [discussed later in Fig. 5.6]. The rich and unique energy dispersions frequently appear, especially for those near the Fermi level. Most of the energy levels are repelled away from $E_F = 0$ by V_z, where certain energy dispersions might be rather weak within a finite k_y-range. Moreover, the oscillatory valence and conduction subbands, the left- and right-hand pairs of oscillatory subbands, being very close to the Fermi level, linearly intersect near the K point. This crossing phenomenon results in the creation of holes and electrons simultaneously. That is, two pairs of the linearly crossing subbands occurs near E_F, thus leading to a special van Hove singularity in density of states. The semimetallic property is in great contrast with the semiconducting behavior in the gated bilayer AB stacking.

As for valence and conduction wave functions, the electric potential and geometric asymmetry make them become highly complicated, as clearly displayed in Figs. 5.5(a)–5.5(h). The V_z-enhanced abnormal standing waves exhibit the irregular features in the oscillatory forms, amplitudes, zero-point numbers, and relations among four sublattices. In general, there are no analytic sine/cosine waves suitable for such random spatial distributions. The amplitudes could appear at any AB/DW/BA/DW regions in the absence of a concise rule. The number of zero points does not grow with state energies monotonously; furthermore, it might be identical or different on the top and bottom layers. For example, the v_1 (v_1') energy subband at the K point, as shown in Fig. 5.5(a) [Fig. 5.5(c)], presents the 4- and 4-zero-point (4- and 6-zero-point) subenvelope functions on the first and second layers, respectively. Of course, a simple linear combination of the (A^1, A^2)- and (B^1, B^2)-dependent tight-binding functions is thoroughly absent, while it almost remains similar for the (A^1, B^1) and (A^2, B^2) sublattices. Apparently, the above-mentioned features of Bloch wave functions are expected to induce very complex optical excitation channels.

Specifically, the subenvelope functions of the Fermi-momentum states, which are situated in the linearly intersecting valence and conduction subbands [Fig. 5.2(c)], are worthy of a closer examination. There exist four Fermi momenta, namely, k_{F1}, k_{F2}, k_{F3}, and k_{F4}, in which the former (latter) two consists of a pair due to the valence (conduction) hole (electron) states. Apparently, their wave functions exhibit the unusual phenomenon, as clearly displayed in Figs. 5.6(a)–5.6(d). That is to say, they are mostly localized in/near a certain region of DW, e.g., the k_{F1}/k_{F4} (k_{F2}/k_{F3}) state corresponding to the second (first) domain. Moreover, the localization behavior is accompanied with a plateau-structure density of states across E_F [discussed in Fig. 5.7]. Apparently, this result suggests that the metallic transport properties might come to exist along the y-direction of a narrow DW. Such a phenomenon is similar to those in metallic/armchair carbon nanotubes [11,14]. The high-resolution STS measurements could directly verify the significant characteristics of electronic energy spectra and wavefunctions under the different gate voltages.

The significant characteristics of electronic structures are directly revealed as various van Hove singularities in the density of states, as clearly displayed in Figs. 5.7(a)–5.7(c). A pristine system has a finite, but low magnitude at E_F [Fig. 5.7(a)], indicating the semi-metallic property. A very weak band overlap leads to a pair of very close shoulder structures across the Fermi level, being consistent with the first-principles calculations [15]. Also another two well separated shoulder structures, which comes

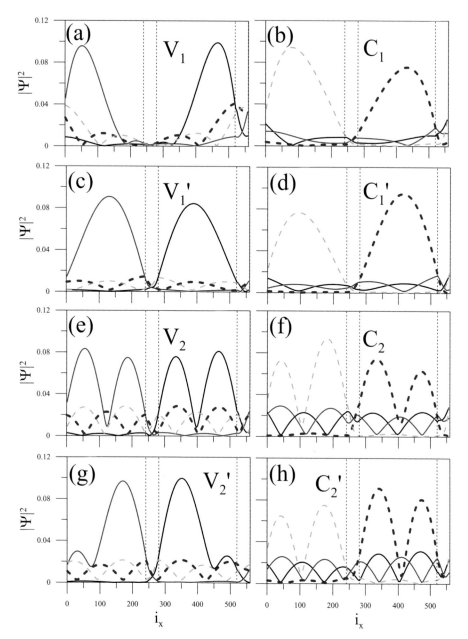

FIGURE 5.5 The stacking- and voltage-modulated subenvelope functions on the four sublattices at $V_z = 0.1$ eV for the K point in the energy subbands: (a) v_1, (b) c_1, (c) v_1', (d) c_1', (e) v_2, (f) c_2, (g) v_2', and (h) c_2'.

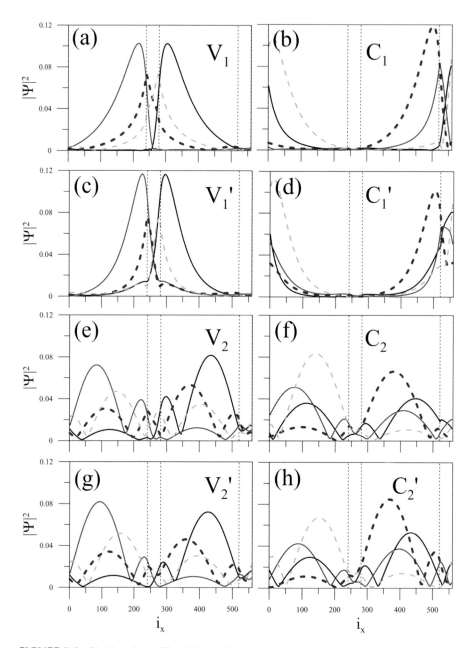

FIGURE 5.6 Similar plot as Fig. 5.5, but displayed under for the four momentum states in the linearly intersecting energy subbands.

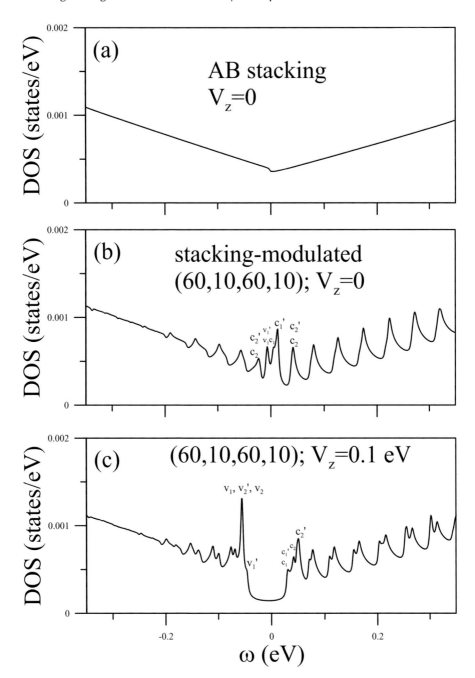

FIGURE 5.7 The significant density of states for bilayer graphene systems under (a) a normal AB stacking, (b) a periodical structure of AB/DW/BA/DW, and (c) a geometric modulation at $V_z = 0.1$ eV.

from the valence and conduction bands of the second pair, appear at the deeper/higher energies, ~ -0.43 eV and 0.45 eV. Such structures are due to the extreme points (the local minima and maxima) of 2D energy bands in the energy-wave-vector space (details in Refs). On the other side, the geometry modulation can create a quasi-1D system and thus the dimension-diversified features [Figs. 5.7(b) and 5.7(c)]. There are a lot of asymmetric pronounced peaks divergent in the square root form, mainly owing to 1D parabolic energy subbands. They could be further classified into single- and double-peak structures, in which most of vHSs belong to the latter. The coexistence of the composite structures obviously corresponds to a nonuniform state splitting [Figs. 5.2(b) and 5.2(c)]. A finite density of states [DOS] at E_F in the geometry-modulated system shows the semimetallic behavior, being accompanied with a pair of rather strong peak structures at $\sim \pm 0.01$ eV. The latter are closely related to the weakly dispersive energy bands near E_F. Under a gate voltage, the DOS near E_F becomes a plateau structure. Apparently, it is induced by the linear energy dispersions across E_F, as shown in Fig. 5.2(c). Furthermore, two very prominent peaks appear at $\sim \pm 0.1$ eV, mainly due to the weak dispersions generated by the gate voltage. It should be noticed that the E_F-dependent densities of states are comparable under the geometric and electric-field modulations. That is, these manipulation methods cannot generate the unusual transitions among the semimetallic, metallic, and semiconducting behaviors. However, they have created the important differences in the low-energy van Hove singularities, especially for those across the Fermi level.

The geometry- and gate-modulated bilayer graphene systems exhibit the rich and unique absorption spectra. The joint density of states, which is proportional to the number of vertical optical excitations, is very different from one another among three kinds of bilayer systems, as clearly revealed in the inset of Fig. 5.8. A pristine system is finite at $\omega = 0$ (the dashed black curve), reflecting the parabolic band-edge states at E_F [Fig. 5.2(a)]. It is featureless below $\omega = 0.41$ eV, and then has a shoulder structure there due to the 2D first/second valence band and second/first conduction band (details in [Ref. [6]]). Joint density of states [JDOS] at $\omega = 0$ remains almost the same in the asymmetric bilayer systems (the dashed red curve), while there exist a plenty of asymmetric peaks in the square root form. Such peaks are very prominent, when the valence and conduction band-edge states correspond to the same wave vector, such as, (v_1, c_1, v_2, c_2) around K and (v_2, c_3) at $k_y a \approx 2.06$. However, some structures are weak but observable because of the non-vertical relation between them. Specifically, the magnitude of JDOS and the number of peak structures are greatly reduced under the effect of gate voltage, as shown in the inset of Fig. 5.8. Such results mainly arise from the optical excitations of band-edge/non-band-edge valence states to non-band-edge/band-edge conduction ones [Fig. 5.2(c)].

In addition to van Hove singularities in joint density of states, the available optical transition channels, being dominated by the characteristics of subenvelope functions, will codetermine 1D absorption peaks. The AB-stacked bilayer graphene does not show any absorption structures at lower frequency ($\omega < 0.3$ eV), as clearly illustrated by the dashed black solid curve in Fig. 5.8. Both the optical gap and the energy gap are zero. The same result is revealed in the geometry- and gate-manipulated bilayer systems. Apparently, the geometric modulation displays the significant absorption peaks (all the solid curves), only coming from the band-edge states of valence

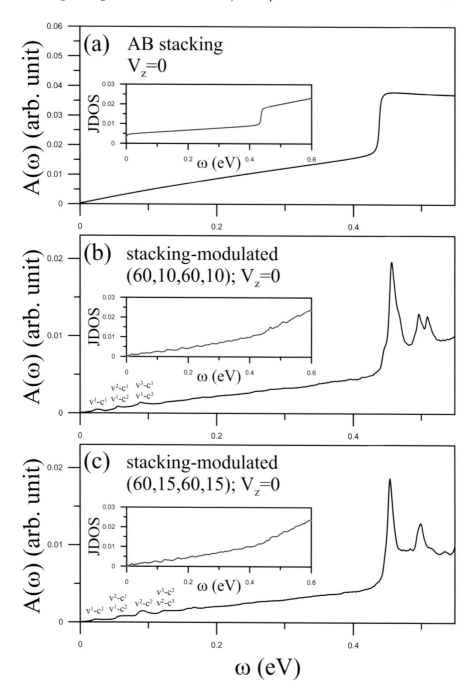

FIGURE 5.8 The optical absorption spectra for pristine and stacking-modulated bilayer graphene systems under various domain-wall widths. Also shown in the inset are the typical joint densities of states.

and conduction subbands with distinct index numbers. The v_n to c_n (v'_n to c'_n) optical excitations, which are due to the same pair of valence and conduction bands, are forbidden. The main mechanism is the linear symmetry/antisymmetric superposition of the (A^1, A^2) and (B^1, B^2) sublattices, as discussed in Figs. 5.4(a)–5.4(h). The first absorption peak, corresponding to (v'_1, c_2) at $k_y a = 2.06$, comes to exist at 0.049 eV. Furthermore, the second, third, fourth, and fifth significant absorption channels are, respectively, related to (v_2, c_5), (v_6, c_3), (v_2, c_7) and (v_7, c_3). Of course, there exist many weak absorption structures in between them, and certain very strong peaks being closely related to multichannel optical excitations [Fig. 5.8]. Apparently, there are no specific optical selection rules, being thoroughly different the edge-dependent ones in 1D graphene nanoribbons [11]. The main reason is that the 1D quantum confinement can create the well-behaved standing waves [11, 14] [Fig. 5.8].

The intensity, frequency, number, and form of optical special absorption structures are very sensitive to the changes in the width of DW and the perpendicular electric-field strength. In general, a simple relation between these optical features and the modulation width is absent at lower ωs [Fig. 5.8]. However, the red-shift phenomenon in the increase of width appears at the larger ones. Apparently, the first absorption structure might be replaced by another excitation channel during the variation of DW width. On the other side, the spectral absorption functions for a specific geometry-modulated bilayer graphene clearly present a regular variation under various gate voltages [Fig. 5.9]. That is to say, the reduced intensity and the enhanced number of absorption structures occur in the increment of V_z. This directly reflects the gradual changes in energy dispersions, band-edge states, and wave functions with the gate voltages. It should be noticed that the band-edges states near the Fermi level might exhibit a sharp change, strongly affecting the threshold optical excitations. Certain channels become apparent during the variation of V_z, as indicated in Fig. 5.9.

The periodical boundary condition in the asymmetry-enriched bilayer graphene is responsible for the rich 1D electronic and optical properties, being in sharp contrast with 2D behaviors in sliding [9] and twisted [16] systems. The latter two exhibit two pairs/more pairs of 2D energy bands composed of the $2p_z$ orbitals, as clearly displayed along the high-symmetry points of the first Brillouin zone. The wave-vector-dependent energy dispersions belong to the linear, parabolic oscillatory and partially flat forms, in which the first ones are the well-known vertical/non-vertical Dirac cones in the AA/AA′ stacking. Their band-edge states, which are the critical points in the energy-wave-vector space, respectively, correspond to the Dirac points, the extreme and saddle points, the highly-degenerate states and the effective 1D constant-energy loops. They present the following vHSs in density of states: the V-shape structures, the shoulder structures & logarithmically symmetric peaks, delta-function-like peaks, and square root asymmetric peaks from 1D band edges. These five kinds of special structures are further revealed in the optical absorption spectra. An electric field could create optical gaps/energy gaps in most of bilayer graphene systems.

The experimental measurements could verify the predicted band structures, densities of states, and optical absorption spectra. The high-resolution angle-resolved photoemission spectroscopy [ARPES; details in [17–22]] is the only experimental instrument able to directly examine the wave-vector-dependent occupied electronic states. The measured results have confirmed the feature-rich band structures of

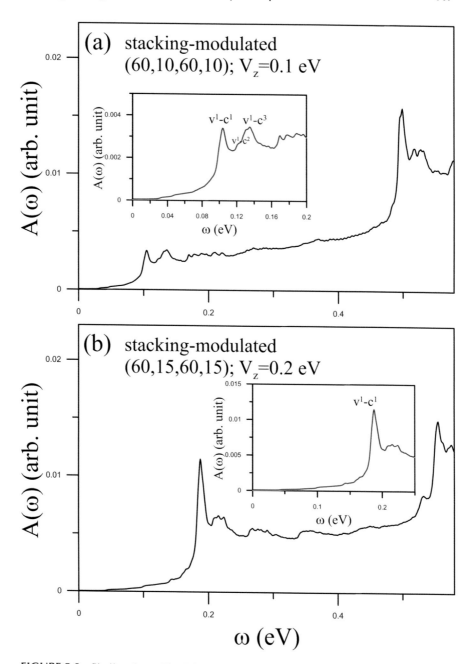

FIGURE 5.9 Similar plot as Fig. 5.8, but only shown for stacking modulated systems in the presence of distinct gate voltages. The inset clearly illustrates the detailed absorption structures for typical cases.

carbon-related sp^2-bonding systems. Graphene nanoribbons are identified to possess 1D parabolic energy subbands centered at the high-symmetry point, accompanied by an energy gap and nonuniform energy spacings [17]. Recently, a lot of ARPES measurements are conducted on few-layer graphenes, covering the proof on the linear Dirac cone in the monolayer system [18–20], two low-lying parabolic valence bands in bilayer AB stacking [18], the coexistent linear and parabolic dispersions in symmetry-destroyed bilayer systems [21], one linear and two parabolic bands in trilayer ABA stacking [18, 19], the partially flat and sombrero-shaped and linear bands in tri-layer ABC stacking [18,19]. The Bernal (AB-stacked) graphite possesses the 3D band structure, with the bilayer- and monolayer-like energy dispersions, respectively, at $k_z = 0$ and 1 (K and H points in the 3D first Brillouin zone) [22]. The ARPES examinations on the geometry- and electric-field-modulated bilayer graphene systems could provide the unusual band structures, such as, the splitting of electronic states, parabolic/oscillatory/linear energy dispersions near the Fermi level, wave-vector-dependent, band-edge states, and semi-metallic properties. These directly reflect the composite effects due to the irregular stacking/the complicated hopping integrals and Coulomb potentials.

Up to date, four kinds of optical spectroscopies, absorption, transmission, reflection, and Raman scattering spectroscopies, are frequently utilized to accurately explore vertical optical excitations. Concerning the AB-stacked bilayer graphene, their measurements have successfully identified the ~ 0.3-eV shoulder structure under zero field [8], the V_z-created semimetal-semiconductor transition and two low-frequency asymmetric peaks [8], the two rather strong π-electronic absorption peaks at the middle frequency, specific magneto-optical selection rule for the first group of Landau levels [23], and linear magnetic-field-strength dependence of the inter-Landau-level excitation energies [23]. Similar verifications conducted on trilayer ABA stacking cover one shoulder at ~ 0.5 eV, the gapless behavior unaffected by gate voltage, the V_z-induced low-frequency multi-peak structures, several π-electronic absorption peaks, and monolayer- and bilayer-like inter-Landau-level absorption frequencies [24]. Moreover, the identified spectral features in trilayer ABC stacking are two low-frequency characteristic peaks and gap opening under an electric field [24]. The above-mentioned optical spectroscopies are worthy of thoroughly examining the vanishing optical gaps in stacking-modulated bilayer graphenes, prominent asymmetric absorption peaks, absence of selection rule, forbidden optical excitations associated with the linear relations in the (A^1, A^2) & (B^1, B^2) sublattices, and drastic changes/the regular variations in absorption structures due to the modulation of DW width/gate voltage.

5.2 SIGNIFICANT LANDAU SUBBANDS

The nonuniform magnetic quantization in stacking-modulated bilayer graphene systems is fully explored by the generalized tight-binding model. The various interlayer atomic interactions and the magnetic field are included in the calculation without the low-energy perturbation. The quasi-1D Landau subbands, which are created by the periodical AB/DW/BA structure, exhibit the partially flat and oscillatory dispersions with or without the anticrossing phenomena. The greatly reduced state degeneracy and extra band-edge states lead to more van Hove singularities in the wider energy ranges. The well-behaved, perturbed, and seriously distorted wave functions appear at the specific regions associated with the stacking configurations. The close relations among the geometric symmetries, the band-edge states, and the spatial probability distributions are identified from the delicate analyses.

The generalized tight-binding model is delicately developed for a full understanding of the nonuniform magnetic quantization due to the stacking-modulated configuration [similar calculations in Sec. 3.1]. Most importantly, there exists a very strong competition between the geometric modulation and magnetic field. This method is reliable for the complicated energy bands with the oscillatory dispersions and the multi-constant loops. The non-homogeneous interlayer and interlayer hopping integrals and the magnetic-field Peierls phases effects are simultaneously included in the diagonalization of the giant magnetic Hamiltonian matrix. The ratio of the periods, which is related to the Peierls phase and the stacking modulation, is characterized by the number of supercell (N_{sc}). For example, concerning (60,10,60,10) stacking-modulated bilayer graphene with a period of 140, the Hamiltonian is a $3,360 \times 3,360$ Hermitian matrix with $N_{sc} = 6$ under $B_z = 95$ T, in which the number of independent magnetic matrix elements is about 560. The magneto-electronic properties, energy dispersions, density of states, and magnetic wave functions are investigated in detail. Their close relations are directly linked together by the detailed analysis. A lot of oscillatory Landau subbands are predicted to initiate from the Fermi level, in great contrast with the dispersionless Landau levels in the well-stacked graphene systems [7,12]. The theoretical predictions on the significant characteristics of quasi-1D Landau subbands could be verified by the high-resolution STS measurements. The typical methods in generating many Landau subbands and the important differences among them are also discussed.

Apparently, the magneto-electronic energy spectra of bilayer graphene systems display the rich features, strongly depending on the stacking configurations. A pristine bilayer AB stacking has two groups of valence and conduction Landau levels under the magnetic quantization of the π-bonding electronic states. The first group is initiated from the Fermi level, as indicated by the red points in Fig. 5.10. The well-behaved levels, which are similar to electronic states of an oscillator, are highly degenerate in the reduced first Brillouin zone of the 2D (k_x, k_y)-space. Their quantum numbers are characterized by the zero points of the oscillatory distributions in the dominating sublattice; furthermore, they have a normal ordering according to $n^c = 1/n^v = 0, 2, 3 \ldots$ etc. A very small energy spacing between the $n^v = 0$ and $n^c = 1$ Landau levels is band gap of ~ 20 meV across $E_F = 0$ under $B_z = 95$ T. The dispersionless Landau levels dramatically changes into the quasi-1D Landau subbands with the partially flat and oscillatory k_y-dispersions. The band-edge states, directly determining the significant density of states, correspond to the dispersionless k_y states, and the locally extreme states (the local minima and maxima). When the magneto-electronic states present the almost dispersionless behavior within a certain k_y-range, they could be regarded as the quasi-Landau-level states. This is also identified from the well-behaved magnetic wave functions [discussed later in Fig. 5.11]. For each (k_x, k_y) state, the fourfold degenerate Landau-level states are replaced by the doubly degenerate Landau-subband ones, because the x-direction associated with stacking modulation has been specified. Apparently, the asymmetry of electron and

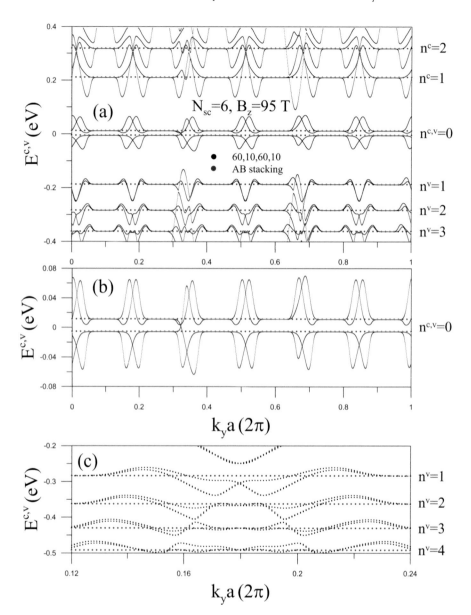

FIGURE 5.10 (a) The low-lying Landau subbands of the stacking-modulated bilayer graphene under $B_z = 95$ T and $N_{sc} = 6$, and the enlarged energy dispersions (b) near the Fermi level and (c) for certain anti-crossing phenomena.

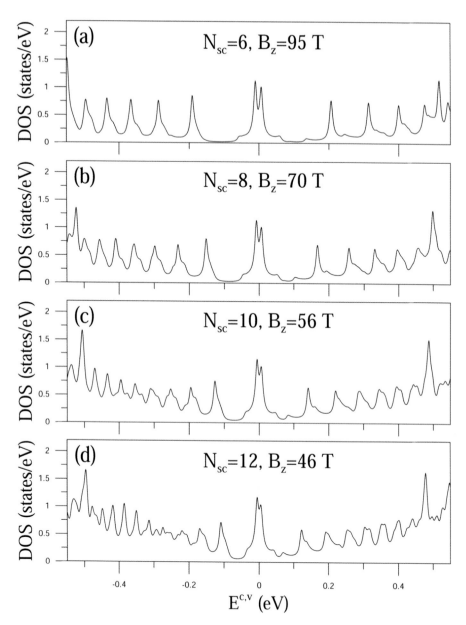

FIGURE 5.11 The low-energy density of states for the various domain-wall widths: (a) $N_{sc} =$ 6, (b) 8, (c) 10, and (d) 12.

hole states is greatly enhanced by the stacking modulation, being attributed to the nonuniform intralayer and interlayer hopping integrals. The oscillation width grows quickly during the increment of $|E^{c,v}|$, in which the anticrossing phenomena appear and thus create more band-edge states. At higher/deeper state energies, the geometry modulation shows the amplified effects, mainly owing to the larger radii of the magnetic cyclotron motions. That is to say, the crossings and anticrossings of Landau subbands come to exist very frequently. However, this unusual phenomenon is very difficult to observe in a composite external field (discuss later in this chapter).

Many van Hove singularities of quasi-1D appear as the special structures in density of states, in which they mainly originate from the various band-edge states. The partially flat quasi-Landau-level and the 1D parabolic extreme states, respectively, generate the delta-function-like peaks and the square root-form asymmetric peaks, as apparently indicated in Fig. 5.11. A well-behaved AB stacking only presents the regular symmetric peaks [7, 12]. The height of each peak is related to the Landau-level (k_x, k_y)-degeneracy. Two neighboring peaks across the Fermi level, being due to the $n^v = 0$ and $n^c = 1$ Landau levels of the first group, represent the most important characteristics. The peak positions, the Landau-level energies, clearly display a redshift under the decrease of n_{sc} (the magnetic-field strength). In general, the stacking modulation results in the greatly reduced peak intensities, the significant shifts of the peak positions, and many extra asymmetric peaks in the extended energy ranges. Obviously, these drastic changes directly reflect the main features of quasi-1D Landau subbands, the finite k_y-ranges, their energy differences with those of Landau levels, and a plenty of local maxima and minima in the oscillatory band dispersions. The modulation effects are relatively easily observed in the presence of a wider DW/a weaker magnetic field. For example, the density of states under $n_{sc} = 6$ [Fig. 5.11(a)] is more seriously deviated from that of a normal stacking, compared with that of $n_{sc} = 12$ [Fig. 5.11(d)].

The homogeneous and non-homogeneous magnetic quantizations, respectively, present in the normal and stacking-modulated AB bilayer graphene systems. The former exhibits a lot of fully degenerate Landau level in the reduced first Brillouin zone [details in Sec. 3.1]. Within the chosen gauge, each k_x, k_y-state Landau level has four fold degeneracy except for the freedom degree of spin configuration, in which the magnetic wave functions are localized at 1/6, 2/6, 4/6, and 5/6 of an B_z-enlarged unit cell under a zero wave vector. Specifically, the magneto-electronic energy spectrum of bilayer AB stacking, corresponding to the magnetically quantized states of the first pair of valence and conduction bands, is characterized by the normal ordering of $n^v = 0, 2, 3...$ ($n^c = 1, 2, 3, ...$), etc. It is well known that the dominating sublattice depends on the localization center, and its zero point determines the magnetic quantum number. For example, such Landau levels are defined by the dominant B^1 sublattice under the 4/6 center, as clearly illustrated in Fig. 5.12 for $n^c = 1$. Generally speaking, the quantum-mode differences for A^1 & A^2, B^1 & B^2, and A^1 & B^1/A^2 & B^2 are, respectively, 0, ± 2, and ± 1 [details in Ref. [7]]. This directly reflects the intrinsic geometric symmetries, i.e., the equivalence of two sublattices on the same layers, and the different chemical environments of them.

The nonuniform intralayer and interlayer hopping integrals have dramatically changed the main features of magnetic wave functions; that is, Landau subbands

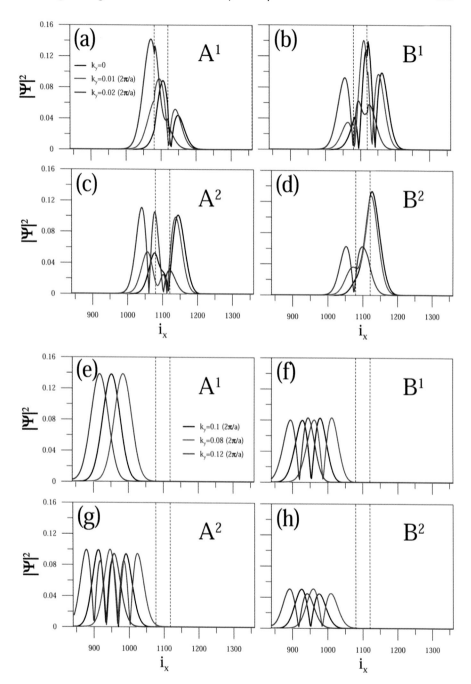

FIGURE 5.12 The subenvelope functions for the $n^c = 1$ Landau subband, with the localization center near 4/6, on four sublattices at smaller wave vectors: (a) B^1, (b) A^1, (c) A^2 & (d) B^2, and under larger ones: (e) B^1, (f) A^1, (g) A^2 & (h) B^2.

are quite different from Landau levels in the spatial distributions. The state splitting, as clearly shown in Fig. 5.10(a), clearly indicates that the neighboring Landau subbands present the 4/6 and 2/6 localization centers, in which the former is chosen for a model study. For example, the subenvelope functions of the $n^c = 1$ Landau subband are very sensitive to the change of k_y [the various curves in Figs. 5.12(a)–5.12(h)]. The $k_y = 0.1$ Landau-subband state (the black curves), even corresponding to the partially flat energy dispersions, exhibits the obvious changes in the amplitudes and modes of magnetic subenvelope functions. The dominating B^1 sublattice only reduces its amplitude [Fig. 5.12(f)], while it remains the one-zero-point antisymmetric distribution under the 4/6 localization center well separated from the DW. On the other side, the A^1-, A^2-, and B^2-related subenvelope functions, respectively, display the drastic transformations: (I) the great enhancement of amplitude and the zero-point change from 0 to 1, especially for the DW region [Fig. 5.12(e)], (II) the variation of the well-behaved symmetric distribution without zero points into the oscillatory one [Fig. 5.12(g)] and (III) the emergence of the one-zero point subenvelope function [Fig. 5.12(h)]. The above-mentioned unusual behaviors quickly grow as k_y increases from zero to 0.12, since the magnetic localization center approaches, enters, and then crosses the first stacking-modulation region. That is, the enhanced competition between the magnetic field and stacking modulation is responsible for the strong k_y-dependence on the main features of the sublattice-dependent subenvelope functions.

For k_y from 0 to 0.05, the $n^c = 1$ Landau subband has a weak parabolic energy dispersion [the black curves in Fig. 5.10(b)], being different from the normal dispersionless ones. Apparently, the dramatic changes, as clearly indicated in Figs. 5.12(a)–5.12(d), cover the obvious deviation of localization center from the DW region, the significant transformation between the B^1- and A^2-sublattice dominance, the drastic variation of the oscillating modes in these two sublattices, and the reduced contributions in the A^1 & B^2 sublattices. According to the subenvelope functions of the whole k_y-dependent Landau subband states, there are no simple relations between two any sublattices (discussed earlier in the pristine case), e.g., the absence of the same zero-point number for A^1 and A^2 sublattices. That is to say, the hexagonal symmetry and the AB stacking symmetry are thoroughly broken by the geometric modulation thoroughly the nonuniform intralayer and interlayer hopping integrals. It should be noticed that the four sublattices would have the comparable amplitudes in the increase of energy/quantum number. In addition, the effective-mass approximation is not unsuitable in solving the unusual magnetic subenvelope functions, since each Landau-subband state might be the superposition of various Landau-level modes.

The anticrossings of two distinct Landau-subband states frequently appear in the magneto-electronic spectra, as shown in Fig. 5.10(a). The specific anticrossing, which is due to the $n^v = 1 \sim n^v = 4$ Landau subbands at smaller wave vectors $k_y \sim 0.02 \rightarrow 0.04$ and $E^c \sim -0.2 \rightarrow -0.5$ eV, is chosen for a clear illustration [the red curves in Fig. 5.10(c)]. As shown in Fig. 5-13, when wave vectors are well separated from the anticrossing center, the former and the latter are approximately characterized by the well-behaved zero points of the dominating B^1 sublattice. During the variation of wave vector from $k_y = 0.02$ to 0.04, the magnetic subenvelope functions present the gradual alternation, the drastic change corresponding to the highly

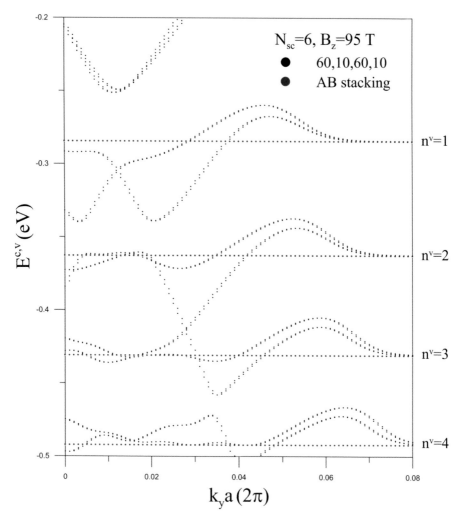

FIGURE 5.13 The specific anti-crossing due to the $n^v = 1 \sim n^v = 4$ Landau subbands at smaller wave vectors and $E^v \sim -0.2 \sim -0.5$ eV for the (a) upper and (b) lower branches.

distorted oscillations across the anticrossing center, and the dramatic transformation changing into another mode. The fourth and the third Landau-subbands states (the upper and lower branches), respectively, become the three- and four-zero-point oscillation modes after the unusual anticrossing. It can be deduced that each Landau-subband state consists of the major and minor modes in the non-anticrossing regions, where the latter might belong to $n^{c,v} \pm 1, \pm 2, \ldots$ etc. Apparently, this leads to a unique phenomenon that all the Landau subbands anticross one another in the k_y-dependent energy spectrum [Fig. 5.10(a)].

In addition to the stacking modulation in few-layer graphenes, the 1D graphene nanoribbons and 3D graphites also exhibit the quasi-1D Landau subbands in the presence of a uniform perpendicular magnetic field. However, their key features are quite different from one another, e.g., the B_z-dependent energy spectra, van Hove singularities, and magnetic wave functions. While the ribbon width is much longer than the magnetic length, many 1D Landau subbands are initiated near the Fermi level, in which each one consists of dispersionless quasi-Landau-level and parabolic energy dispersion [details in Ref. [12]]. Their densities of states appear in the delta-function-like forms. The almost well-behaved wave functions of quasi-Landau-levels are localized about the ribbon center; furthermore, the standing waves come to exist for the other states. These characteristics directly reflect the strong competition among the finite-size confinement, edge structures, and magnetic field. On the other side, the periodic interlayer hopping integrals play critical roles in the AA-, AB-, and ABC-stacked graphites, respectively, displaying one pair [12], two pairs [12], and one pair [12] of 3D valence and conduction bands. The 3D magnetically quantized states form many Landau subbands which are composed of 2D dispersionless Landau levels and 1D k_z-dependent parabolic energy dispersions. The number of Landau-subband groups is, respectively, one, two, and one for simple hexagonal, Bernal and rhombohedral graphites. Some Landau subbands in AA-stacked graphite cross the Fermi level, and their bandwidths are ~ 1 eV, mainly owing to the strongest interlayer atomic interactions in three graphitic systems. Such 1D Landau subbands generate a plenty of square root asymmetric peaks in density of states, as observed those for Bernal and rhombohedral graphites. The Landau subbands of AB-stacked graphite possess band widths of about 0.2 eV, and only two of them intersect with E_F [25]. The Landau-subband energy spectrum, corresponding to the K and H points ($k_z = 0$ and 1), is almost asymmetric and symmetric about E_F, respectively. The ABC-stacked graphite presents the Landau-subband bandwidth of <10 meV, and only one Landau level is located at E_F [26]. In general, the magnetic wave functions of Landau-subband states are similar to those of the well-behaved Landau levels.

The third method in creating the quasi-1D landau subbands is to introduce the specific composite field, the superposition of a uniform perpendicular magnetic field and the spatially modulated magnetic field/electric field [27]. The generalized tight-binding is also suitable for fully exploring the nonuniform quantization phenomena, while the external fields are commensurate to each other. For example, such composite fields in monolayer graphene are predicted to exhibit the lower-degeneracy Landau subbands with the strong energy dispersions and high anisotropy. The 1D characteristics further lead to a plenty of square root-form asymmetric peaks in density of states. Apparently, the magnetic wave functions present the seriously distorted spatial distribution and even a dramatic transformation of oscillation modes, depending on the strength and period of a spatially modulated field. This field has very strong effects on the main features of Landau subbands, covering the enhancement in dimensionality, reduction of state degeneracy, variation of energy dispersions, generation of band-edge states, and the changes in the center, width, phase, and symmetric/antisymmetric distribution of the localized quantum modes. However, the anti-crossing phenomena are thoroughly absent. In addition, the above-mentioned quasi-1D magneto-electronic states have been expected to display the rich and unique absorption spectra, e.g., the drastic changes of magneto-optical selection rules [27].

The anticrossing phenomena of magneto-electronic states come to exist in the wave-vector- [27], magnetic-field- [28], and electric-field-dependent energy spectra [29], where the first, second, and third types frequently appear in Landau sub-bands, Landau levels, and both. They mainly come from the similar physical pictures/ the identical Van-Born theorem [30], being assisted by the critical mechanisms: the special stacking configurations [e.g., the non-AA configurations; Ref. [12]], the multi-orbital hybridizations [31], the nonuniform chemical environments [31], and the significant spin-orbital couplings [31]. However, the specific relation of quantum numbers (the quantum-number difference of Δn) between the main and side modes might be different in layered condensed-matter systems with various stacking configurations. For example, only the AB-stacked graphite [25], but not the AA- [12] and ABC-stacked [26] systems, present the frequently Landau-subband anticrossings between the first and second groups during the variation of k_z, with $\Delta n = 3$ [details in Ref. [12]]. Apparently, the k_z-decomposed wave vectors play important roles on the interlayer atomic interactions and thus the unusual behaviors, e.g., the bilayer- and monolayer-like hopping integrals at $k_z = 0$ and π, respectively [12,25]. The similar relations are also revealed in the well-stacked graphene systems for the B_z-induced energy spectra, such as, the AB [25], ABC [26], and AAB stackings [32]. In addition, the undefined Landau levels, corresponding to non-well-behaved sliding bilayer systems [9], exhibit the continuous intergroup anticrossings. Specifically, the various gate voltages could be utilized to manipulate the frequent anticrossings of Landau levels as a result of the splitting Landau levels. Such electric fields, which are, respectively, cooperated with the specific interlayer hopping integrals [layered graphenes; Refs. [27–29]], the important spin-orbital interactions [germanene and tinene; Ref. [31]], and the complicated intralayer and interlayer hopping integrals [bilayer phosphorene; Ref. [31]], present quantum number differences: $\Delta n = 3l, \pm 1$ and l.

The measurements of STM, as thoroughly discussed in Sec. 2.1, are very powerful in exploring the van Hove singularities due to the band-edge states and the metallic/semiconducting/semimetallic behaviors, They have successfully identified diverse electronic properties in graphene nanoribbons [11], carbon nanotubes [12], few-layer graphene systems [12, 31], and graphite [12]. The focuses of the STS examinations on the geometry-modulated and gated bilayer graphenes should cover the square root asymmetric peaks, the major double-peak structures and the minor single-peak ones, and the significant density of states at the Fermi level. Furthermore, a pair of very close shoulder structures just across E_F, a finite DOS at E_F accompanied by the prominent valence and conduction asymmetric peaks, and a sufficiently long plateau crossing E_F, can distinguish the distinct effects of AB configuration, stacking modulations, and nonuniform hopping integrals & Coulomb potentials, respectively. Moreover, such STS measurements are very reliable for identifying the uniform and nonuniform magnetic quantization in electronic energy spectra of layered graphene systems. As to the former, the measured tunneling differential conductance directly reflects the structure, energy, number, and degeneracy of the Landau-level delta-function-like peaks. Part of the theoretical predictions on the Landau-level energy spectra have been verified by the experimental measurements, e.g., the $\sqrt{B_z}$-dependent Landau-level energy of monolayer graphene [12, 31], the linear B_z-dependence in AB-stacked bilayer graphene [12,31], the coexistent square root and linear B_z-dependences in trilayer ABA stacking [12,31], and the 2D and 3D characteristics of the Landau subbands in Bernal graphite [12, 25]. The experimental examinations on the energy range and number of prominent asymmetric and symmetric peaks of the stacking-modulated bilayer graphene systems can clearly illustrate the significant characteristics of the oscillatory quasi-1D Landau subbands, namely, the partially flat and parabolic dispersions, the oscillation width, and the subband anticrossing phenomena.

The energy-fixed STS measurements can directly map the spatial probability distributions of wave functions in the presence/absence of a uniform magnetic (electric) field. Up to now, they have verified the rich and unique electronic states in graphene-related systems. For example, the well-behaved standing waves, with the specific zero points, could survive in finite-length carbon nanotubes [33]. The topological edge states are identified to be created by the AB-BA DW of bilayer graphene systems [3]. Moreover, the normal Landau-level wave functions, being similar to those of an oscillator, frequently appear in the well-stacked few-layer graphene systems, such as, the $n^{c,v} = 0$, and ± 1 Landau levels for the monolayer and AB-stacked bilayer graphenes [34]. The higher-n [25], undefined [9], and anticrossing Landau levels [35] are worthy of the further experimental examinations. The similar STS measurements are available in examining the current predictions: (I) the drastic changes in the main features of standing waves due to the stacking modulation, (II) the gate-voltage-induced localized states within the DWs, directly linking with a plateau-structure density of states across the Fermi level, and (III) the nonuniform magnetic quantization, establishing the direct relations between the band-edge state energies and the distribution range of the regular or irregular magnetic wave functions.

REFERENCES

1. Ju L, Shi Z, Nair N, Lv Y, Jin C, Velasco Jr J, et al. 2015 Topological valley transport at bilayer graphene domain walls *Nature* **520** 650.
2. Butz B, Dolle C, Niekiel F, Weber K, Waldmann D, Weber H B, et al. 2014 Dislocations in bilayer graphene *Nature* **505** 533.
3. Yin L J, Jiang H, Qiao J B, and He L 2016 Direct imaging of topological edge states at a bilayer graphene domain wall *Nat. Commun.* **7** 11760.
4. Jiang L, Wang S, Shi Z, Jin C, Utama M I B, Zhao S, et al. 2018 Manipulation of domain-wall solitons in bi- and trilayer graphene *Nat. Nanotech.* **13** 204.
5. Vaezi A, Liang Y, Ngai D H, Yang L, and Kim E A 2013 Topological edge states at a tilt boundary in gated multilayer graphene *Phys. Rev. X* **3** 021018.
6. Huang B L, Chuu C P, and Lin M F 2019 Asymmetry-enriched electronic and optical properties of bilayer graphene *Sci. Rep.* **9** 859.
7. Lin C Y, Do T N, Huang Y K, and Lin M F 2017 Electronic and optical properties of graphene in magnetic and electric fields *IOP Concise Physics*. San Raefel, CA, USA: Morgan & Claypool Publishers.
8. Zhang Y, Tang T T, Girit C, Hao Z, Martin M C, Zettl A, et al. 2009 Direct observation of a widely tunable bandgap in bilayer graphene *Nature* **459** 820.
9. Huang Y K, Chen S C, Ho Y H, Lin C Y, and Lin M F 2014 Feature-rich magnetic quantization in sliding bilayer graphenes *Sci. Rep.* **4** 7509.
10. Wang Z F, Liu F, and Chou M Y 2012 Fractal Landau-level spectra in twisted bilayer graphene *Nano Lett.* **12** 3833.

11. Chung H C, Chang C P, Lin C Y, and Lin M F 2016 Electronic and optical properties of graphene nanoribbons in external fields *Phys. Chem. Chem. Phys.* **18** 7573.

12. Lin C Y, Chen R B, Ho Y H, and Lin M F 2018 Electronic and optical properties of graphite-related systems. Boca Raton, FL, USA: CRC Press.

13. Oostinga J B, Heersche H B, Liu X, Morpurgo A F, and Vandersypen L M 2008 Gate-induced insulating state in bilayer graphene devices *Nature Materials* **7** 151.

14. Lin C Y, Chen S C, Wu J Y, and Lin M F 2012 Curvature effects on magnetoelectronic properties of nanographene ribbons *J. Phys. Soc. Jpn.* **81** 064719.

15. Aoki M and Amawashi H 2007 Dependence of band structures on stacking and field in layered graphene *Solid State Communications* **142** 123.

16. Moon P and Koshino M 2013 Optical absorption in twisted bilayer graphene *Phys. Rev. B* **87** 205404.

17. Ruffieux P, Cai J, Plumb N C, Patthey L, Prezzi D, Ferretti A, et al. 2012 Electronic structure of atomically precise graphene nanoribbons *ACS Nano* **6** 6930.

18. Bao C, Yao W, Wang E, Chen C, Avila J, Asensio M C, et al. 2017 Stacking-dependent electronic structure of trilayer graphene resolved by nanospot angle-resolved photoemission spectroscopy *Nano Lett.* **17** 1564.

19. Knox K R, Wang S, Morgante A, Cvetko D, Locatelli A, Mentes T O, et al. 2008 Spectromicroscopy of single and multilayer graphene supported by a weakly interacting substrate *Phys. Rev. B* **78** 201408.

20. Chang C P 2011 Exact solution of the spectrum and magneto-optics of multilayer hexagonal graphene *J. Appl. Phys.* **110** 013725.

21. Kim K S, Walter A L, Moreschini L, Seyller T, Horn K, Rotenberg E, et al. 2013 Coexisting massive and massless Dirac fermions in symmetry-broken bilayer graphene *Nat. Mater.* **12** 887.

22. Sugawara K, Sato T, Souma S, Takahashi T, and Suematsu H 2006 Fermi surface and edge-localized states in graphite studied by high-resolution angle-resolved photoemission spectroscopy *Phys. Rev. B* **73** 045124.

23. Orlita M, Faugeras C, Plochocka P, Neugebauer P, Martinez G, Maude DK, et al. 2008 Approaching the Dirac point in high-mobility multilayer epitaxial graphene *Phys. Rev. Lett.* **101** 267601.

24. Lui C H, Li Z, Mak K F, Cappelluti E, and Heinz T F 2011 Observation of an electrically tunable band gap in trilayer graphene *Nat. Phys.* **7** 944.

25. Ho Y H, Wang J, Chiu Y H, Lin M F, and Su W P 2011 Characterization of Landau subbands in graphite: A tight-binding study *Phys. Rev. B* **83** 121201.

26. Ho C H, Chang C P, and Lin M F 2015 Optical magnetoplasmons in rhombohedral graphite with a three-dimensional Dirac cone structure *J. Phys.: Condens. Matter* **27** 125602.

27. Ou Y C, Chiu Y H, Yang P H, and Lin M F 2014 The selection rule of graphene in a composite magnetic field *Optics Express* **22** 7473.

28. Ou Y C, Sheu J K, Chiu Y H, Chen R B, and Lin M F 2011 Influence of modulated fields on the Landau level properties of graphene *Phys. Rev. B* **83** 195405.

29. Ou Y C, Chiu Y H, Lu J M, Su W P, and Lin M F 2013 Electric modulation effect on magneto-optical spectrum of monolayer graphene *Comput. Phys. Commun.* **184** 1821.

30. Neumann J and Wigner E 1929 *Phys. Z.* **30** 467.

31. Chen S C, Wu J Y, Lin C Y, and Lin M F 2017 Theory of magnetoelectric properties of 2D systems *IOP Concise Physics*. San Raefel, CA, USA: Morgan & Claypool Publishers.

32. Do T N, Lin C Y, Lin Y P, Shih P H, and Lin M F 2015 Configuration-enriched magnetoelectronic spectra of AAB-stacked trilayer graphene *Carbon* **94** 619.

33. Venema L C, Wildoer J W G, Janssen J W, Tans S J, Tuinstra H L J T, Kouwenhoven L P, et al. 1999 Imaging electron wave functions of quantized energy levels in carbon nanotubes *Science* **283** 52.
34. Matsui T, Kambara H, Niimi Y, Tagami K, Tsukada M, and Fukuyama H 2005 STS observations of Landau levels at graphite surfaces *Phys. Rev. Lett.* **94** 226403.
35. Lin C Y, Wu J Y, Chiu Y H, and Lin M F 2014 Stacking-dependent magneto-electronic properties in multilayer graphenes *Phys. Rev. B* **90** 205434.

6 AA-Bottom-Top Bilayer Silicene Systems

Po-Hsin Shih,[a] *Chiun-Yan Lin,*[a] *Thi-Nga Do,*[c,d]
Jhao-Ying Wu,[b] *Shih-Yang Lin,*[e]
Ching-Hong Ho,[b] *Ming-Fa Lin*[a,f,g]

[a] Department of Physics, National Cheng Kung University, Tainan 701, Taiwan

[b] Center of General Studies, National Kaohsiung University of Science and Technology, Kaohsiung 811, Taiwan

[c] Laboratory of Magnetism and Magnetic Materials, Advanced Institute of Materials Science, Ton Duc Thang University, Ho Chi Minh City, Vietnam

[d] Faculty of Applied Sciences, Ton Duc Thang University, Ho Chi Minh City, Vietnam

[e] Department of Physics, National Chung Cheng University, Chiayi 621, Taiwan

[f] Quantum Topology Center, National Cheng Kung University, Tainan 701, Taiwan

[g] Hierarchical Green-Energy Materials Research Center, National Cheng Kung University, Tainan, Taiwan

CONTENTS

Few-layer silicene systems, which are purely made of silicon atoms through the dominating sp^2 bondings and the weak, but significant sp^3 ones, have stirred many researches. Up to now, they are successfully synthesized by the epitaxial growth on the different substrate surfaces. Monolayer silicene systems are generated on Ag(111) [1, 2], Ir(111) [3], and ZrB$_2$(0001) [4], with the 4×4, $\sqrt{3} \times \sqrt{3}$ and 2×2 unit cells, respectively. The high-resolution measurements of STM and low-energy electron diffraction clearly identify the buckled single-layer honeycomb lattice, in which it might present an enlarged unit cell as a result of the strong correlation with substrate (the non-negligible orbital hybridization between silicon atoms and substrate ones). Specifically, the Dirac-cone structure, with a graphene-like Fermi velocity, is directly confirmed from the angle-resolved photoemission spectroscopy [ARPES] experiments [1, 2] in the presence of a small spin-orbital coupling ($\sim$ 2 meV). Bilayer silicene systems, corresponding to the AB stacking configurations, are revealed in the experimental growth [details in Chap. 7]; furthermore, there exist the AB-bt and AB-bb stable bucklings [5]. However, no experimental evidence has been done for the AA-stacked bilayer silicene systems. In addition, similar planar structure is formed in bilayer graphene [5].

Up to now, a lot of theoretical studies are conducted on few-layer silicene systems without and with adatom chemisorptions/guest atom substitutions through the first-principles calculations [6–12], the generalized tight-binding model [13, 14], and the effective-mass approximation [15], especially for the essential properties of monolayer and bilayer. It is well known that the first and second methods are suitable in studying the optimal geometries and the magnetic quantizations, respectively. A single-layer silicene is predicted to have a very narrow gap [$\sim$ 5 meV; Ref. 6], mainly owing to a quite weak, but important spin-orbital coupling. When the intrinsic interactions further combine with the gate voltages and magnetic fields, the destruction of $z = 0$-plane mirror symmetry and the periodical Peierls phases can induce the spin-dominated splitting, the dramatic change of band gaps [15], and the highly degenerate Landau levels [15], respectively. Such electronic properties are predicted to greatly diversify Coulomb excitations [Ref. 6; similar results for germanene in Chap. 10] and induce inter-Landau-level magnetoplasmon modes/electron-gas-like ones. It should be noticed that the VASP calculations are frequently utilized to explore the obvious chemical modifications due to the significant multi-orbital hybridizations in Si-Si, Si-X, and X-X bonds (X for adatoms), e.g., the creations of metallic or semiconducting behaviors [16, 17]. But for bilayer silicene systems with the non-negligible bucklings, the largely enhanced spin-orbital interactions and the complex interlayer hopping integrals make the phenomenological models more difficult to solve the magnetic-field-dominated fundamental properties. For example, after the delicate analysis and detailed fittings, it is almost impossible to get a set of reliable parameters of the generalized tight-binding model, being suitable for the low-lying two pairs of valence and conduction bands. The main reason lies in that the first-principle results on AA and AB stackings [6–12] have shown the very complicated energy dispersions for bilayer silicene systems, such as, the non-monotonous energy dispersions and the irregular valleys at the non-high-symmetry points. Apparently, the free carrier densities in these pristine materials are expected to be very sensitive to the buckling and stacking configurations [e.g., AA-bt, AA-bb, AB-bt, & AB-bb].

The AA-bt silicene systems in the presence of electric and magnetic fields are very suitable for exploring the diversified magnetic quantization phenomena. The generalized tight-binding model, being combined with the dynamic Kubo formula, is further developed to study the essential electronic and optical properties, especially for the main features of magneto-electronic states. The delicate calculations and analyses cover the band properties across the Fermi level (semimetals or zero-gap/finite-gap semiconductors), energy dispersions (linear or parabolic/monotonous or non-monotonous relations), critical points in energy-wave-vector space (band-edge states: minima, maxima, and saddle points), distinct valleys (initiation of electronic states), special structures of van Hove singularities in density of states, optical absorption spectra, significant gate-voltage- and magnetic-field-dependences, categories/subgroups of valence and conduction Landau levels, and possible magneto-optical excitations. Very interestingly, how to unify the valley-enriched magnetic quantizations is the main studying focus. This is done through the detailed examinations on the spatial oscillation modes of magnetic wave functions, in which the complex combined effects of different subgroups need to be clarified from the distinct stable/metastable valleys. The important differences between an AA-bt bilayer silicene and a monolayer silicene/graphene in the essential physical properties are thoroughly investigated through the viewpoint of valley structures in electronic energy spectra.

6.1 ESSENTIAL ELECTRONIC AND OPTICAL PROPERTIES

The tight-binding mode and the gradient approximation are, respectively, utilized to investigate electronic properties and optical absorption spectra. A bilayer AA-bt silicene, as clearly shown in Figs. 6.1(a) and 6.1(b), present the same and opposite bucklings for the (A^1, A^2) and (B^1, B^2) sublattices. There are four Si atoms in a primitive unit cell. The low-energy Hamiltonian, which is built from the four tight-binding functions of Si-$3p_z$ orbitals in the absence of spin-orbital couplings, is suitable in exploring the essential physical properties. It is expressed by

$$H = \sum_{i,l} U_i^l c_i^{\dagger l} c_i^l + \sum_{\langle i,j \rangle, l, l'} \gamma_{ij}^{ll'} c_i^{\dagger l} c_j^{l'}. \tag{6.1}$$

In this notation, $c_i^{\dagger l}$ (c_i^l) can create (destroy) an electronic state at the i-th site of the l-th layer. The Coulomb potential energy, $U_i^l(A^l, B^l)$, depends on the heights of buckled sublattices, mainly owing to the applied gate voltage. The term covers the intralayer and interlayer hopping integrals, $\gamma_{ij}^{ll'}$. The former, mainly arising

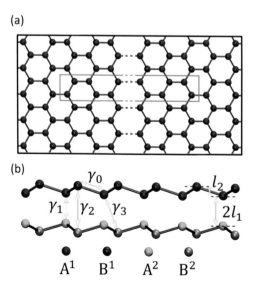

FIGURE 6.1 The geometric structure of AA-bt bilayer silicene system: (a) top view with a uniform magnetic field, and (b) side view including the significant hopping integrals.

from the nearest-neighbor interactions in the (A^l, B^l) sublattices, is $\gamma_0 = 1.004$ eV. Furthermore, the latter possess the similar interlayer atomic interactions; that is, $\gamma_1 = -2.110$ eV, $\gamma_2 = -1.041$ eV, and $\gamma_3 = 0.035$ eV, respectively, correspond to the (A^1, A^2), (B^1, B^2), and (A^1, B^2) sublattices [Fig. 6.1(b)]. It should be noticed that γ_1/γ_2 is much higher/comparable to γ_0. This result and the high stacking symmetry might be responsible for the negligible spin-orbital couplings.

An AA-bt bilayer silicene exhibits the unusual electronic properties, in sharp contrast with those of a single-layer silicene [2] and an AA-stacked bilayer graphene [18], such as, energy bands and density of states. This system, as clearly shown in Fig. 6.2(a), has two pairs of valence and conduction bands due to the Si-$3p_z$-orbital π bondings, in which they are somewhat asymmetric about the Fermi level of $E_F = 0$. Apparently, it presents the semimetallic behavior with zero gap and a finite density of states at E_F [details in Fig. 6.4(a)]. The second pair is situated at deeper/higher energy ranges of $|E^{c,v}| \geq 2$ eV; therefore, the low-energy essential properties are expected to be dominated by the first pair. Very importantly, the stable/non-stable valleys are formed from the electronic states near the high-symmetry points of the hexagonal first Brillouin zone [Fig. 6.2(b)], i.e., there exist the M, K and Γ valleys [Figs. 6.2(c), 6.2(d), and 6.2(e)], as measured from $E_F = 0$ with conduction/valence band-edge state energies, respectively, ~ 0.5 eV, 1.19 eV; 1.43 eV/-0.51 eV, -1.21 eV; -1.50 eV. They are expected to be closely related to the magnetic quantizations of the initial Landau levels [discussed later in Figs. 6.7–6.10]. Furthermore, the first one and the other two, respectively, belong to the saddle point and the local extreme point, thus leading to the different van Hove singularities, absorption structures, and magnetic quantization behaviors. Apparently, the Dirac-cone band structure is absent, since the K/K$'$ valleys are not responsible for the low-lying electronic states closest to the Fermi level. That is to say, the upward conduction/downward valence Dirac cone initiated from the K/K$'$ points is dramatically changed into the concave-downward conduction/concave-upward valence valleys. This is the most important difference between monolayer and bilayer silicene systems (between AA-stacked bilayer graphene and silicene). Such a result might be due to the more complex and strong hopping integrals [the large γ_1 and γ_2 in Fig. 6.1(b)]. It should be noticed that the constant-energy loops of $E^c \sim 4$ meV and $E^v \sim -4$ meV are almost merged together by the concave-downward conduction valley and concave-upward valence one, and a very narrow energy spacing between conduction and valence ones is ~ 8 meV [inset in Fig. 6.4(a)]. Very interestingly, the Dirac-cone-like valence and conduction valleys thoroughly disappear in the AA-bt bilayer silicene; therefore, the low-lying Landau levels do not correspond to the initial magneto-electronic states [discussed later in Fig. 6.10].

The Bloch wave functions of AA-bt bilayer silicene, which consist of the Si-$3p_z$-orbital tight-binding functions on the $(A^1, B^1)/(A^2, B^2)$ sublattices, strongly depend on the wave vectors, as clearly indicated in Fig. 6.2(f) by the solid and dashed blue curves, respectively. For example, electronic states of the low-lying conduction band near the K point only have the identical B^1 and B^2 components (the vanishing A^1 and A^2 ones). Obviously, this illustrates the equal distribution probability on the B^1 & B^2 (A^1 & A^2) sublattices. mainly owing to the same (x, y)-plane projection (chemical environment). And then along the KM→MΓ

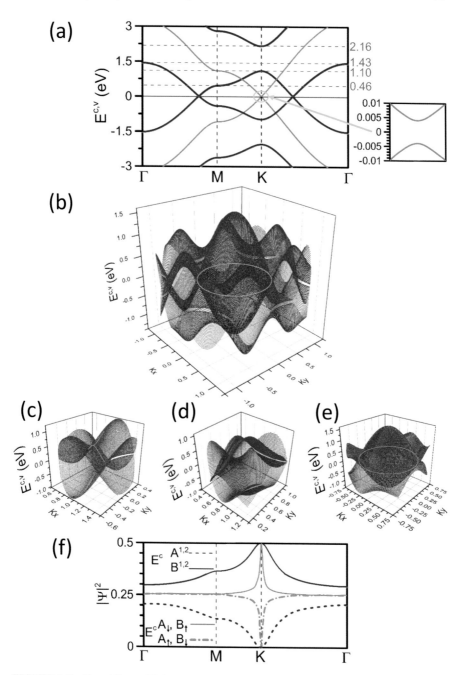

FIGURE 6.2 For a bilayer AB-bt stacking, (a) its 2D band structure along the high symmetry points, (b) the whole 3D low-energy electronic spectrum & those near the (c) M, (d) K, (e) Γ points, and (f) the wave-vector-dependent wave functions [the solid and dashed blue curves of the A^1 and B^1 sublattices, respectively]. Also shown in (a) and (f) are those of monolayer systems by the green curves.

directions, the B^l- and A^l-sublattice probabilities, respectively, present the variations of $0.5 \rightarrow 0.375 \rightarrow 0.265$ and $0 \rightarrow 0.125 \rightarrow 0.2355$, being quite different from those in monolayer silicene (the solid and dashed green curves). The B^l-sublattice dominance is also revealed in the shallower valence band; therefore, the equivalence of the intralayer (A^l, B^l) sublattices is thoroughly broken by the very strong hopping integrals of γ_1 and γ_2 [Fig. 6.1(b)]. It should be noticed that the higher/deeper conduction/valence band is dominated by the A^l sublattice, but not the B^l one. On the other hand, monolayer silicene possesses a significant spin-orbital coupling [6.2], so the $(A_\uparrow, B_\uparrow)/(A_\uparrow, B_\downarrow)$ sublattices exhibit the same probabilities during the variation of wave vector, as shown in Fig. 6.2(f) by the green curves. This result obviously indicates their equivalence under the honeycomb lattice/hexagonal symmetry.

From all the predicted band structures, the conduction/valence states are initiated from the Γ- and K-point valleys, in which the latter cover the saddle M-point one. Furthermore, their energy spectra present the monotonous dispersions, the strong anisotropy, and the conduction and valence constant-energy loops without band overlap. Apparently, the first pair of conduction and valence bands of bilayer silicene sharply contrast with those in monolayer silicene/graphene [1, 2, 19]. The latter presents a similar concave-downward/concave-upward parabolic conduction/valence valley which is initiated from the Γ point at the higher/deeper energy [the green curve in Fig. 6.2(a)]. This structure is directly transformed into an upward Dirac cone/a downward one along the direction of ΓK, or it first becomes a saddle form and then a pair of Dirac cones near the Fermi level along the path of ΓM$\rightarrow$K. The K-point conduction and valence valleys, being slightly separated by a very narrow gap of $E_g \sim 8$ meV, belong to very stable structures, as clearly illustrated by the initial magnetic quantization (discussed later).

The low-lying band structures are very sensitive to the change of gate voltage, as clearly display in Figs. 6.3(a)–6.3(d). The zero-gap property across the Fermi level is thoroughly altered by the various V_zs. The distinct Coulomb potentials on the four sublattices of (A^1, B^1, A^2, B^2), as well as the significant interlayer hopping integrals, are responsible for the energy separations of the highest occupied valence state and the lowest unoccupied conduction one along the KΓ and MΓ directions. Specifically, the smaller energy spacing in between the K and Γ points is band gap, in which its magnitude gradually grows as the gate voltage increases. Very interestingly, such critical factors also lead to the strong modification on energy dispersions which are initiated from the Γ, K, and M valleys, e.g., the great enhancement of the second band-edge state energy. When V_z is sufficiently high [0.4 eV in Fig. 6.3(d)], the K-point conduction/valence state is higher/deeper than the Γ-point one. On the other hand, the spin-configuration-dominated state degeneracy remains doubly degenerate. This result sharply contrasts with the spin-split valence and conduction bands in monolayer slicene [1, 2], since the latter are strongly affected by the combined effects of the spin-orbital coupling and the difference of Coulomb potential. Furthermore, the spin-split states can create dramatic variations of energy gaps $[E_g \sim 0 \rightarrow E_g \neq 0 \rightarrow E_g = 0 \rightarrow E_g \neq 0$ during the increment of gate voltage [20]]. As for the theoretical predictions on zero-field and gate-voltage-dependent band structures, the high-resolution angle-resolved measurements of photoemission spectroscopies [21] are available in examining the unusual three valence valleys due to

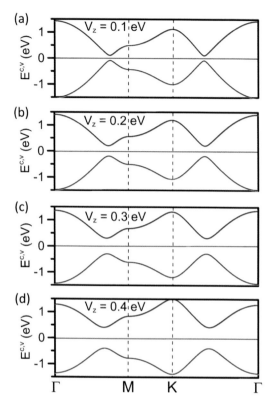

FIGURE 6.3 Electronic structures under the various gate voltages: (a) V_z=0.1 eV, (b) 0.2 eV, (c) 0.3 eV, and (d) 0.4 eV.

the specific buckling, AA stacking configuration, and external Coulomb potentials, and the important differences between bilayer and monolayer silicene systems. That is, such experimental examinations could provide the full information the rich valley structures due to the quite large interlayer hopping integrals. From the theoretical points of view, the effective-mass approximation is not suitable dealing with the zero-file electronic properties and the magnetic quantizations as a result of the absence of stable valleys near the Fermi level and the energy overlap of K and Γ valleys. This further illustrates that only the generalized tight-binding model is reliable in fully exploring the diverse quantization phenomena.

There are three kinds of band-edge states, and so do the van Hove singularities in the density of states [Figs. 6.4(a)–6.4(e)]. The special structures cover the asymmetric peaks in the square root divergent form, shoulders (discontinuous structures), and logarithmically divergent peaks. A pair of temple-like cusp structures, as clearly shown in the inset of Fig. 6.4(a), crosses the Fermi level with a very narrow energy spacing ($\sim$ 8 meV). They arise from the special conduction/valence constant-energy loops in the energy-wave vector space [the green circle in Fig. 6.2(b)], since such energy bands could be regarded as the one-dimensional parabolic dispersions. With the increasing energy, the first/second pair of prominent symmetric peaks, which correspond to the

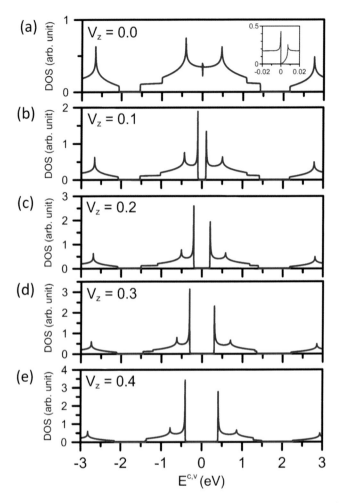

FIGURE 6.4 The van Hove singularities in density of states under (a) the pristine case, (b) $V_z = 0.1$ eV, (c) 0.2 eV, (d) 0.3 eV, and (e) 0.4 eV.

2D saddle M-point [Fig. 6.2(a)], appear at 0.50 eV & −0.52 eV/2.60 eV & −2.65 eV. Moreover, the K and Γ valleys, with the extreme points, create the specific shoulder structures, respectively, at (1.10 eV, −1.15 eV)/(2.16 eV, −2.20 eV) and (1.43 eV, −1.5 eV). Very interestingly, the external gate voltage [Figs. 6.4(b)–6.4(e)] creates the dramatic transformations, the creation of band gap, the greatly enhanced asymmetric peaks, and even the merged shoulder structures due to the K and Γ valleys [e.g., $V_z = 0.4$ eV in Fig. 36.4(f)]. The former two results mainly originate from the separated conduction and valence constant-energy loops in Figs. 6.3(a)–6.3(d), where their curvatures decline in the increment of gate voltage. The above-mentioned rich forms of 2D van Hove singularities, directly reflecting the unusual energy bands in AA-bt bilayer silicene [Fig. 6.2], could be verified through the high-resolution scanning tunneling spectroscopy [STS] measurements [details in Sec. 2.1]. Such examinations are useful in understanding the combined effects of buckling structures and sublattice-dependent Coulomb potentials. Most importantly, the density of states is rather high near the M point, so that it would be very difficult to generate the highly degenerate Landau levels from this valley, as observed in monolayer silicene/graphene [22,23]. That is to say, the M-point valley is not stable for the magnetic quantization, or the Landau levels are not initiated from the unstable valley.

Obviously, the optical absorption structures directly reflect the main features of energy bands and wave functions [Figs. 6.2(a)–6.2(e)], being sensitive to the changes in gate voltages. Both AA-bt bilayer silicene and monolayer system, as clearly shown in Figs. 6.5(a)/6.5(c) and 6.5(b)/6.5(d), present the unique vertical excitation spectra during the valence-conduction transitions under the same wave vector at zero temperature. The occupied and unoccupied (valence and conduction) band-edge states in the former, which respectively, correspond to the constant-energy loops near the Fermi level, the saddle M-point structures, the K valleys and the Γ ones, satisfy the conservation of momentum; therefore, the joint density of states of optical excitations is created by the similar van Hove singularities [Fig. 6.4(a)]. If the dipole matrix elements due to such critical points are almost vanishing [finite], there are no (exist) special structures in the optical spectral functions. As a result, only the first pair of valence and conduction energy bands in a bilayer silicene [the solid blue curve in Fig. 6.5(a) at $V_z = 0$] exhibit the weak, but observable asymmetric peak in the square root form ($\sim$ 8 meV; inset). However, the prominent peak with the logarithmic symmetry ($\sim$ 1.00 eV), the shoulder structure ($\sim$ 2.20 eV) and the similar one ($\sim$ 2.92 eV) hardly survive (discussed in the following paragraph). The first optical absorption structure determines a threshold excitation frequency (an optical gap identical to an energy gap); furthermore, its frequencies and intensities are greatly enhanced by the increasing gate voltages, as clearly shown in Fig. 6.5(c). The initial asymmetric peak represents the strongest vertical optical excitations at any gate voltages, since the other higher-energy absorption structures reveal the negligible intensities. On the other hand, a single-layer silicene displays two significant absorption structures within $\omega < 3$ eV [the sold blue curve in Fig. 6.5(b) at $V_z = 0$], the weak shoulder [$\sim$ 6.00 meV in inset due to the slight separation of Dirac cones in Fig. 6.2(a)] and the strong logarithmic peak ($\sim$ 2.30 eV arising from the saddle M-point). With the application of V_z [other curves in Fig. 6.5(d)], two neighboring shoulder structures/one prominent peak appear/occurs at the higher absorption frequencies. The unusual former structures originate from the low-lying spin-spilt electronic states are induced by the cooperation of gate voltage and spin-orbital coupling. It should be noticed that the inversion symmetry about the $z = 0$ plane is broken in monolayer silicene, but not for the AA-bt system. The intensities/frequencies of special absorption spectra rapidly grow/vary during the increase of gate voltage.

The negligible absorption intensities, which are induced by the electronic states near the K-, M-, and Γ-band-edge ones, deserve a closer examination. They come from the corresponding dipole matrix elements, being mainly determined by the first derivative of Hamiltonian matrix elements about k_x/k_y and the four subenvelope functions of the initial and final states [the theoretical details in Sec. 3.2]. The former are closely related to ($\gamma_0, \gamma_1, \gamma_2, \gamma_3$) the ($A^1/A^2$, B^1/B^2), (A^1, A^2), (B^1, B^2), (A^1/B^1,

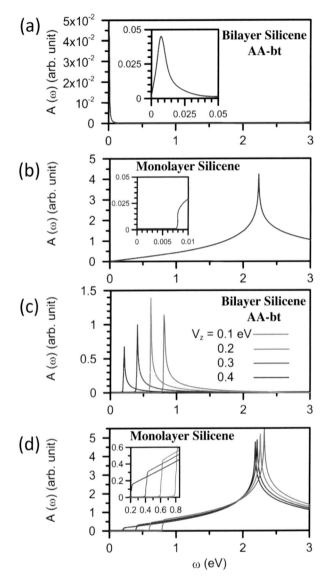

FIGURE 6.5 The optical absorption spectra at $V_z = 0/V_z=0.1$, 0.2, 0.3; 0.4 eV's for (a)/(c) bilayer and (b)/(d) monolayer systems. The insets in (a) and (b) show the initial absorption structures.

B^2/A^2) atomic interactions in Fig. 6.1(b)], in which the very large hopping integrals, the second and third terms are independent of wave vectors, i.e., they do not contribute to the optical transitions. Furthermore, both occupied valence states and unoccupied conductions. Respectively, possess the symmetric and antisymmetric superpositions of subenvelope functions on A^l and B^l sublattices. This is responsible for the vanishing contributions due to the first and fourth hopping integrals. The featureless

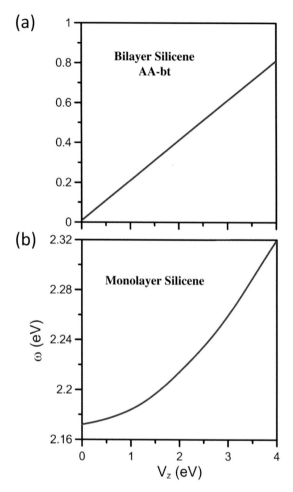

FIGURE 6.6 The gate-voltage-dependent vertical transition frequencies of prominent absorption structures due to the first pair of valence and conduction bands in (a) bilayer and (b) single-layer silicenes.

absorption spectrum is hardly affected by the external gate voltages. On the other hand, the constant-energy valence and conduction loops, being far away from these valley structures, exhibit the V_z-created prominent absorption peak.

The gate-voltage-dependent optical threshold absorption structure can provide the full informations on the direct experimental examination. As to the AA-bt bilayer system, the optical gap in Fig. 6.6(a), which is related to the constant-energy valence and conduction loops [Fig. 6.2(b)], quickly grow in the increment of gate voltage. Furthermore, its absorption intensity also behaves the monotonous enhancement. On the other hand, a monolayer silicene exhibits the V_z-doubled absorption frequencies in Fig. 6.6(b). The dependence of optical gap on V_z is non-monotonous; that is, it first reduces, vanishes at a critical gate voltage ($\sim$ 6 meV in the inset), and then gradually enlarges with the increasing gate voltage. The above-mentioned theoretical predictions on the main features of absorption spectra could be directly examined by the high-resolution optical spectroscopies [details in Sec. 2.2]. It is very useful in understanding the unusual valley structures due to the geometric symmetries, the intralayer & interlayer hopping integrals, the spin-orbital couplings, and the gate voltages.

6.2 DIVERSE CHARACTERISTICS OF MAGNETO-ELECTRONIC STATES

The generalized tight-binding model [details in Sec. 3.1] is very efficient and reliable in fully exploring the unusual magnetic quantization in an AA-bt bilayer silicene. By the detailed calculations, the independent magnetic Hamiltonian matrix elements include:

$$\langle A_m^1|H|A_m^1\rangle = E_z l_1, \tag{6.2}$$

$$\langle B_m^1|H|B_m^1\rangle = E_z(l_1 + l_2), \tag{6.3}$$

$$\langle A_m^2|H|A_m^2\rangle = -E_z l_1, \tag{6.4}$$

$$\langle B_m^2|H|B_m^2\rangle = -E_z(l_1 + l_2), \tag{6.5}$$

$$\langle A_m^1|H|B_m^1\rangle = \langle A_m^2|H|B_m^2\rangle = \gamma_0(QP_2 + Q^*P_3), \tag{6.6}$$

$$\langle A_m^1|H|B_m^2\rangle = \langle A_m^2|H|B_m^1\rangle = \gamma_3(QP_2 + Q^*P_3), \tag{6.7}$$

$$\langle A_m^1|H|A_m^2\rangle = \gamma_1, \tag{6.8}$$

$$\langle B_m^1|H|B_m^2\rangle = \gamma_2, \tag{6.9}$$

$$\langle B_m^1|H|A_{m+1}^1\rangle = \langle B_m^2|H|A_{m+1}^2\rangle = \gamma_0 P_1, \tag{6.10}$$

$$\langle B_m^1|H|A_{m+1}^2\rangle = \langle B_m^2|H|A_{m+1}^1\rangle = \gamma_3 P_1, \tag{6.11}$$

in which $l_1 = 1.245$ Å and $l_2 = 0.68$ Å, respectively, represent the half of interlayer distance and the intralayer spacing [Fig. 6.1(b)]. The related terms are:

$$P_1 = exp[ik_x b], \tag{6.12}$$

$$P_2 = exp[i(k_x b/2 + k_y a/2)], \tag{6.13}$$

$$P_3 = exp[i(k_x b/2 - k_y a/2)], \tag{6.14}$$

$$Q = exp[i\pi\psi(j - 1 + 1/6)], \tag{6.15}$$

where j indicates the lattice site. $a = 3.83$ Å and $b = 2.21$ Å are the lattice constant and the bond length, respectively. Generally speaking, the magnetic Hamiltonian is a quite huge Hermitian matrix. For example, it is a ~13,000 × 13,000 Hermitian matrix with real elements for ($k_x = 0, k_y = 0$) state at $B_z = 20$ T.

The Γ- and K-point conduction/valence valleys [Fig. 6.2] account for the rich and unique magnetic quantizations arising from the first pair of energy bands. The valley-dependent diverse phenomena cover the Landau-level state degeneracy, the different localization centers, the significant differences among the localized oscillation modes on the distinct sublattices, the well-behaved & perturbed Landau levels, the complex magnetic-field dependences in the composite energy spectra, and the non-crossing & crossing behaviors. Specifically, the constant-energy loops near the Fermi level are closely related to the higher and lower valleys; therefore, their magneto-electronic states and energy spectra are expected to be very complicated. That is, it might be quite difficult to directly identify the low-energy Landau levels from the high-resolution STS measurements [24–27]. Of course, a lot of observable van Hove singularities, which correspond to the prominent delta-function-like peaks with the sufficiently large energy spacings at the sufficiently high/deep energies, could provide the experimental examinations. In addition, the highly degenerate Landau levels due to the second pair of energy bands are partially discussed.

The first investigations are conducted on the magneto-electronic states of an AA-bt bilayer silicene, being initiated from the Γ-point top/bottom valley. Under a uniform perpendicular magnetic field of $B_z = 20$ T, the initial conduction and valence Landau levels, respectively, appear at 1.427 eV and −1.539 eV. Apparently, the well-behaved magnetic subenvelope functions on the four sublattices of (A^1, B^1, A^2, B^2), as clearly shown in Figs. 6.7(a)–6.7(b) and 6.7(c)–6.7(d), are, respectively, localized about 1/2 and 2/2 of the B_z-enlarged unit cell. That is to say, each specific (k_x, k_y)-state possesses the fourfold degeneracy, covering the spin degree of freedom and the direction of magnetic field. The similar Landau-level state degeneracy is revealed in that of monolayer silicene/graphene [Fig. 6.11(c); Refs. 22 and 23]. Furthermore, four subenvelope functions have similar oscillation modes, in which both A^1 & A^2/B^1 & B^2 sublattices are fully identical. These results directly reflect the equivalence of the intralayer (A_i, B_i) sublattices and the same chemical environment for the interlayer (A^1, A^2) & (B^1, B^2) sublattices. The specific oscillation mode is purely dominated by the B^1 or B^2 sublattices, corresponding to the 1/2 and 2/2 localization centers, could serve as the quantum number. Such behavior is consistent with the zero-field wave-vector-dependent wave functions [Fig. 6.2(f)]. The $n_Γ^{c,v}$ Landau-level subgroup displays a roughly uniform energy spectrum [discussed later in Figs. 6.12(b) and 6.12(c)], mainly owing to the isotropically parabolic energy dispersion near the Γ point [Fig. 6.2(a)]. Apparently, it behaves like a 2D electron gas [28].

The Landau levels, which are magnetically quantized from electronic states near the K/K′ points, exhibit the valley-diversified phenomena [Fig. 6.8]. The magnetic subenvelope functions have four localization centers at 1/6, 2/6, 4/6, and 5/6 of an enlarged unit cell, where the first and third spatial distributions (the second and fourth ones) are identical. Their spatially oscillations might become slightly distorted under the significant interlayer hopping integrals [$γ_1$ and $γ_2$ in Fig. 6.1(b)]; therefore, the K-induced Landau levels belong to the perturbed ones. There exists one zero-point difference between the intralayer A^l and B^l sublattices, another illustration of the almost equivalent A^l and B^l sublattices [Figs. 6.1(a) and 6.1(b)]. The fully dominant

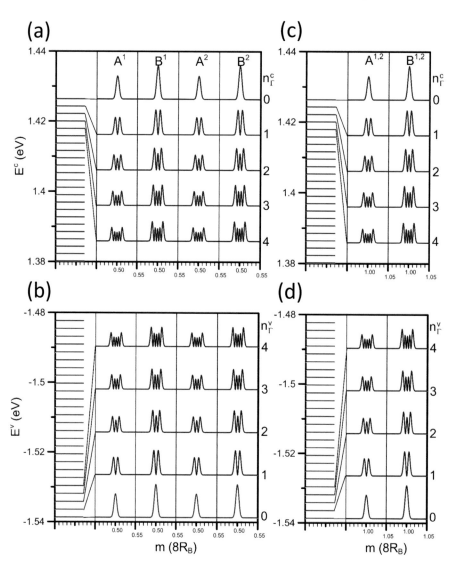

FIGURE 6.7 At $B_z = 50$ T, the conduction and valence Landau levels and wave functions of the AA-bt bilayer silicene within the energy ranges: (a) 1.49 eV $\leq E^c \leq$ 1.43 eV, and (b) -1.54 eV $\leq E^c \leq -1.51$ eV. They, being localized about 1/2, are initiated from the Γ valley. Also shown in (c) and (d) are those of the 2/2 localization center.

B^l sublattices about the (1/6, 4/6]/[2/6, 5/6) localization centers possess the zero-point number higher/less than that of the A^l sublattices by one, as clearly displayed in Figs. 6.8(c) & 6.8(d). As a result, the quantum numbers of the degenerate Landau levels are characterized by the oscillation modes of B^1/B^2 sublattice. In general, each magneto-electronic state in the reduced first Brillouin zone is eightfold degenerate, covering the degree of freedom in spin configurations, magnetic-field directions and equivalent valleys. However, the extra Landau levels, the initial ones under the

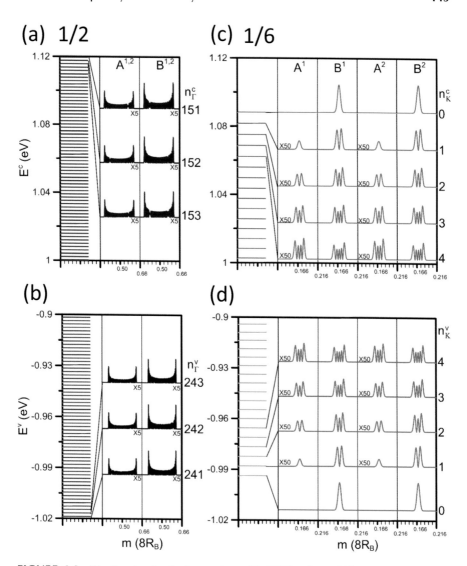

FIGURE 6.8 The Landau levels due to the stable K/K' valley within the energy ranges. (c) 1.02 eV $\leq E^c \leq$ 1.10 eV and (d) −0.99 eV $\leq E^v \leq$ −0.94 eV. The composite magneto-electronic states also cover those arising from the Γ valley in (a) and (b).

(2/6, 5/6) localization centers [the $n_K^c = 0$ and $n_K^v = 0$ Landau levels in Figs. 6.8(c) and 6.8(d), respectively], exhibit the unique fourfold degeneracy. The main reason is that such magneto-electronic states only originate from the B^l sublattices, regardless of the A^l ones [Fig. 6.2(f)]. Very interestingly, the K- and Γ-point-induced Landau-level subgroups only present a direct composition in the magnet-electronic energy spectra, as clearly shown in Figs. 6.8(a)-6.8(d) [also sees Fig. 6.12(c)], since these two stable valleys are independent of each other [Fig. 6.2]. The Landau-level anti-crossings between them are expected to be absent in the magnetic-field- and gate-voltage

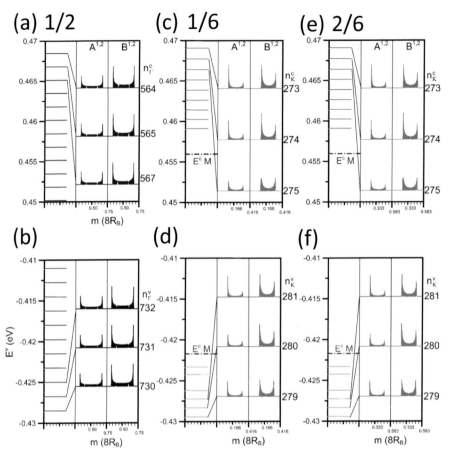

FIGURE 6.9 (a)–(f) Similar plot as Fig. 6.7, but displayed in 0.40 eV $\leq E^c \leq$ 0.60 eV due to the M valley.

dependences. In addition, the higher-/deeper-energy Landau levels, which are magnetically quantized from the second pair energy bands near the K/K′ valleys [Fig. 6.2(a)], exhibit the normal behaviors, i.e., the well-behaved Landau wave functions, with the specific oscillation modes, represent the initial ones [discussed latter in Fig. 6.12(a)].

With the further decrease/increase of conduction/valence Landau-level energies, the Γ- and K-valley-induced magneto-electronic states present the direct composition in the magnetic-field-dependent energy spectra and the well-separated wave functions, such as, the Landau levels within the energy range covering the saddle M-point [Figs. 6.9(a)–6.9(f)], or near the Fermi level [Fig. 6.10]. Apparently, the former remain two degenerate localization centers of 1/2 & 2/2, as observed under the initial case [Figs. 6.7(a)–6.7(d)]. The very wide oscillation distributions, which are ∼ 35% of a unit cell in Figs. 6.9(a) and 6.9(b), obviously indicate the quite large quantum numbers. Such results further illustrate the monotonous variation of the stable Γ valley along the various directions [Figs. 6.2(b) and 6.2(e)]. Specifically, the latter

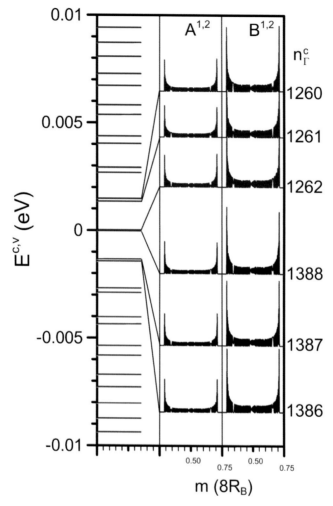

FIGURE 6.10   Similar plot as Fig. 6.7, but depicted under -0.1 eV $\leq E^{c,v} \leq 0.10$ eV across the Fermi level.

exhibit the negligible state splittings; that is, the (1/6, 2/6, 4/6, 5/6) Landau levels are identical to one another in energy spectra and spatial distributions [Figs. 9(c)–9(f)]. Their quantum numbers are much smaller than those of the former, mainly owing to the higher density of states in the K-related valleys [DOS in Fig. 6.4]. The magnetic quantization also reflects the fact that the M-point saddle structure belongs to the K/K′ valley.

Very interestingly, it is quite difficult to characterize the magneto-electronic states across the fermi level, as clearly illustrated in Fig. 6.10. There exist a lot of Landau levels, with rather narrow energy spacings. Their spatial distributions present the rather strong and wide oscillations, leading to the meaningless modes. Moreover, it is very difficult to present the well-behaved magnetic-field-dependent energy spectrum [too complicated to analyze the B_z-dependence in Fig. 6.12(e)]. These results directly reflect the facts that the constant-energy conduction/valence loop possesses a square root divergent density of states [inset in Fig. 6.5(a)], and this structure is due to the cooperation of the stable K- and Γ-point valleys. Such unique magnetic quantization phenomenon is never observed in the other condensed-matter systems according to the previous theoretical and experimental studies [13, 22, 23].

Apparently, the low-energy magneto-electronic states in the AA-bt bilayer silicene are responsible for the other essential physical properties, such as, the delta-function-like van Hove singularities, magneto-optical absorption spectra/selection rules, quantum Hall transports, and Inter-Landau-level dampings and magnetoplasmon modes. However, the Landau-level energy spacings are very narrow, being even lower than the thermal energy of 10 K [~ 1 meV]. The critical factors in experimental verifications of the valley-enriched magnetic quantizations should rely on them, but not the regular or irregular spatial distributions. For example, a plenty of prominent symmetric peaks in density of states near the Fermi level could be examined by the high-resolution STS measurements [details in Sec. 2.1]. The theoretical calculations, which are conducted on the optical, transport and Coulomb excitation properties, might have the high barriers in the numerical technique and the delicate analysis as a result of too many neighboring Landau levels. Whether there exist the concise physical pictures is worthy of a systematic investigation.

An AA-bt bilayer system [Figs. 6.7–6.10] sharply contrasts with monolayer silicence/graphene in the main features of magneto-electronic properties [Fig. 6.11; Refs. 22 and 23]. Apparently, the latter has the conduction and valence Dirac cones covering the Fermi level near the K/K' point [the green curve in Fig. 6.2(a)]. The stable valleys can create the highly degenerate Landau levels with the eightfold degeneracy for each (k_x, k_y) state in the reduced first Brillouin zone, in which two localization centers of (1/6, 4/6)/(2/6, 5/6) are dominated by the B/A sublattice [Fig. 6.11(a)]. The spin degree of freedom is also degenerate even in the presence of spin-orbital interactions. With the further increase/decrease of conduction/valence state energy, the Landau levels near the saddle M-point show a weak splitting behavior [Fig. 6.11(b)]. Furthermore, the degenerate Landau levels of four localization centers are replaced by the four-split ones with the almost random distributions in a B_z-enlarged unit cell; that is, they cannot be well characterized by the non-well-behaved oscillation modes. And then, the doubly degenerate magneto-electronic states come to exist [Fig. 6.11(c)], when their energies are close to that of the Γ point. Such Landau levels possess two different localization centers of 1/2 and 2/2, and they present the initial Landau levels of smaller quantum numbers near the Γ-point energies. The above-mentioned features of magnetic quantizations reflect the direct linking among the K-point Dirac cones, the saddle M-points, and the parabolic Γ valley [discussed earlier in Fig. 6.2]. On the other side, the K-point/Γ-point valleys of the AA-bt bilayer silicene display any parabolic dispersions, occurs at the higher/deeper energy, directly combine with the Γ-valley/K-valley Landau-level spectrum, and the B^l-sublattice dominance under localization centers. These two independent valleys only create the very high quantum numbers of Landau levels within the energy ranges of the M point and the Fermi level [Figs. 6.9 and 6.10]. Apparently, the stable valley structures, which are determined by the intrinsic interactions and lattice symmetries, are responsible for the diversified magnetic properties.

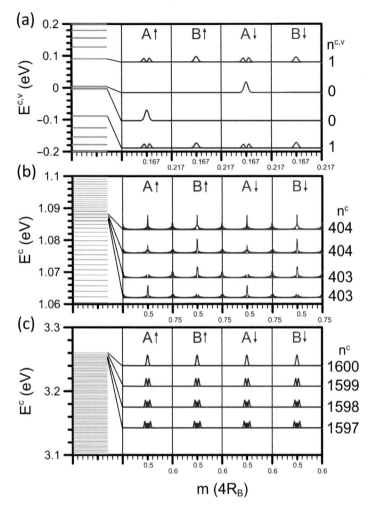

FIGURE 6.11 The Landau-level energies and wave functions of monolayer silicene at the different energy ranges: (a) $0 \leq E^c \leq 0.1\gamma_0$, (b) $0.95\gamma_0 \leq E^c \leq 1.05\gamma_0$, and (c) $2.90\gamma_0 \leq E^c \leq 3.0\gamma_0$.

The magnetic-field- and gate-voltage-dependent energy spectra [Figs. 6.12 and 6.13] are very useful in thoroughly understanding the magnetic quantization phenomena and providing the full informations on the experimental examinations. For a bilayer AA-bt silicene, the linear B_z-dependence, as obviously displayed in Fig. 6.12(a), is revealed in the Landau-level energy spectrum of the second conduction/valence band. This result further illustrates the stable K/K' valleys for such electronic structure [Fig. 6.2(a)]. However, two independent Landau-level subgroups in Figs. 6.12(b)–6.12(d), being closely related to the first conduction/valence, exhibit the composite energy spectrum with/without only the frequent crossings/anti-crossings. It is very difficult to reach a final conclusion on the B_z-dependent behaviors for each

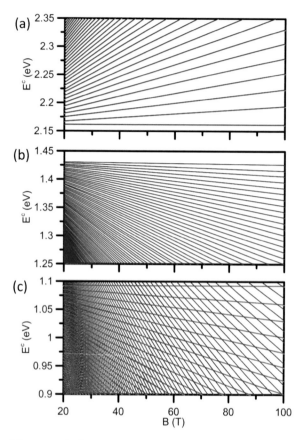

FIGURE 6.12   The magnetic-field-dependent Landau-level energy spectrum, being plotted in the different regions: (a) 2.15 eV $\leq$ E^c $\leq$ 2.35 eV, (b) 1.25 eV $\leq$ E^c $\leq$ 1.45 eV, (c) 0.90 eV $\leq$ E^c $\leq$ 1.1.10 eV, (d) 0.40 eV $\leq$ E^c $\leq$ 0.6 eV, and (e) −0.1 eV $\leq$ $E^{c,v}$ $\leq$ 0.10 eV.

Landa-level subgroup. Moreover, there are too many low-energy magneto-electronic across the Fermi level to distinguish the Γ- and K-valley-induced energy spectra [Fig. 6.12(e)]. On the other side, a monolayer silicene shows the initial Landau levels near $E_F = 0$ [Fig. 6.13(a)], in which the B_z-induced energy spectrum is close to that of the linear Dirac cone [23]. Furthermore, gate voltages in Fig. 6.13(b) can create the spin-up- and spin-down-dominated Landau levels, the semiconductor-semimetal transition at a critical value, and the frequent crossings/anti-crossings. It should be noticed that the V_z-related anti-crossing behaviors are relatively easy to be observed in groups-IV systems with the larger spin-orbital couplings and the weaker hopping integrals (the smaller Fermi velocities), such as, monolayer germanene and tinene [29, 30]. The predicted Landau-level energy spectra could be verified from the experimental measurements of STS, optical spectroscopies, and Hall apparatus [details in Secs. 2.1–2.3].

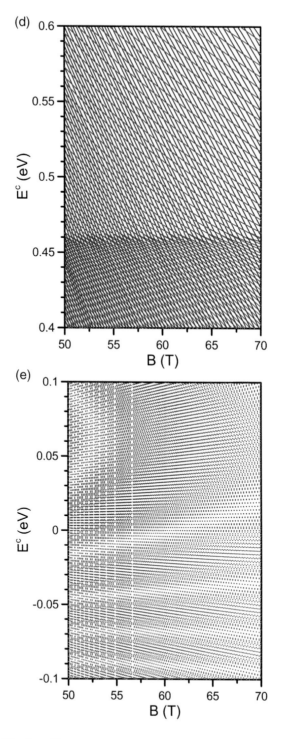

FIGURE 6.12 (*Continued*).

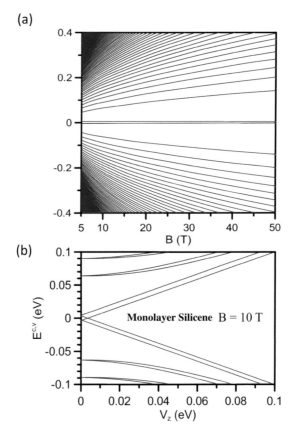

FIGURE 6.13 The magnetic-field- and gate-voltage-related magneto-electronic energy spectra for monolayer silicene.

REFERENCES

1. Vogt P, Padova P D, Quaresima C, Avila J, Frantzeskakis E, Asensio M C, et al. 2012 Silicene: Compelling experimental evidence for graphenelike two-dimensional silicon *Phys. Rev. Lett.* **108** 155501.
2. Feng B, Ding Z, Meng S, Yao Y, He X, Cheng P, et al. 2012 Evidence of silicene in honeycomb structures of silicon on Ag(111) *Nano Lett.* **12 (7)** 3507.
3. Meng L, Wang Y, Zhang L, Du S, Wu R, Li L, et al. 2013 Buckled silicene formation on Ir(111) *Nano Lett.* **13 (2)** 685.
4. Fleurence A, Friedlein R, Ozaki T, Kawai H, Wang Y, and Yamada-Takamura Y 2012 Experimental evidence for epitaxial silicene on diboride thin films *Phys. Rev. Lett.* **108** 245501.
5. Yaokawa R, Ohsuna T, Morishita T, Hayasaka Y, Spencer M J S, and Nakano H 2016 Monolayer-to-bilayer transformation of silicenes and their structural analysis *Nat. Comm.* **7** 10657.
6. Mohan B, Kumar A, and Ahluwalia P K 2013 A first principle calculation of electronic and dielectric properties of electrically gated low-buckled mono and bilayer silicene *Physica E* **53** 233.

7. Liu H S, Han N, and Zhao J J 2014 Band gap opening in bilayer silicene by alkali metal intercalation *J. Phys.: Condens. Matter* **26** 475303.
8. Liu J J and Zhang W Q 2013 Bilayer silicene with an electrically-tunable wide band gap *RSC Adv.* **3** 21943.
9. Liuy F, Liuy C C, Wu K H, Yang F, and Yao Y G 2013 $d + d'$ Chiral superconductivity in bilayer silicene *Phys. Rev. Lett.* **111** 066804.
10. Padilha J E and Pontes R B 2015 Free-standing bilayer silicene: The effect of stacking order on the structural, electronic, and transport properties *J. Phys. Chem. C* **119** 3818.
11. Fu H X, Zhang J, Ding Z J, Li H, and Menga S 2014 Stacking-dependent electronic structure of bilayer silicene *Appl. Phys. Lett.* **104** 131904.
12. Wang X Q and Wu Z G 2017 Intrinsic magnetism and spontaneous band gap opening in bilayer silicene and germanene *Phys. Chem. Chem. Phys.* **19** 2148.
13. Do T N, Shih P H, Gumbs G, Huang D, Chiu C W, and Lin M F 2018 Diverse magnetic quantization in bilayer silicene *Phys. Rev. B* **97** 125416.
14. Do T N, Gumbs G, Shih P H, Huang D, Chiu C W, Chen C Y, et al. 2019 Peculiar optical properties of bilayer silicene under the influence of external electric and magnetic fields *Scientific Reports* **9** 624.
15. Tabert C J and Nicol E J 2013 Magneto-optical conductivity of silicene and other buckled honeycomb lattices *Phys. Rev. B* **88** 085434.
16. Lin X and Ni J 2012 Much stronger binding of metal adatoms to silicene than to graphene: A first-principles study *Phys. Rev. B* **86** 075440.
17. Sahin H and Peeters F M 2013 Adsorption of alkali, alkaline-earth, and 3d transition metal atoms on silicene *Phys. Rev. B* **87** 085423.
18. Rakhmanov A L, Rozhkov A V, Sboychakov A O, and Nori F 2012 Instabilities of the AA-stacked graphene bilayer *Phys. Rev. Lett.* **109** 206801.
19. Sprinkle M, Siegel D, Hu Y, Hicks J, Tejeda A, Taleb-Ibrahimi A, et al. 2009 First direct observation of a nearly ideal graphene band structure *Phys. Rev. Lett.* **103** 226803.
20. Ni Z, Liu Q, Tang K, Zheng J, Zhou J, Qin R, et al. 2012 Tunable bandgap in silicene and germanene *Nano Lett.* **12 (1)** 113.
21. De Padova P, Vogt P, Resta A, Avila J, Razado-Colambo I, Quaresima C, et al. 2013 Evidence of Dirac fermions in multilayer silicene *Appl. Phys. Lett.* **102** 163106.
22. Yin L J, Li S Y, Qiao J B, Nie J C, and He L 2015 Landau quantization in graphene monolayer, bernal bilayer, and bernal trilayer on graphite surface *Phys. Rev. B* **91** 115405.
23. Tabert C J and Nicol E J 2013 Magneto-optical conductivity of silicene and other buckled honeycomb lattices *Phys. Rev. B* **88** 085434.
24. Jhang S H, Craciun M F, Schmidmeier S, Tokumitsu S, Russo S, Yamamoto M, et al. 2011 Stacking-order dependent transport properties of trilayer graphene *Phys. Rev. B* **84 (16)** 161408.
25. Miller D L, Kubista K D, Rutter G M, Ruan M, and de Heer W A 2009 Observing the quantization of zero mass carriers in graphene *Science* **324 (5929)** 924.
26. Rutter G M, Jung S, Klimov N N, Newell D B, Zhitenev N B, and Stroscio J A 2011 Microscopic polarization in bilayer graphene *Nature Physics* **7 (8)** 649.
27. Li G and Andrei E Y 2007 Observation of Landau levels of dirac fermions in graphite *Nature Physics* **3 (9)** 623.
28. Kallin C and Halperin B I 1984 Excitations from a filled Landau level in the two-dimensional electron gas *Phys. Rev. B* **30** 5655.
29. Groves S H, Pidgeon C R, and Feinleib 1966 Infrared magnetoelectroreflectance in Ge, GaSb, and InSb *Phys. Rev. Lett.* **17** 643.
30. Chen S C, Wu C L, Wu J Y, and Lin M F 2016 Magnetic quantization of sp^3 bonding in monolayer gray tin 2016 *Phys. Rev. B* **94** 045410.

7 AB-Bottom-Top Bilayer Silicene

Thi-Nga Do,[c,d] *Chiun-Yan Lin,*[a] *Jhao-Ying Wu,*[b]
Po-Hsin Shih,[a] *Shih-Yang Lin,*[e]
Ching-Hong Ho,[b] *Ming-Fa Lin*[a,f,g]

[a] Department of Physics, National Cheng Kung University,
Tainan 701, Taiwan
[b] Center of General Studies, National Kaohsiung University of
Science and Technology, Kaohsiung 811, Taiwan
[c] Laboratory of Magnetism and Magnetic Materials, Advanced
Institute of Materials Science, Ton Duc Thang University,
Ho Chi Minh City, Vietnam
[d] Faculty of Applied Sciences, Ton Duc Thang University,
Ho Chi Minh City, Vietnam
[e] Department of Physics, National Chung Cheng University,
Chiayi 621, Taiwan
[f] Quantum Topology Center, National Cheng Kung University,
Tainan 701, Taiwan
[g] Hierarchical Green-Energy Materials Research Center,
National Cheng Kung University, Tainan, Taiwan

CONTENTS

Two-dimensional materials, such as the IV- and V-group layered structures [1–23], have become the mainstream condensed-matter systems since the discovery of graphene in 2004 through the mechanical exfoliation. Their rich geometric properties cover the nano-scaled thicknesses, specific lattice symmetry, planar or buckled structure, and unique stacking configuration. Such systems are identified/predicted to exhibit the diverse physical properties [1–16] with many potential device applications [17–23]. Most importantly, the unusual Hamiltonians include the complex effects arising from the significant orbital hybridizations in chemical bonds, spin-orbital couplings, magnetic fields, electric fields, and interlayer atomic interactions. How to solve them becomes one of the basic tasks in solid-state physics today. The studying focuses are the magnetic- and electric-field-modulated electronic and optical properties in AB-stacked bilayer silicene.

Recently, few-layer silicene systems, with the buckled honeycomb lattices, have been successfully synthesized on Ag(111), Ir(111), and $ZrBi_2$ surfaces [24, 25]. According to the accurate first-principles calculations [26–32], most of stacking-dependent structures belong to the meta-stable systems. Both AB and AA stackings, which are defined by the (x, y)-plane projections, display bottom-top (bt) and bottom-bottom (bb) configurations on the (x, z) plane [30]. Up to date, the AB-bt and AB-bb configurations have been confirmed from the delicate measurements of high-angle annular dark field scanning transmission electron microscopy [33]. The geometric symmetry, the intralayer and interlayer atomic interactions, and layer-related spin-orbital couplings are expected to dominate the low-energy fundamental physical properties. For example, monolayer silicene presents a slightly separated Dirac cone with a narrow direct band gap ($E_g \sim 10$ meV) in the presence of spin-orbital interactions [24]. The low-lying band structures in bilayer silicene become very sensitive to changes in stacking configurations, such as the stacking-induced indirect gap in AB-bt system [26–32] and semimetal in AA-bb one [26, 30, 31]. The former, with the lowest ground state energy, is very suitable in illustrating the rich magnetic quantization phenomena.

The low-energy electronic properties of monolayer silicene are mainly determined by the outer $3p_z$ orbitals, similar to graphene systems [details in Sec. 6.1]. The perturbation approximation of the 4×4 Hamiltonian could be made around the high-symmetry point [the K/K′ valleys], and then the magnetic quantization only follows in a straightforward way. The Landau-level energies are identified to be associated with the energy gap and Fermi velocity analytically [34]. These magneto-electronic spectra look similar to those of monolayer graphene as their magnitudes become much larger than E_g. It should be noticed that the magneto-electronic states remain doubly degenerate for the spin degree of freedom even in the presence of spin-orbital coupling [34]. However, the effective-mass approximation becomes too cumbersome for bilayer silicene with unusual band structures. On the other hand, the generalized tight-binding model has been developed for solving the various Hamiltonians in distinct condensed-matter systems. The theoretical framework is based on the subenvelope functions of distinct sublattices (Fig. 7.1(d)), in which all the intrinsic interactions and the external fields can be taken into consideration simultaneously [details in Sec. 3.1; Ref. 35]. Specifically, the magnetically quantized energy spectra, wave functions, and inter-Landau-level transitions could be evaluated very efficiently through the exact diagonalization method even for a very large Hamiltonian matrix with complex elements.

The generalized tight-binding model [35], being directly combined with the dynamic Kubo formula under the gradient approximation [36], is very suitable for fully exploring the low-energy electronic and optical properties of AB-bt silicene in the presence of uniform magnetic $[B_z\hat{z}]$ and electric $[E_z\hat{z}]$ fields. First, the intrinsic interactions in the tight-binding model are obtained from the well fitting with the first-principles calculations [26–32]. And then, the main features of the magnetically quantized Landau levels, energy spectra and spatially oscillating distributions, are thoroughly examined, especially for the composite effects arising from intrinsic

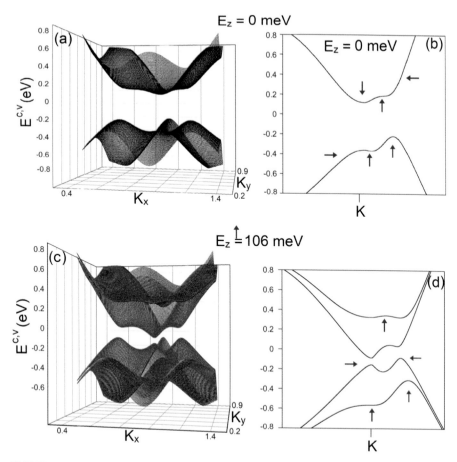

FIGURE 7.1 (a)–(d) For the AB-bt bilayer silicene, the first pair of conduction and valence energy bands in the 3D and 2D forms under the various electric-field strengths; (a) & (b) $E_z = 0$, (c) & (d) $E_z = 106$ meV/Å, (e) & (f) $E_z = 124$ meV/Å, and (g) & (h) $E_z = 153$ meV/Å.

interactions and external fields. This work clearly demonstrates that the magneto-electronic states are characterized by the dominating (B^1 and B^2) sublattices and spin-dependent configurations, leading to four subgroups of conduction/valence Landau-level states. This Landau-level degeneracy splitting will effectively reduce both the impurity and phonon scatterings and result in enhanced mobilities at the same time. The unique Landau levels are directly reflected in the magneto-optical conductivities with a lot of single, double, and twin nonuniform delta-function-like absorption peaks. The sublattice- and spin-dependent Landau-level energies are confirmed from those of the van Hove singularities in density of states, for which the B_z-created energy splitting behaviors could be verified through the high-resolution scanning tunneling spectroscopy [STS] measurements [details in Sec. 2.1]. Specifically, the Landau-level energies and wave functions are predicted to be easily modulated by an electric field, leading to the frequent crossing and anti-crossing phenomena in the E_z- and B_z-dependent energy spectra. Therefore, the use of bilayer silicene, in comparison

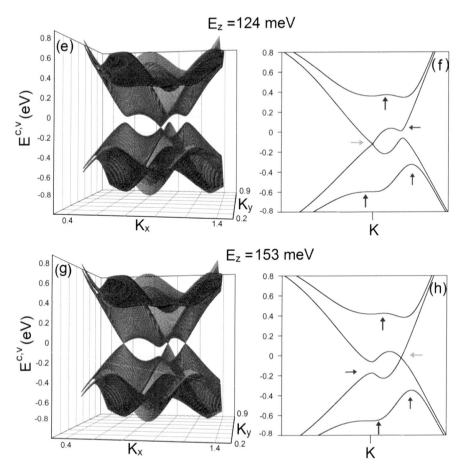

FIGURE 7.1 (e)–(h) For the AB-bt bilayer silicene, the first pair of conduction and valence energy bands in the 3D and 2D forms under the various electric-field strengths; (a) & (b) $E_z = 0$, (c) & (d) $E_z = 106$ meV/Å, (e) & (f) $E_z = 124$ meV/Å, and (g) & (h) $E_z = 153$ meV/Å.

with monolayer silicene and bilayer graphene, has brought in new opportunities for the gate-controlled quantum conductivities, which is expected to be very useful for novel designs of *Si*-based nano-electronic devices [37–39].

7.1 RICH ELECTRONIC AND OPTICAL PROPERTIES UNDER AN ELECTRIC FIELD

For an AB-bt bilayer silicene, the buckled honeycomb lattice, the intralayer & interlayer hopping integrals, and the layer-dependent spin-orbital interactions are fully discussed in Sec. 3.1. The main features of the low-lying energy bands will be delicately examined from the single-orbital Hamiltonian, covering the direct or indirect band gap, the distinct valleys in wavevector-energy space, the various energy dispersions, the critical points/van Hove singularities, and the dramatic variation with the external electric field. Moreover, the E_z-enriched optical absorption spectra are calculated from the gradient approximation of the Hamiltonian matrix elements [details in Sec. 3.2]. The absorption structures and the threshold absorption frequencies are explored in detail.

An AB-bt bilayer silicene presents the feature-rich electronic properties, mainly owing to its buckled hexagonal lattice, complex intralayer and inter-layer hopping integrals, and significant layer-dependent spin-orbital interactions. Apparently, there exist two pairs of conduction and valence bands. This work is focused on the magnetic quantization of the low-lying energy bands, as clearly displayed in Figs. 7.1(a) and 7.1(b). The conduction and valence bands clearly show an asymmetric energy spectrum about the Fermi level, strong energy dispersions, a highly anisotropic behavior, and a spin-dependent double degeneracy [the spin-up- and spin-down-dominated degenerate states discussed with respect to Figs. 7.2(b)–7.2(e)]. The conduction-band valley is initiated from the K point, exhibits a special shoulder-like structure/a partially flat dispersion along the KΓ direction within the range of 0.2 eV $< E^c <$ 0.22 eV, and then grows quickly in the further increase of wave vector. On the other side, the valence states are built from the T point between the K and Γ points; furthermore, the unusual energy spectrum, with an extreme K point, is revealed along the TK direction at $E^v \sim -0.33$ eV. There is a noticeable indirect gap of 0.3 eV, which is determined by the highest occupied state at the T point and the lowest unoccupied state at the K point. This result sharply contrasts with the zero-gap band structures of bilayer graphene systems [40]. These special properties leave footprints in the diversified magneto-electronic properties discussed later. The calculated band structure is consistent with that obtained from the first-principles result [32]. The low-lying unusual energy dispersions near valence bands near the T and K valleys and an indirect band gap are worthy of the high-resolution angle-resolved photoemission spectroscopy (ARPES) examinations [41].

The state probabilities due to the subenvelope functions on the distinct sublattice of $A^{1,2}$ and $B^{1,2}$, with the spin-up and spin-down configurations (↑ and ↓), could provide the full information about their dominance in the magnetic quantization. The doubly degenerate states have the identical wave functions under interchange of $(B^1_\uparrow, B^1_\downarrow, A^1_\uparrow, A^1_\downarrow)$ and $(B^2_\uparrow, B^2_\downarrow, A^2_\uparrow, A^2_\uparrow)$, in which each one exhibits very strong sublattice, spin and wave vector dependence. It should be noticed that the specific interchange relation is broken by a gate voltage across bilayer silicene. The conduction states are dominated by the $B^1_\uparrow$ sublattice, especially for the full dominance at the K point [the solid blue curve in Fig. 7.2(a)]. The $B^2_\uparrow$ sublattice also makes certain important contributions to the K-valley states [the solid purple curve in Fig. 7.2(b)]. Concerning the T-valley valence states, the $B^1_\downarrow$ sublattice shows strong dominance [the red curve in Fig. 7.2(c)], being accompanied with the partial contribution from the $B^2_\downarrow$ sublattice [the green curve in Fig. 7.2(d)]. The $A^{1,2}$ sublattices do not show the dominating features; therefore, the dominant $B^{1,2}$ sublattices are expected to determine largely the quantum modes of the magnetic Landau-level states. Apparently, the sublattice equivalence between $A^1/A^1/B^1$ and $B^1/A^2/B^2$, as observed in the AB-stacked bilayer graphene [40], is thoroughly absent; that is, there are no simple relations about the oscillation modes of the magnetic subenvelope functions on the four sublattices.

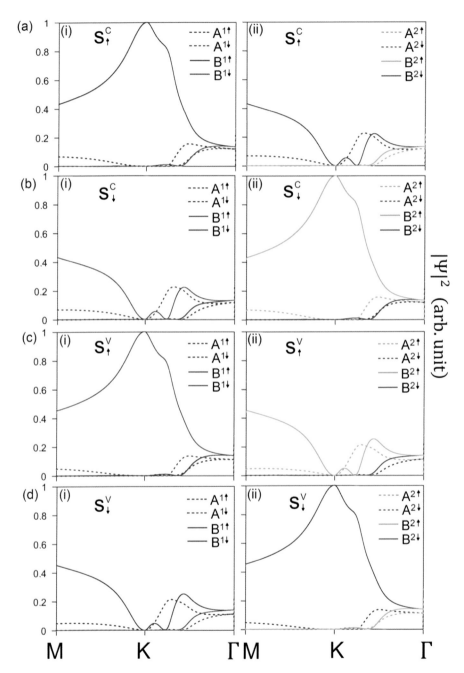

FIGURE 7.2 The sublattice-decomposed state probabilities for the (a) & (b) conduction-band and (c) & (d) valence-band states, clearly indicating the strong A/B sublattice-, layer-, and spin-dependent behaviors.

A uniform perpendicular electric field can dramatically change the electronic properties of bilayer silicene, mainly owing to the breaking of the $z = 0$ mirror symmetry and thus the absence of equivalence between any two sublattices. This field also destroys the spin-dependent state degeneracy and alters considerably the energy gap, leading to a drastic transformation of the electronic energy spectrum. It is noticed that the effects due to an external electric field are revealed through the distinct Coulomb potentials on each sublattice being determined by the atom heights. Such a field does not directly change the various hopping integrals and spin-orbital couplings. Each conduction/valence band is split into a pair of energy subbands, as denoted by $S^c_{1,2}$ and $S^v_{1,2}$ in Figs. 7.1(c)–7.1(h) under the different electric fields. The first and second conduction (valence) subbands are, respectively, characterized by the dominating $B^2_\uparrow$ and $B^1_\uparrow$ ($B^1_\downarrow$ and $B^2_\downarrow$) sublattices, where the former is relatively close to the Fermi level. This feature will be magnified in the E_z-enriched Landau-level energy spectra, as discussed below. Most importantly, the sizable band gap is easily tuned by the external electric field. With the increasing electric-field strength, both K-valley S^c_1 and T-valley S^v_1 energy subbands simultaneously approach to the Fermi level, while the opposite is true for the S^c_2 and S^v_2 ones. Apparently, the band gap is reduced and then vanishes at the first critical field [$E_z = 106$ meV/Å in Figs. 7.1(c) and 7.1(d)]; that is, the parabolic S^c_1 and S^v_1 subbands start to overlap there. In a further increase of E_z, a linear Dirac-cone structure initiated from the K point comes to exist at the second critical electric field ($E_z = 124$) meV/Å in Figs. 7.1(e) and 7.1(f)]. Very interestingly, the similar band structure, which appears at the T point, is created at the third critical one [$E_z = 153$ meV/Å in Figs. 7.1(g) and 7.1(h)]. The E_z-enriched band structures in Figs. 7.1(a)–7.1(h) present the diverse critical points in the energy-wave-vector space: the extremal states of parabolic bands (red arrows), the constant-energy loops/the partial energy dispersions along the specific direction (blue arrows), and the linear Dirac-cone structures (green arrows), in which the second ones could be regarded as the 1D parabolic dispersions or the the partially flat bands. Obviously, they are expected to diversify the magnetic quantization phenomena. The gate-voltage-modulated electronic properties make bilayer silicene extremely useful for electronic device applications [37–39], where they require the high-resolution ARPES verifications [41].

Van Hove singularities in density of states directly reflect the critical points of the unusual energy bands, as clearly illustrated in Figs. 7.3(a)–7.3(d) under the various electric-field strengths. At zero field, a large band gap, which is characterized by a pair of valence and conduction shoulders across the Fermi level, is obviously revealed in the zero-field density of states in Fig. 7.3(a). However, a finite but very low density of states appears, when an electric field reaches among the first, second, and third critical ones [Figs. 7.3(b)–7.3(d)]. These results indicate the semimetallic behaviors. Moreover, there exist three kinds of special structures: shoulders, square-root form asymmetric peaks, and V-shape forms near the Fermi level, respectively, corresponding to the band-edge states of parabolic dispersions, the constant-energy loops/the partially flat energy bands along the specific directions, and the linear Dirac cones. To fully comprehend the complicated relations among the intralayer & interlayer hopping integrals, the layer-dependent spin-orbital couplings, and the electric field, the E_z-enriched band gaps and distinct van Hove singularities/energy dispersions could be directly verified from the high-resolution STS measurements [details

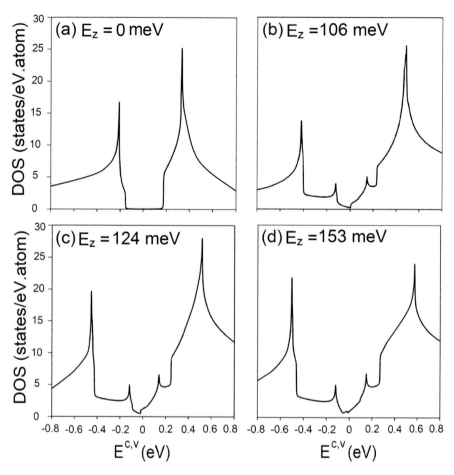

FIGURE 7.3 The low-energy density of states for the AB-bt bilayer silicene under the electric-field strengths: (a) $E_z = 0$, (b) $E_z = 106$ meV/Å, (c) $E_z = 124$ meV/Å, and (d) $E_z = 153$ meV/Å.

in Sec. 2.1]. Compared with AB bilayer graphene [40], a sufficiently high electric field can create an energy gap, the extra parabolic band-edge states, and the oscillatory energy dispersions with the constant-energy loops. The significant differences lie in the E_z-independent state degeneracy or the E_z-induced splitting, and the semimetal-semiconductor transition or the inverse one.

The optical transitions of AB-bt bilayer silicene exhibit the electric-field-enriched absorption spectra, in a great contrast to those of monolayer system [42]. Under a vanishing electric field, three special absorption structures appear in the optical spectrum, as shown by the black curve in Fig. 7.4(a). The first two present the shoulder forms, in which the former and the latter are, respectively, associated with the band-edge states with the parabolic dispersions in the valence and conduction bands near the K and T points [purple and red arrows in Figs. 7.4(a) and 7.4(b)]. The third structure, the antisymmetric peak [a green arrow], is due to the weak energy dispersion/the

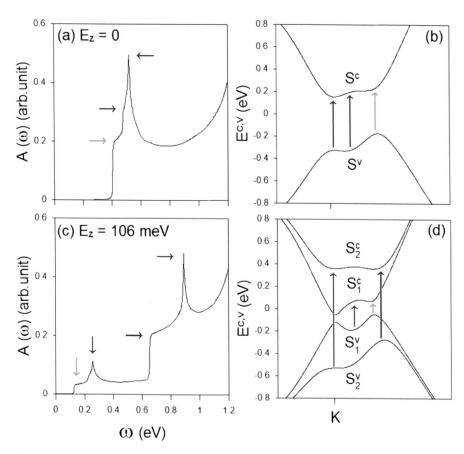

FIGURE 7.4 (a)–(d) The low-energy optical absorption spectra and the significant excitation channels due to the critical points under the various electric-field strengths. (a) & (b) $E_z = 0$, (c) & (d) $E_z = 106$ meV/Å, (e) & (f) $E_z = 124$ meV/Å, and (g) & (h) $E_z = 153$ meV/Å.

constant-energy loop between the K and T points. All of them belong to the specific van Hove singularities in the joint density of states in the presence of vertical excitations. An electric field makes the low-energy absorption spectra become more complicated as a result of the spin-split energy bands. Two distinct absorption regions come to exist, since the vertical transitions are allowed only for the almost identical spin configuration of valence and conduction bands. $S_1^v \rightarrow S_1^c$ and $S_2^v \rightarrow S_2^c$, respectively, correspond to the lower- and higher-frequency absorption regions. The pairs of the shoulder and asymmetric peak appear simultaneously, when the electric field is not too strong, such as the red curve in Fig. 7.4(c) for $E_z \leq 106$ meV/Å. The lowest-frequency threshold shoulder disappears for the higher electric field, e.g., $E_z = 124$ meV/Å and $E_z = 153$ meV/Å, respectively, in Figs. 7.4(e) and 7.4(g). The main mechanism is that the band-edge states of the first valence/conduction band become unoccupied/occupied near the T/K point [Figs. 7.4(f) and 7.4(h)]; that is, they do not make any contributions. Apparently, a finite optical gap occurs even for a semimetallic systems, mainly owing to the higher/lower Dirac point than the Fermi

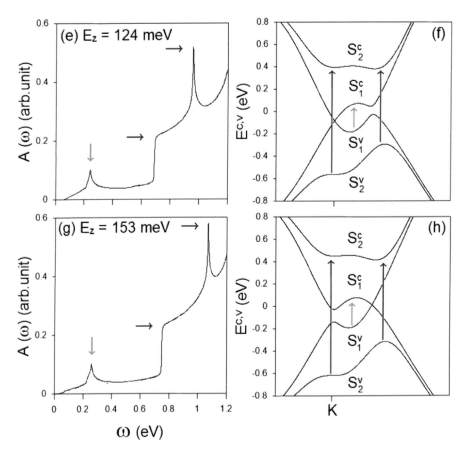

FIGURE 7.4 (e)–(h) The low-energy optical absorption spectra and the significant excitation channels due to the critical points under the various electric-field strengths. (a) & (b) $E_z = 0$, (c) & (d) $E_z = 106 \,\mathrm{meV/\mathring{A}}$, (e) & (f) $E_z = 124 \,\mathrm{meV/\mathring{A}}$, and (g) & (h) $E_z = 153 \,\mathrm{meV/\mathring{A}}$.

level. The predicted electric-field-enriched optical spectra could be verified from the high-resolution infrared reflection/absorption/transmission spectroscopies [43–48].

7.2 UNUSUAL MAGNETO-ELECTRONIC STATES

Based on the complicated band structure of AB-bt bilayer silicene in Fig. 7.1, the magnetic quantization can only be solved by the generalized tight-binding model. This directly reflects the fact that the highly buckled structure, the large interlayer atomic interactions, the layer-dependent spin-orbital couplings, and the external fields need to be taken into account in the Hamiltonian diagonalization simultaneously. That is, all the Hamiltonian matrix elements are complex and non-negligible/significant, where they require the high-resolution ARPES verifications [41]. On the other hand, the effective-mass model is almost impossible in dealing with the magnetic quantization through the low energy expansion, mainly owing to the distinct valleys, strong anisotropy, and weakly energy dispersions of asymmetric conduction and valence bands. Even when the low-energy conduction K valley and the valence T one are analytically obtained, it is not accurate to make the magnetic quantization separately, which is inconsistent with the basic requirement of quantum statistics (the Fermi-Dirac distribution for fermions).

The low-energy magneto-electronic properties are thoroughly investigated through the magnetic Hamiltonian [details in Sec. 3.1]. The diverse magnetic quantization phenomena are delicately identified from the main features of Landau levels. It covers the state degeneracy within the reduced first Brillouin zone, the subenvelope functions on the layer- and spin-related sublattices, the specific oscillation modes, the specific classification of Landau-level subgroups, the linear or nonlinear/square-root dependence of state energy on field strength, the anti-crossing/crossing/non-crossing behaviors in the field-dependent energy spectra. One of the studying focuses is the dramatic changes in the main features of Landau-level subgroups, such as the separation in the distinct energy ranges, the interchange of the well-behaved and irregular wave functions, and the occupation number of valence and conduction states. The important differences between bilayer AB-bt silicene and AB-stacked graphene are examined in detail.

Obviously, an AB-bt bilayer silicene presents the rich magneto-electronic properties, being thoroughly different from bilayer graphene systems [40]. The buckled honeycomb lattice, complex interlayer atomic interactions, and significant spin-orbital couplings remarkably enrich the main features of Landau levels. The low-lying conduction and valence Landau levels are magnetically quantized from the electronic states near the K and T valleys [Figs. 7.1(a) and 7.1(b)], respectively. The magneto-electronic states are doubly degenerate under the interplay of non-equivalent sublattices, spin-orbital interactions, and magnetic field, while there exist an eight-fold degeneracy in non-stacking-modulated bilayer graphenes [40]. The conduction Landau-level wave functions are centered at 1/6 [4/6] and 2/6 [5/6] of the magnetic unit cell, and the valence ones are localized about the 1/4 [3/4]. Such wave functions are the well-behaved spatial distributions characterized by the $3p_z$- and spin-dependent subenvelope functions on the eight sublattices, as clearly displayed in Fig. 7.5(a)–7.5(e). The four quantum numbers are defined by the zero-points number of the spatial probability distributions in the dominant sublattices. Furthermore, their detailed information are very useful in fully understanding the magneto-optical selection rules due to the inter-Landau-level vertical transitions. In principle, the Landau-level states can be classified as four distinct subgroups [$n_{\uparrow 1}^c$, $n_{\downarrow 1}^c$, $n_{\uparrow 2}^c$, and $n_{\downarrow 2}^c$], based on the sublattice- and spin-dominated wave functions [blue, red, purple, and green lines in Figs. 7.5(a) and 7.5(d)]. Four Landau-level subgroups possess the usual orderings of state energy and energy spacing only within a finite energy range; that is, such properties, respectively, grow and decline with the increase/decrease of E^c/E^v. The Landau-level splitting is induced by the above-mentioned critical mechanisms; furthermore, the split energy is strongly dependent on the magnetic field strength (discussed later).

Since the low-energy valence and conduction Landau levels, respectively, corresponding to electronic states initiated from the T and K points, have different localization centers, their vertical magneto-optical transitions are forbidden. The higher conduction and deeper valence Landau levels, with large zero-point numbers, play

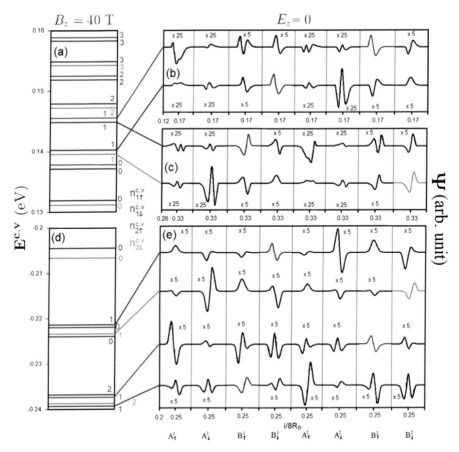

FIGURE 7.5 At $B_z = 40$ T, (a)/(d) the conduction/valence Landau-level energy spectrum and (b) & (c)/(e) conduction/valence subenvelope functions on the spin-split eight different sublattices.

critical roles in the magneto-optical excitations; therefore, such magneto-electronic states deserve a closer observation. The former and the latter are, respectively, quantized from the electronic states near the T and K valleys [the shoulder-like energy bands in Fig. 7.1(b)]. This is thoroughly different from the magnetic quantization in bilayer graphene systems [40]. The magneto-electronic energy spectrum and spatial distributions of the higher conduction Landau levels are clearly illustrated in Figs. 7.6(a)-7.6(c). Obviously, such states exhibit the non-symmetric/non-antisymmetric and non-well-behaved spatial distributions; that is, their oscillation behaviors should belong to the perturbed Landau levels with the major and minor modes [3]. For the higher conduction/deeper valence Landau states, their contribution widths are wider and even the effective width covers two localization centers of (1/6, 1/4) & (2/6, 1/4). As a result, the low-lying conduction/valence and the deeper valence/higher conduction Landau levels will have a significant overlap in the spatial probability distributions, leading to the critical excitation channels in the magneto-optical threshold excitation. The theoretical predictions on the magnetic wave functions, as shown in Figs. 7.5 and 7.6, could be verified by the energy-fixed STS measurements [details in Sec. 2.1].

The B_z-dependent magneto-electronic energy spectrum, which is identified from the unique sublattice- and spin-dominated magnetic wave functions in bilayer AB-bt silicene, plays a critical role in comprehending the diverse quantization phenomena. The four subgroups of conduction/valence Landau levels exhibit the similar magnetic-field dependence, as clearly displayed in Figs. 7.7(a)–7.7(c). As for the unoccupied magneto-electronic states in the range of 165 meV $\leq E^c \leq$ 195 meV, the smaller-n^c Landau-level energies present the monotonic/almost linear B_z-dependence and the normal ordering among four subgroups. However, the abnormal behaviors, i.e., the unusual field dependence and Landau-level anti-crossings, come to exist frequently at higher energies. The $n_{\uparrow 1}^c$ and $n_{\downarrow 2}^c$ Landau levels (blue and green curves) anti-cross with each other within a certain magnetic-field range, and so do the $n_{\downarrow 1}^c$ and $n_{\uparrow 2}^c$ ones (red and purple curves). Such phenomena are illustrated and marked by the pink and yellow circles for the $n^c = 7$ Landau levels. As a result, two kinds of inter-subgroup anti-crossings often appear for the specific quantum modes. Obviously, these anti-crossings indicate that the magnetic subenvelope functions of the perturbed Landau levels consist of the main and side modes, but not a single mode. Furthermore, such localized oscillation modes change substantially during the variation of field strength [discussed later with respect to Fig. 7.8] [48]. Similarly, there exists a simple relationship between the valence Landau-level energies and the field strength within the range of -300 meV $\leq E^v \leq -165$ meV. The deeper-state energy spectrum exhibit two kinds of inter-subgroup Landau level anti-crossings due to the neighboring Landau levels, covering the $n_{\uparrow 1}^v$ & $n_{\downarrow 2}^v + 2$ Landau levels [blue and green curves in Fig. 7.7(c)] and the $n_{\uparrow 2}^v$ & $n_{\downarrow 1}^v + 2$ ones (purple and red curves). Here, the $n^v = 10$ [$n^2 = 12$] Landau levels acquire a comparable side mode of 12 [10], leading to the anti-crossings for [$n_{\uparrow 1}^v = 10$ and $n_{\downarrow 2}^v = 12$] and [$n_{\uparrow 2}^v = 10$ and $n_{\downarrow 1}^v = 12$] Landau levels, as indicated by the blue circle in Fig. 7.7(c). According to the Wigner-von Neuman non-crossing rule, two multi-mode Landau-level states avoid crossing each other, while they simultaneously possess certain identical modes with comparable amplitudes on the same sublattices. The details of Landau-level anti-crossings will be discussed later in Fig. 7.8. The rich and unique magneto-electronic energy spectra are closely related to the magnetic quantization of the unusual conduction and valence bands, such as their anti-crossings corresponding to the partially flat bands/the constant-energy loops for the conduction/valence states near the T/K point [Figs. 7.1(a) and 7.1(b)].

The magneto-electronic properties are greatly diversified by an external electric field. For the zero-gap band structure under the first critical electric field, $E_z = 106$ meV/Å, the magnetic quantization is initiated from the valence and conduction states near the K and T valleys [Figs. 7.1(c) and 7.1(d)]. The electric field has very strong effects on the amplitudes, distribution forms, oscillation modes, localization centers of magnetic subenvelope functions, but not the sublattice dominance. Most importantly, the low-lying valence and conduction Landau levels, as shown in Figs. 7.8(a)–7.8(c), present the same localization center simultaneously, instead of only conduction or valence ones under the vanishing electric field [Fig. 7.2]. In particular, the perturbed valence Landau levels near the 1/6 and 2/6 centers (the conduction ones at 1/4 center;

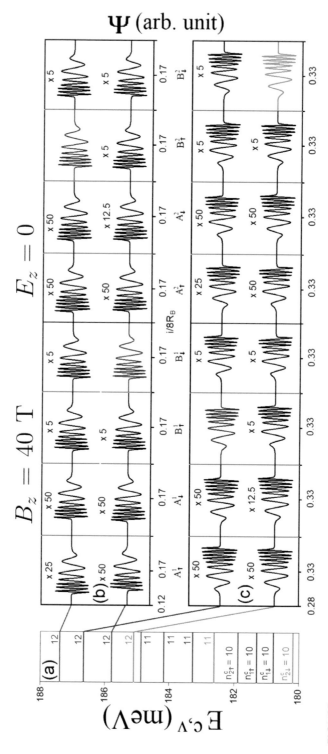

FIGURE 7.6 (a) The higher conduction Landau-level energy spectrum of four subgroups and (b)–(c) magnetic subenvelope functions on eight distinct sublattices for $B_z = 40$ T in the absence of gate voltage.

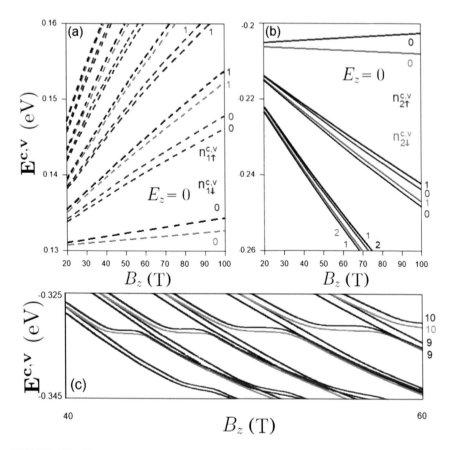

FIGURE 7.7 The magnetic-field-dependent Landau-level energy spectrum: the blue, red, purple, and green curves, respectively, corresponding to the dominant $B_\uparrow^1$, $B_\downarrow^1$, $B_\uparrow^2$, and $B_\downarrow^2$ sublattices of different four subgroups for the (a) conduction and (b) & (c) valence Landau levels.

not shown at higher energies) come into existence, as illustrated by the titled colored lines. Such magneto-electronic states arise from the rather pronounced oscillating band structure in the presence of an electric field [Fig. 7.2(e)]. Four subgroups of Landau levels do not appear together, but are well separated into two lower- and higher/deeper-energy ones. This clearly indicates that subgroup splitting closely associated with the non-equivalence of sublattices is greatly enhanced by the electric field because of the distinct Coulomb site energies. The current work is focused on the former arising from the lower-frequency magneto-optical excitations. For each valley, the two subgroups of low-lying conduction and valence Landau levels correspond to $[n_{\uparrow 2}^c, n_{\downarrow 2}^c]$ and $[n_{\uparrow 1}^v, n_{\downarrow 1}^v]$, respectively. It should be noticed that they are dominated by the different sublattices. The vertical transitions, valence to conduction Landau levels, from the different valleys provide a major contribution to the magneto-optical absorption spectra.

On the other side, a perpendicular electric field might result in the formation of Dirac cones at the K or T valleys, giving rise to the unique Landau-level quantization.

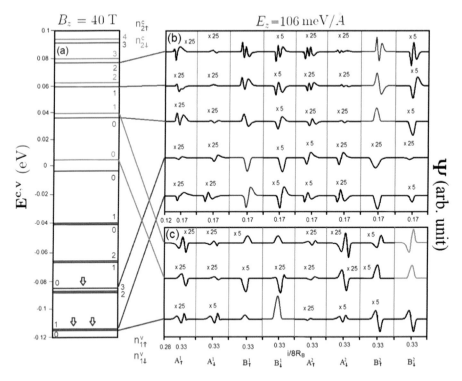

FIGURE 7.8 (a) The low Landau-level energies and (b) & (c) subenvelope functions on eight independent sublattices at $B_z = 40$ T and $E_z = 106$ meV/Å.

For the second critical electric field, $E_z = 124$ meV/Å [Figs. 7.9(a)–7.9(e)], the conduction and valence Landau levels, being initiated from the K and T valleys (the T and K ones) are, respectively, characterized by $n^c_{\downarrow 2}$ and $n^v_{\downarrow 1}$ [$n^c_{\uparrow 2}$ and $n^v_{\uparrow 1}$]. All the magnetic wave functions are slightly distorted in the spatial distribution, as clearly shown in Figs. 7.9(b)–7.9(d). The former is similar to those from the linear Dirac cone [42], their energies approximately agree with $E^{c(v)}_{2(1)} \propto \sqrt{n^c_{\downarrow 2}(n^v_{\uparrow 1})}$. A simple relation is absent for the other subgroups. Specifically, the energy spacing of $n^c_{\downarrow 2} = 0$ and $n^v_{\uparrow 1} = 0$ is finite and gradually grows with the magnetic field strength, having a magnitude of ~ 25 meV at $B_z = 40$ T. The opposite is true for the third critical electric field, $E_z = 153$ meV/Å. That is, the Dirac-cone-like Landau levels are created by the electronic states near the T valley. The sharp contrast between the K and T valleys will be directly reflected in the magneto-optical excitations. Moreover, there exist certain important differences compared to monolayer graphene with a zero energy spacing between the $n^c = 0$ and $n^v = 0$ Landau levels, eight-fold degeneracy, and the same dominant sublattices for the valence and conduction states [42].

The magneto-electronic energy spectra are drastically changed by the electric field, as clearly indicated by the strong B_z-dependence in Figs. 7.10(a)–7.10(c). At $E_z = 0$, Figs. 7.7 and 7.10(a)] have demonstrated four subgroups of Landau levels behave similarly during the variation of B_z, in which the low-lying spectrum exhibits an almost linear B_z-dependence, except for the initial valence $n^v_{\downarrow 1} = 0$ Landau level near

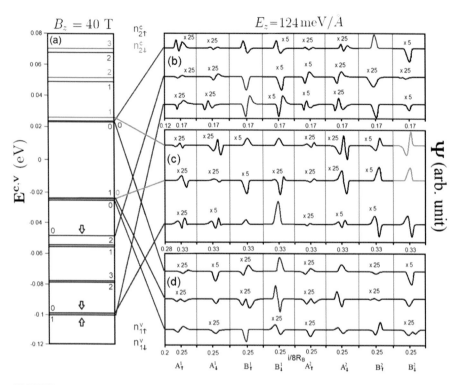

FIGURE 7.9 The magneto-electronic (a) energy spectrum and (b)–(d) wave functions under $B_z = 40$ T and $E_z = 124$ meV/Å.

the T valley. It should be noticed that the onset energy of each Landau-level subgroup is strongly dependent on the electric-field strengths. The effects of composite fields are clearly reflected in the B_z-dependent Landau-level energy spectra under the critical electric fields. The electron-hole asymmetry of the energy spectrum is greatly enhanced at the first critical electric field, $E_z = 106$ meV/Å [Fig. 7.10(c)]. There are three initial Landau levels of $[n^v_{\uparrow 1} = 0$ & $n^c_{\downarrow 2} = 0]$ from the K valley and $[n^v_{\downarrow 1} = 0]$ from the T one which present very weak B_z-dependence due to the band-edge states [Fig. 7.1(d)]. Specifically, the above-mentioned $n^c_{\downarrow 2} = 0$ and $n^v_{\uparrow 1} = 0$ Landau levels determine the Fermi level, being the middle of the nearest occupied and unoccupied states. The energy gap between these Landau levels becomes very narrow and slowly grows with the increasing B_z. It is almost vanishing at a sufficiently low B_z. According to the delicate numerical examinations, the energy spacing and magnetic field strength present a neither simple linear nor square-root relationship. In addition, there are only a few well-behaved conduction Landau levels from the T valley $[n^c_{\uparrow 2}$ & $n^c_{\downarrow 2}]$ [e.g., $n^c_{\uparrow 2} = 0$ Landau level under $B_z \geq 20$ T] and they are located at relatively high energies, compared with those due to the K valley. Apparently, the latter will dominate the threshold magneto-optical excitations.

Also, the magneto-electronic properties can be remarkably diversified by applying an external electric field, as clearly illustrated by the E_z-dependent energy spectra in Figs. 7.11 and 7.12. In the presence of composite electric and magnetic fields,

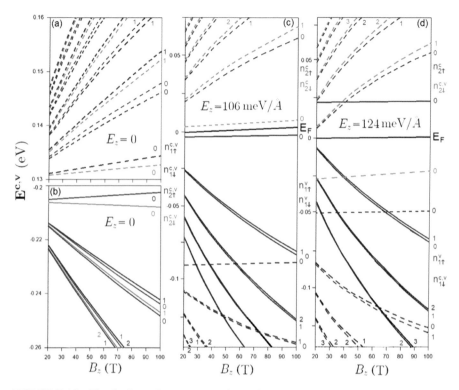

FIGURE 7.10 The B_z-dependent magneto-electronic energy spectra for the electric fields: (a) & (b) $E_z = 0$, (c) $E_z = 106$ meV/Å and (d) 124 meV/Å.

the magneto-electronic state energies, which belong to [$n_{\uparrow 1}^{c,v}$ and $n_{\uparrow 1}^{c,v}$] along with [$n_{\uparrow 2}^{c,v}$ and $n_{\downarrow 2}^{c,v}$] subgroups, respectively, grow or decline in the increase of electric field [Figs. 7.11(a) and 7.12(a) by the blue and red curves and the purple and green curves]. The E_z dependence of these Landau-level subgroups is opposite for the B^1- and B^2-dominated state energies. This clearly indicates the E_z-enhanced sublattice non-equivalence by means of the different Coulomb potential energies. A plenty of Landau-level anti-crossings and crossings appear in the E_z-created energy spectra. The former are induced by the neighboring [$n_{\uparrow 1}^{c,v}$ and $n_{\downarrow 2}^{c,v}$] magneto-electronic states [blue and green curves] and [$n_{\downarrow 1}^{c,v}$ and the $n_{\uparrow 2}^{c,v}$] ones [red and purple curves], in which their oscillation modes gradually change during the variation of E_z. However, such behaviors never happen between the $n_{\uparrow 1}^{c,v}$ and $n_{\downarrow 1}^{c,v}$ subgroups as well as between the $n_{\uparrow 2}^{c,v}$ and $n_{\downarrow 2}^{c,v}$ subgroups. That is to say, the spin-induced Landau-level energy spacing is hardly affected by the electric field.

The frequent Landau-level anti-crossings deserve a closer examination. The E_z-dependent energy spectra [Figs. 7.11(a) and 7.12(a)] clearly show that the inter-subgroup anti-crossings are only associated with two neighboring Landau levels with the quantum-number differences of $\Delta n = 0, \pm 1, \pm 2$. Such electronic transitions are either difficult to observe or getting into the crossing behaviors [for large Δn's]; furthermore, they might disappear at the higher magnetic fields, e.g., the absence of anti-

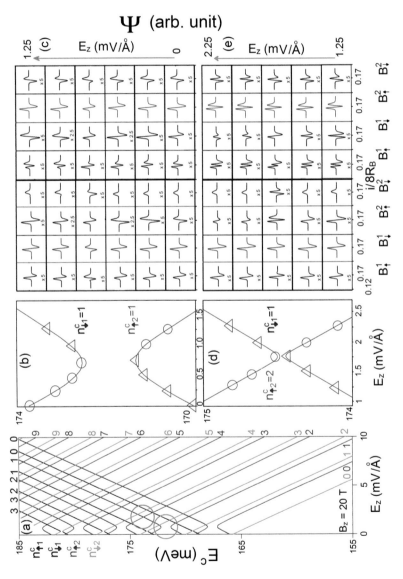

FIGURE 7.11 The electric-field-dependent conduction Landau-level energy spectrum (a) under $B_z = 20$ T, and (b) & (d) two certain anti-crossings of the $[n_{\downarrow 1}^c = 1, n_{\uparrow 2}^c = 1]$ & $[n_{\downarrow 1}^c = 1, n_{\uparrow 2}^c = 2]$ magneto-electronic states, respectively, corresponding to the evolutions of sublattice-related subenvelope functions in (c) and (e).

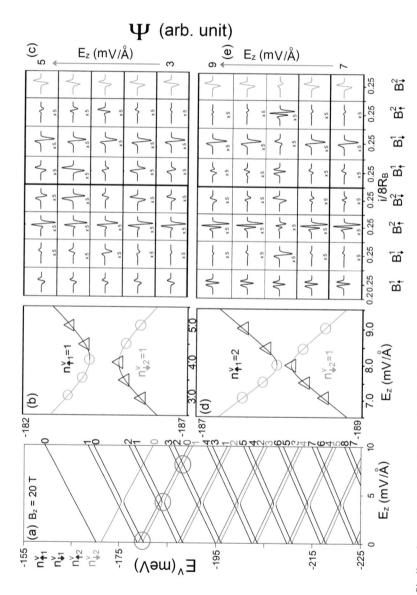

FIGURE 7.12 Similar plot as Fig. 7.11, but shown for (a) valence energy spectrum with (b) & (d) two certain anti-crossings of the $[n^v_{\downarrow 2} = 1, n^v_{\uparrow 1} = 1]$ & $[n^v_{\downarrow 2} = 1, n^v_{\uparrow 1} = 2]$ Landau-level states, respectively, corresponding to the dramatic variations of sublattice-related subenvelope functions in (c) and (e).

crossings at $B_z = 40$ T [49]. The conduction Landau levels anti-cross each other for $[n^c_{\downarrow 1}$ and $n^c_{\uparrow 2}]$, $[n^c_{\downarrow 1}$ and $n^c_{\uparrow 2} + 1]$ as well as $[n^c_{\uparrow 1}$ and $n^c_{\downarrow 2} + 1]$, as shown in Fig. 7.11(a). The similar phenomena are revealed in the magneto-electronic valence states [Fig. 7.12(a)]. The $n^c_{\downarrow 1} = 1$ Landau level is very suitable serving as an example for fully understanding the evolution of sublattice- and spin-dependent envelope functions, when the electric-field strength gradually grows [the red curve in Fig. 7.11(b)]. This quantized state has its first anti-crossing with the $n^c_{\uparrow 2} = 1$ Landau level as the electric field increases initially from zero [the dramatic transformation of the red curve into the purple one]. They remain their oscillation modes for $[n = 1]$ on the $B^1_{\uparrow}$ and $B^2_{\uparrow}$ sublattices, as displayed in Fig. 7.11(c). However, one of two amplitudes [red] gives rise to phase switching during the anti-crossing, e.g., those at $E_z = 0$ and 0.75 mV/Å. Moreover, the magnetic wave functions of the $n^c_{\downarrow 1} = 1$ and $n^c_{\uparrow 2} = 1$ Landau levels on the $B^2_{\uparrow}$ and $B^1_{\downarrow}$ sublattices, respectively, are greatly enhanced near the anti-crossing center and acquire comparable amplitudes commensurate with the dominating ones [the left-hand-side red and black curves; the right-hand-side black and purple ones]. With further increase of E_z, the $n^c_{\downarrow 1} = 1$ and $n^c_{\uparrow 2} = 2$ Landau levels encounter another weak anti-crossing at $E_z \sim 1.75$ mV/Å in Fig. 7.11(d). In this case, the latter one on the $B^2_{\uparrow}$ sublattice is mapped directly onto the former wave function on the $B^2_{\uparrow}$ sublattice and vice versa in Fig. 7.11(e). The above-mentioned Landau-level wave functions are consistent with the Wigner-von Neuman non-crossing rule. Similar anti-crossing behaviors are revealed for the valence Landau levels in Fig. 7.12(a). For example, the $n^v_{\downarrow 2} = 1$ state anti-crosses with the $n^v_{\uparrow 1} = 1$ and $n^v_{\uparrow 1} = 2$ ones at lower and higher electric fields in Figs. 7.12(b) and 7.12(d), respectively, could be identified from the drastic changes of amplitude in Figs. 7.12(c) and 7.12(e), or even the unique transformation of oscillating modes on the $B^1_{\uparrow}$ and $B^2_{\downarrow}$ sublattices. The theoretical predictions on the frequent anti-crossings [Figs. 7.11 and 7.12] could be directly examined by the STS measurements on the magnetic wave functions [details in Sec. 2.1].

The above-mentioned magneto-electronic energy spectra, being sensitive to the strengths of magnetic and electric fields, will behave as a lot of delta-function-like prominent peaks in density of states within a narrow energy range [details in Ref. 49]. In general, there exist many pair-like symmetric peaks corresponding to Figs. 7.5 and 7.7. The state energies of four Landau-level subgroups might present the different orderings for the conduction and valence ones, e.g., the first unoccupied and occupied Landau level due to the distinct subgroups. The initial subgroup, energy, number, and height of prominent symmetric peaks are dramatically changed by the electric field prominent peaks, mainly owing to the complicated crossing and anti-crossing behaviors [Fig. 7.10]. The theoretical predictions on the rich van Hove singularities could be verified by the high-resolution measurements from the gate-voltage-dependent STS [details in Sec. 2.1]; that is, such verifications are very useful in understanding the cooperative/competitive relations among the geometric symmetries.

The AB-bt bilayer silicene sharply contrasts with the AB-stacked bilayer graphene with respect to electronic and magneto-electronic properties. The former and the latter, respectively, belong to an indirect-gap semiconductor and a semimetal. The graphene structure has two pairs of valence and conduction bands, with the monotonic parabolic dispersions, a weak anisotropy and a small overlap near the K valley [40]. Electronic states are independent of the spin configurations in the absence of significant spin-orbital couplings. They remain doubly degenerate for the spin degree of freedom under any external fields. In general, the neighboring electronic states are magnetically quantized into well-behaved Landau levels with the specific nodes in the oscillating probability distributions [40]. For each $[k_x, k_y]$ magneto-electronic state, all the Landau levels are eight-fold degenerate as a result of the equivalent sublattices and spin degeneracy. The conduction/valence Landau levels cannot be classified into four Landau-level subgroups, in which they only present the regular B_z-dependent energy spectra without the anti-crossing behaviors. Moreover, a perpendicular electric field is responsible for a semimetal-semiconductor phase transition, but not the semiconductor-semimetal transition associated with the split energy bands. For silicene, the electric field also leads to a lifting of the degenerate K and K' valleys for the Landau-level states. That is to say, the Landau levels in graphene and silicene systems, respectively, possess four-fold and double degeneracies under the composite electric and magnetic fields. There exist four degenerate Landau levels with the frequent anti-crossings in the E_z-dependent energy spectra. The important differences between bilayer silicene and graphene highlight the main features of the spin- and sublattice-dependent energy bands and Landau levels, directly demonstrating the distinct geometric structures, hopping integrals, and spin-orbital interactions. Specifically, the former could provide the full information about the unusual spintronics and their potential applications [50], such as the spin-dependent quantum Hall conductivities [50]. Apparently, few-layer silicene materials are worthy of a systematic investigation.

7.3   UNIQUE MAGNETO-ABSORPTION SPECTRA

To fully comprehend the diversified magneto-optical phenomena, the generalized tight-binding model is capable of directly linking with the dynamic Kubo formula [details in Sec. 3.2], since the sublattice-related tight-binding functions are consistent with the gradient approximation in calculating the electric dipole moment [36]. Obviously, the unusual valley- and spin-split Landau levels in Sec. 7.2 are expected to present the rich and unique magneto-absorption spectra, such as the total categories of the available magneto-excitation channels/the number of absorption structures within a narrow frequency range, the various absorption structures the distinct forms of the symmetric absorption peaks, the dramatic transformation in the threshold transition frequency/excitation channel during the variation of the electric-field strength, and the existence or absence of magneto-optical selection rules. The above-mentioned critical characteristics require very accurate calculations.

An AB-bt bilayer silicene presents the feature-rich magneto-absorption spectra, directly reflecting the unusual band structures and magneto-electronic properties. The vertical optical transitions among four subgroups of Landau levels lead to a plenty of single, double, and twin delta-function-like prominent absorption peaks with nonuniform intensities, as clearly displayed in Figs. 7.13(a)–7.13(c). Apparently, there exist 4×4 categories of inter-Landau-level optical transitions, covering 4 intra-subgroup (part of the pearks indicated by the arrows in distinct colors) and 12 inter-subgroup ones. By the delicate analysis on the magneto-optical absorption functions [details in Sec. 3.2; Ref. 50], the inter-Landau level transition is available whenever the initial and final states associated with the two sublattices in the large $[t_0, t_1]$ hopping

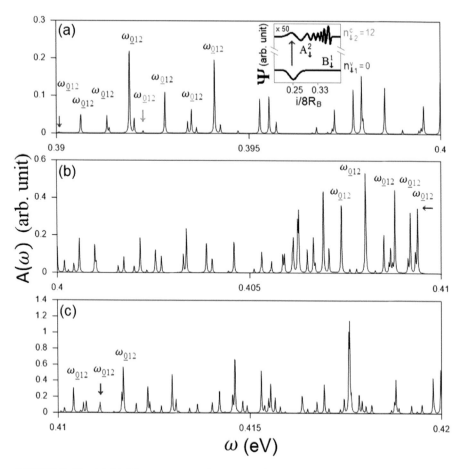

FIGURE 7.13 (a)-(c) The magneto-optical absorption spectrum in AB-bt bilayer silicene under $B_z = 40$ T in the absence of electric field. Also shown in the inset are the magnetic sublattice envelope functions of the initial and final states in the threshold absorption.

integrals [Fig. 3.1] possess the identical oscillation mode. As a result of an indirect energy gap [Figs. 7.1(a) and 7.1(b)], in each category, absorption peaks are generated by the optical transitions closely related to the few- and multi-mode Landau levels under the absence of a specific selection rule. This results in many magneto-absorption peaks within a very narrow-range frequency of ~ 30 meV, never observed in other condensed-matter systems [40]. For a Landau level with a sufficiently large quantum number, the extended and non-well-behaved sublattice envelope function localized at [1/4]/[1/6 & 2/6] along the x-axis could overlap with that of another state at the neighboring localization centers of [1/6 & 2/6]/[1/4]. For example, the spatial amplitude distributions of the $n^v_{\downarrow 1} = 0$ and $n^c_{\downarrow 2} = 12$ Landau levels are illustrated in the inset of Fig. 7.13(a). The former and the latter are, respectively, localized at 1/4 and 1/6 centers which are very close to each other, leading to an obvious overlapping phenomenon between them. This enables the vertical optical transitions between the initial- and final-state Landau levels near the T [K] valley, where the magneto-absorption peaks present the distinct intensities and frequencies, e.g., the 16 absorption structures due to the $[n_{1\downarrow}^{v} = 0, n_{1\uparrow}^{v} = 0, n_{2\uparrow}^{v} = 0, n_{2\downarrow}^{v} = 0]$ and $[n_{1\downarrow}^{c} = 12, n_{1\uparrow}^{c} = 12, n_{2\uparrow}^{c} = 12, n_{2\downarrow}^{c} = 12]$ Landau levels. Such unique magneto-optical property is absent in other well-known 2D systems, e.g., graphene [40], MoS_2 [51], and phosphorene [52]. The threshold absorption frequency, the optical gap, belongs to the intra-subgroup $n_{1\downarrow}^{v} = 0 \rightarrow n_{1\downarrow}^{c} = 12$ transition [the first red arrow in Fig. 7.13(a)]. Its frequency is expected to be dependent on both magnetic and electric fields. The optical gap $[\approx 0.39$ eV for $B_z = 40$ T$]$ is higher than the energy gap $[\approx 0.3$ eV$]$, mainly owing to the forbidden vertical transitions between the low-lying/small-$n^{c,v}$ conduction and valence Landau levels at the different centers. In addition, another 16 excitation categories arising from the final conduction Landau levels come to exist from the K valley, such as those due to the $[n_{1\downarrow}^{v} = 12, n_{1\uparrow}^{v} = 12, n_{2\uparrow}^{v} = 12, n_{2\downarrow}^{v} = 12]$ and $[n_{1\downarrow}^{c} = 0, n_{1\uparrow}^{c} = 0, n_{2\uparrow}^{c} = 0, n_{2\downarrow}^{c} = 0]$ Landau levels. However, they do not contribute to the magneto-optical threshold absorption.

The magneto-optical spectra are greatly diversified under the interplay of electric field, magnetic field, significant sin-orbital couplings, and interlayer hopping integrals. An applied gate voltage can create the inter-Landau-level optical transitions at lower absorption frequencies, and such available excitation channels are characterized by the specific magneto-optical selection rules, e.g., the absorption spectrum at the first critical electric field of $E_z = 106$ meV/Å under the magnetic field of $B_z = 40$ T [Figs. 7.14(a)–7.14(c)]. Most importantly, the magneto-optical excitations, which originate from the initial and final Landau levels with the same localization center [type-I] or the neighboring ones [1/4 & 1/6 (2/6) centers] [type-II], come to exist simultaneously. The type-I absorption peaks are associated with the E_z-induced low-lying well-behaved valence Landau levels near the T [K] valley. They are available only for the initial and final Landau levels with the same spin configuration $[n_{\uparrow 1}^{v} \rightarrow n_{\uparrow 2}^{c}$ and $n_{\downarrow 1}^{v} \rightarrow n_{\downarrow 2}^{c}]$. The threshold absorption, being determined by the $n_{\uparrow 1}^{v} = 0 \rightarrow n_{\uparrow 2}^{c} = 0$ transition, is located at much lower frequency ($\omega_{th} \approx 121$ meV), compared with that at zero electric field [Fig. 7.13(a)]. It should be noticed that, type-I magneto-optical excitations obey the magneto-optical selection rules of $\Delta n = 0$ and ± 1. The main reason is that each Landau-level state possesses a major mode and certain minor modes, referring to Figs. 7.8(b) and 7.8(c). For example, the $n_{\uparrow 1}^{v} = 1$ Landau level consists of a main mode of 1 on the dominating $B_{\uparrow}^{1}$ sublattice and the side modes of 0 and 2 on the other sublattices. As a result, there are effective inter-Landau-level optical transitions of $n_{\uparrow 1}^{v} = 1 \rightarrow n_{\uparrow 2}^{c} = 0, n_{\uparrow 1}^{v} = 1 \rightarrow n_{\uparrow 2}^{c} = 1$, and $n_{\uparrow 1}^{v} = 1 \rightarrow n_{\uparrow 2}^{c} = 2$, as marked by the red arrows in Figs. 7.14(b) and 7.14(c). As for type-II absorption peaks, there are excitation channels between $n = 0$ and 12 Landau level, similar to those in the absence of an electric Figs. 7.13(a)–7.13(c); that is, they are associated with the low-lying well-behaved valence Landau level and irregular conduction one near the T valley. The above-mentioned absorption peaks, which include both type-I and type-II ones, belong to 5 excitation categories in the low-frequency range, but not 16 ones as in the absence of E_z. The useful channels only cover the $n_{\uparrow 1}^{v} \rightarrow n_{\uparrow 2}^{c}$, $n_{\uparrow 1}^{v} \rightarrow n_{\downarrow 2}^{c}, n_{\uparrow 1}^{v} \rightarrow n_{\downarrow 1}^{c}, n_{\downarrow 1}^{v} \rightarrow n_{\uparrow 2}^{c}$, and $n_{\downarrow 1}^{v} \rightarrow n_{\downarrow 2}^{c}$ magneto-optical transitions.

It is worth considering the magneto-optical absorption spectrum under the second critical electric field $E_z = 124$ meV/Å$]$, when a slightly distorted a Dirac cone

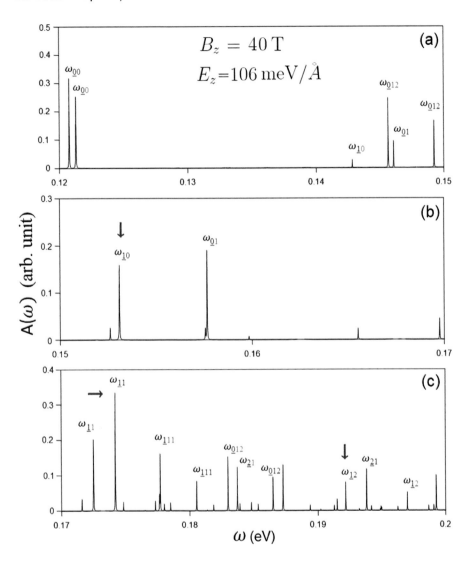

FIGURE 7.14   Similar plot as Fig. 7.12, but shown at $B_z = 40$ T and $E_z = 106$ meV/Å.

is formed at the K valley [Figs. 7.1(e) and 7.1(f)]. Both type-I and type-II magneto-optical excitations, as clearly shown in Figs. 7.15(a)–7.15(c), come to exist frequently in the absorption spectra. The former satisfy the optical selection rules of $\Delta n = 0$ and ± 1, while the opposite is true for the latter. Especially, the formation of the almost linearly intersecting band structure induce an unusual phenomenon, in which the specific Dirac point lies below the Fermi level [Fig. 7.1(f)]. As a result, the occupation of some landau levels near this point is very sensitive to their energies, directly determining the threshold magneto-optical excitation [Figs. 7.15(a)–7.15(c)]. The intra-subgroup excitation channels of $n_{\downarrow 2}^{c} = 0 \rightarrow n_{\downarrow 2}^{c} = 1$ (red arrow) and $n_{\downarrow 1}^{v} = 1 \rightarrow n_{\downarrow 1}^{v} = 0$ (green arrow) comes into existence because the conduction $n_{\downarrow 2}^{c} = 0$ Landau level near the K valley and $n_{\downarrow 1}^v = 0$ one from the T valley become the occupied and unoccupied states, respectively. The threshold absorption peak, closely related to the former, is present at a rather low frequency [$\omega_{th} \sim 50$ meV under $B_z = 40$ T]. Furthermore, there exists a double peak due to the two peaks of [$n_{\uparrow 1}^v = 0 \rightarrow n_{\uparrow 2}^c = 0$ & $n_{\downarrow 1}^v = 0 \rightarrow n_{\downarrow 2}^c = 0$] merging simultaneously in the magneto-absorption spectrum, as indicated by the blue arrow in Fig. 7.15(a). The crossing behavior between two initial/final Landau levels is responsible for this double peak. In general, type-I magneto-optical excitations are

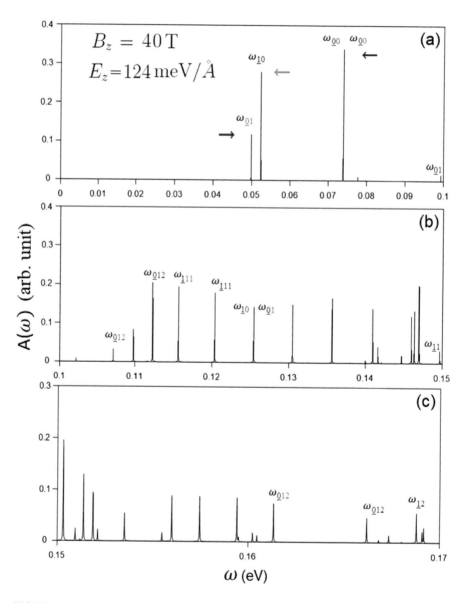

FIGURE 7.15 Similar plot as Fig. 7.12, but displayed at $B_z = 40$ T and $E_z = 124$ meV/Å.

mainly revealed near the K valley, except for $n_{\downarrow 1}^v = 1 \rightarrow n_{\downarrow 1}^v = 0$ [green arrow] near the T valley. Similar magneto-optical properties could also be observed for the third critical electric field [$E_z = 153$ meV/Å], in which the Dirac point within the T valley is situated above the Fermi level [Fig. 7.1(h)].

The optical gap is strongly affected by the strengths of the electric and magnetic fields, as demonstrated in Figs. 7.16(a)–7.16(c). With the increase of the electric field up to the first critical field ($E_z = 106$ meV/Å), the threshold absorption frequency, being determined by the first shoulder-like spectral structure from the K and T valleys [Figs. 7.4(a) and 7.4(c)], gradually declines from around 0.4 eV, as clearly illustrated in Fig. 7.16(a). Right after this field, such structure is absent since the conduction/valence band-edge state near the Fermi level becomes occupied/unoccupied for the K/T valley. Consequently, the optical gap, which is associated with the excitation of electronic states near the K valley, continues to decrease. A further increase in the electric field, which is higher than the second critical one ($E_z = 124$ meV/Å), changes the threshold excitation to be near the T valley and slightly lowers the optical gap. As a result, the variation of absorption gap with electric field presents three kinds of E_z-dependence. Regarding the magneto-optical threshold peak in the absence of an electric field, its frequency is characterized by the type-II absorption peak of $n_{\downarrow 1}^v = 0 \rightarrow n_{\downarrow 1}^c = 12$ [red arrow in Fig. 7.13(a)]. The optical gap monotonously grows with the increment of magnetic field, as clearly shown in Fig. 7.16(b). This is consistent with the B_z-dependent Landau-level energies, in which the conduction/valence state energies gradually rise/declines with the increasing B_z [Fig. 7.10(a)]. Under the composite fields, the threshold frequency has an inverse relation with the electric field for a fixed magnetic field, as demonstrated in Fig. 7.16(c) at $B_z = 40$ T. With an increase of E_z from zero up to 153 meV/Å, the optical gap in general decreases. For $E_z < 106$ meV/Å where the band gap is a finite value, the threshold magneto-absorption is determined by the type-II $n_{\downarrow 1}^v = 0 \rightarrow n_{\downarrow 1}^c = 12$ excitation channel; the optical gap monotonically decreases with E_z. After that, the threshold peak relates to the type-I excitation channel of $n_{\downarrow 2}^c = 0 \rightarrow n_{\downarrow 2}^c = 1$ at the K valley [red arrow in Fig. 7.15(a)]; the optical gap decreases more quickly (from the green arrow to the purple one). As for E_z beyond the second critical one, the threshold peak corresponds to the type-I excitation channels at the T point and its frequency decreases more slowly. Apparently, the spatial distributions of sublattice envelope functions and the Fermi-Dirac occupation number play critical roles in determining the threshold absorption channel/structure/frequency.

There exist certain important differences between bilayer AB-bt silicene and AB-stacked graphene in optical/magneto-optical properties [40], mainly owing to the distinct geometric structures and intrinsic interactions. The latter is a semimetal with monotonous energy dispersions, so the optical gap is vanishing except for a sufficiently large electric field. Generally speaking, this graphene system has the same gap in electronic and optical properties even with the E_z-dependence. Its all Landau levels possess eight-fold degeneracy without sublattice non-equivalence and spin splitting, as a result of the absence of the buckled structure, very large interlayer hopping integrals, and important spin-orbital couplings. Their localization centers are only present at 1/6 [4/6] and 2/6 [5/6], but disappear at 1/4 [3/4]. Such features are hardly affected by an electric field. However, the state degeneracy is reduced to

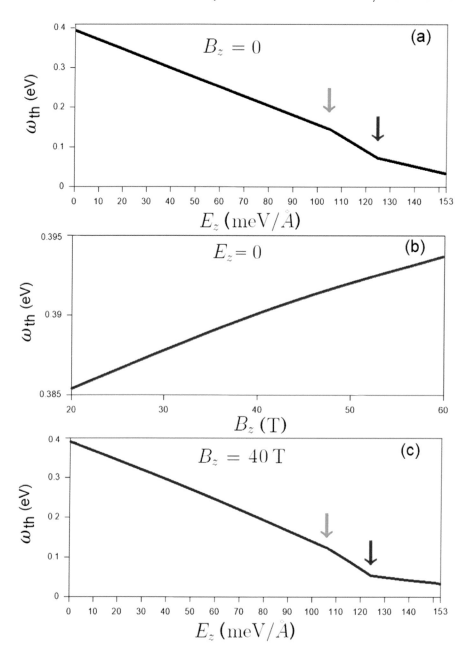

FIGURE 7.16 The optical threshold absorption frequencies with the significant electric- and magnetic-field dependences, respectively, under (a)/(c) B_z=0/40 T and (b) E_z = 0.

half in the presence of E_z, since the inversion symmetry/bi-sublattice equivalence is broken by the layer-dependent Coulomb potential site energies. The sublattice-dominated/sublattice- and spin-dominated Landau-level energy spectra exhibit diverse behaviors, i.e., anti-crossing, crossing & non-crossing behaviors during the variation of B_z or E_z. Considering the first group of valence and conduction Landau levels, the available magneto-excitation category/categories is/are one/two in the absence/presence of E_z [40]. Obviously, the magneto-absorption peaks are well described by the specific selection rule of $\Delta n = \pm 1$ except that the extra $\Delta n = 0$ & 2 rules might appear under a perpendicular electric field. Their quantum numbers are much smaller than those in bilayer silicene. The threshold channel is only associated with the small quantum-number Landau levels [40]; that is, it is mainly determined by $n^v = 0/1/2$ and $n^c = 1/0/1$.

Generally speaking, absorption [43, 44], transmission [44, 45, 47, 56], and reflection spectroscopies [44, 48] are the most efficient techniques in exploring the essential optical excitations of emergent materials [details in Sec. 3.2]. They effectively serve as analytical tools for the characterization of optical properties, when the experimental measurements are taken on the fraction associated with the adsorbed, transmitted, or reflected light by a sample within a significant frequency range. A broadband light source is utilized and done through a tungsten halogen lamp with a wide range for modulation intensity and frequency [45, 48]. Three kinds of optical spectroscopies are very useful in examining the stacking- and E_z-enriched vertical excitation spectra of AB-bt bilayer silicene, e.g., form, intensity, number, and frequency of shoulder and asymmetric-peak absorption structures. Magnetic quantization phenomena of low-dimensional systems could be investigated using magneto-optical spectroscopies [45, 46, 54–56]. The external magnetic field is achieved by a superconducting magnet [45, 46, 56] and semi-destructive single-turn coil [54, 55] with the available field strength below 80 T. The rich and unique magneto-optical spectra in bilayer AB-bt silicene are worthy of further experimental examinations, covering the 16 excitation categories, diverse absorption structures, B_z- and E_z-induced transition channels/threshold frequency, many absorption peaks within a very narrow frequency range, and the absence/presence of specific magneto-optical selection rules. Such measurements could provide rather useful informations about the buckled structure, stacking configuration/strong interlayer hopping integrals, and significant layer-dependent spin-orbital couplings.

REFERENCES

1. Bolotin K I, Sikes K J, Jiang Z, Klima M, Fudenberg G, Hone J, et al. 2008 Ultrahigh electron mobility in suspended graphene *Solid State Communications* **146** 351.
2. Do T N, Lin C Y, Lin Y P, Shih P H, and Lin M F 2015 Configuration-enriched magneto-electronic spectra of AAB-stacked trilayer graphene *Carbon* **94** 619.
3. Huang Y K, Chen S C, Ho Y H, Lin C Y, and Lin M F 2014 Feature-rich magnetic quantization in sliding bilayer graphenes *Sci. Rep.* **4** 7509.
4. Padilha J E and Pontes R B 2015 Free-standing bilayer silicene: the effect of stacking order on the structural, electronic, and transport properties *J. Phys. Chem. C* **119** 3818.
5. Wu J Y, Chen S C, Gumbs G, and Lin M F 2016 Feature-rich electronic excitations of silicene in external fields *Phys. Rev. B* **94** 205427.

6. Borensztein Y, Prevot G, and Masson L 2014 Large differences in the optical properties of a single layer of Si on Ag(110) compared to silicene *Phys. Rev. B* **89** 245410.

7. Derivaz M, Dentel D, Stephan R, Hanf M, Mehdaoui A, Sonnet P, et al. 2015 Continuous Germanene Layer on Al(111) *Nano Lett.* **15** 2510.

8. Shih P H, Chiu Y H, Wu J Y, Shyu F L, and Lin M F 2017 Coulomb excitations of monolayer germanene *Sci. Rep.* **7** 40600.

9. Zhu F F, Chen W J, Xu Y, Gao C L, Guan D D, Liu C H, et al 2015 Epitaxial growth of two-dimensional stanene *Nature Mater.* **14** 1020.

10. Chen R B, Chen S C, Chiu C W, and Lin M F 2017 Optical properties of monolayer tinene in electric fields *Sci. Rep.* **7** 1849.

11. Liu H, Neal A T, Zhu Z, Luo Z, Xu X, Tomnek D, et al. 2014 Phosphorene: An unexplored 2D semiconductor with a high hole mobility *ACS Nano* **8** 4033.

12. Berman O, Gumbs G, and Kezerashvili R 2017 Bose-Einstein condensation and super-fluidity of dipolar excitons in a phosphorene double layer *Phys. Rev. B* **96** 014505.

13. Wu J Y, Chen S C, Do T N, Su W P, Gumbs G, and Lin M F 2018 The diverse magneto-optical selection rules in bilayer black phosphorus *Sci. Rep.* **8** 13303.

14. Ares P, Palacios J J, Abellán G, Gómez-Herrero J, and Zamora F 2018 Recent progress on antimonene: a new bidimensional material *Adv. Mater.* **30** 1703771.

15. Reis F, Li G, Dudy L, Bauernfeind M, Glass S, Hanke W, et al. 2017 Bismuthene on a SiC substrate: a candidate for a high-temperature quantum spin Hall material *Science* **357** 287.

16. Chen R B, Jang D J, Lin M C, and Lin M F 2018 Optical properties of monolayer bis-muthene in electric fields *Optics Lett.* **43** 6089.

17. Avouris P and Dimitrakopoulos C 2012 Graphene: synthesis and applications *Materials Today* **15** 3.

18. Randviir E P, Brownson D A C and Banks C E 2014 A decade of graphene research: production, applications and outlook *Materials Today.* **17 (9)** 426.

19. Xia F, Mueller T, Lin Y M, Valdes-Garcia A, and Avouris P 2009 Ultrafast graphene photodetector *Nature Nano.* **4** 839.

20. Mueller T, Xia F, and Avouris P 2010 Graphene photodetectors for high-speed optical communications *Nature Photon* **4** 297.

21. Echtermeyer T J, Britnell L, Jasnos P K, Lombardo A, Gorbachev R V, Grigorenko A N, et al 2011 Strong plasmonic enhancement of photovoltage in graphene *Nature Commun.* **2** 458.

22. Liu M, Yin X, Ulin-Avila E, Geng B, Zentgraf T, Ju L, et al. 2011 A graphene-based broadband optical modulator *Nature* **64** 474.

23. Bao Q L, Zhang H, Wang Y, Ni Z, Yan Y L, Shen Z X, et al. 2009 Atomic-layer graphene as a saturable absorber for ultrafast pulsed lasers *Adv. Funct. Mater.* **19** 3077.

24. Vogt P, Padova P D, Quaresima C, Avila J, Frantzeskakis E, Asensio M C, et al 2012 Silicene: compelling experimental evidence for graphenelike two-dimensional silicon *Phys. Rev. Lett.* **108** 155501.

25. T Li, Cinquanta E, Chiappe D, Grazianetti C, Fanciulli M, Dubey M, et al. 2015 Silicene field-effect transistors operating at room temperature *Nat. Nanotech.* **10** 227.

26. Mohan B, Kumar A, and Ahluwalia P K 2013 A first principle calculation of electronic and dielectric properties of electrically gated low-buckled mono and bilayer silicene *Physica E* **53** 233.

27. Liu H S, Han N, and Zhao J J 2014 Band gap opening in bilayer silicene by alkali metal intercalation *J. Phys.: Condens. Matter* **26** 475303.

28. Liu J J and Zhang W Q 2013 Bilayer silicene with an electrically-tunable wide band gap *RSC Adv.* **3** 21943.

29. Liuy F, Liuy C C, Wu K H, Yang F, and Yao Y G 2013 $d + d'$ chiral superconductivity in bilayer silicene *Phys. Rev. Lett.* **111** 066804.
30. Padilha J E and Pontes R B 2015 Free-standing bilayer silicene: the effect of stacking order on the structural, electronic, and transport properties *J. Phys. Chem. C* **119** 3818.
31. Fu H X, Zhang J, Ding Z J, Li H, and Menga S 2014 Stacking-dependent electronic structure of bilayer silicene *Appl. Phys. Lett.* **104** 131904.
32. Wang X Q and Wu Z G 2017 Intrinsic magnetism and spontaneous band gap opening in bilayer silicene and germanene *Phys. Chem.Chem. Phys.* **19** 2148.
33. Yaokawa R, Ohsuna T, Morishita T, Hayasaka Y, Spencer M J S, and Nakano H 2016 Monolayer-to-bilayer transformation of silicenes and their structural analysis *Nat. Comm.* **7** 10657.
34. Tabert C J and Nicol E J 2013 Magneto-optical conductivity of silicene and other buckled honeycomb lattices *Phys. Rev. B* **88** 085434.
35. Lin C Y, Wu J Y, Ou Y J, Chiu Y H and Lin M F 2015 Magneto-electronic properties of multilayer graphenes *Phys. Chem. Chem. Phys.* **17** 26008.
36. Lin M F and Shung K W K 1994 Plasmons and optical properties of carbon nanotubes *Phys. Rev. B* **50** 17744.
37. Kamal C, Chakrabarti A, Banerjee A, and Deb S K 2013 Silicene beyond mono-layers— different stacking configurations and their properties *J. Phys.: Condens. Matter* **25** 085508.
38. Zhang Y, Tang T, Girit C, Hao Z, Martin M, Zettl A, et al. 2009 Direct observation of a widely tunable bandgap in bilayer graphene *Nature* **459** 820.
39. Oostinga J B, Heersche H B, Liu X, Morpurgo A F, Vandersypen L M K 2008 Gate-induced insulating state in bilayer graphene devices *Nature Mater.* **7** 151.
40. Lin C Y, Do T N, Huang Y K, and Lin M F 2017 Optical properties of graphene in magnetic and electric fields *IOP Publishing* 2053–2563, ISBN: 978-0-7503-1566-1.
41. Vogt P, Padova P D, Quaresima C, Avila J, Frantzeskakis E, Asensio M C, et al. 2012 Silicene: compelling experimental evidence for graphenelike two-dimensional silicon *Phys. Rev. Lett.* **108** 155501.
42. Borensztein Y, Prevot G, and Masson L 2014 Large differences in the optical properties of a single layer of Si on Ag(110) compared to silicene *Phys. Rev. B* **89** 245410.
43. Mak K F, Shan J, and Heinz T F 2010 Electronic Structure of Few-Layer Graphene: Experimental Demonstration of Strong Dependence on Stacking Sequence *Phys. Rev. Lett.* **104** 176404.
44. Li Z Q, Henriksen E A, Jiang Z, Hao Z, Martin M C, Kim P, et al. 2009 Band structure asymmetry of bilayer graphene revealed by infrared spectroscopy *Phys. Rev. Lett.* **102** 037403.
45. Jiang Z, Henriksen E A, Tung L C, Wang Y J, Schwartz M E, Han M Y, et al. 2007 Infrared spectroscopy of Landau levels of graphene *Phys. Rev. Lett.* **98** 197403.
46. Plochocka P, Faugeras C, Orlita M, Sadowski M L, Martinez G, Potemski M, et al. 2008 High-energy limit of massless Dirac fermions in multilayer graphene using magneto-optical transmission spectroscopy *Phys. Rev. Lett.* **100** 087401.
47. Mak K F, Lui C H, Shan J, and Heinz T F 2009 Observation of an electric-field-induced band gap in bilayer graphene by infrared spectroscopy *Phys. Rev. Lett.* **102** 256405.
48. Mak K F, Sfeir M Y, Wu Y, Lui C H, Misewich J A, and Heinz T F 2008 Measurement of the optical conductivity of graphene *Phys. Rev. Lett.* **101** 196405.
49. Do T N, Shih P H, Gumbs G, Huang D, Chiu C W, and Lin M F 2018 Diverse magnetic quantization in bilayer silicene *Phys. Rev. B* **97** 125416.
50. Wang Y Y, Quhe R G, Yu D P, and Jing L 2015 Silicene spintronics—a concise review *Chinese Phys. B* **24** 8.

51. Do T N, Gumbs G, Shih P H, Huang D, Chiu C W, Chen C Y, et al. 2019 Peculiar optical properties of bilayer silicene under the influence of external electric and magnetic fields *Sci. Rep.* **9** 624.

52. Rose F, Goerbig M O, and Piéon F 2013 Spin- and valley-dependent magneto-optical properties of *MoS₂ Phys. Rev. B* **88** 125438.

53. Wu J Y, Chen S C, Do T N, Su W P, Gumbs G, and Lin M F 2018 The diverse magneto-optical selection rules in bilayer black phosphorus *Sci. Rep.* **8** 13303.

54. Plochocka P, Solane P Y, Nicholas R J, Schneider J M, Piot B A, Maude D K, et al. 2012 Origin of electron-hole asymmetry in graphite and graphene *Phys. Rev. B* **85** 245410.

55. Nicholas R J, Solane P Y, and Portugall O 2013 Ultrahigh magnetic field study of layer split bands in graphite *Phys. Rev. Lett.* **111** 096802.

56. Orlita M, Faugeras C, Borysiuk J, Baranowski J M, Strupinski W, Sprinkle M, et al. 2011 Magneto-optics of bilayer inclusions in multilayered epitaxial graphene on the carbon face of SiC *Phys. Rev. B* **83** 125302.

8 Si-Doped Graphene Systems

Po-Hsin Shih,[a] Chiun-Yan Lin,[a] Thi-Nga Do,[c,d]
Jhao-Ying Wu,[b] Shih-Yang Lin,[e]
Ching-Hong Ho,[b] Ming-Fa Lin[a,f,g]

[a] Department of Physics, National Cheng Kung University, Tainan 701, Taiwan
[b] Center of General Studies, National Kaohsiung University of Science and Technology, Kaohsiung 811, Taiwan
[c] Laboratory of Magnetism and Magnetic Materials, Advanced Institute of Materials Science, Ton Duc Thang University, Ho Chi Minh City, Vietnam
[d] Faculty of Applied Sciences, Ton Duc Thang University, Ho Chi Minh City, Vietnam
[e] Department of Physics, National Chung Cheng University, Chiayi 621, Taiwan
[f] Quantum Topology Center, National Cheng Kung University, Tainan 701, Taiwan
[g] Hierarchical Green-Energy Materials Research Center, National Cheng Kung University, Tainan, Taiwan

CONTENTS

The essential properties are greatly modified by generating the substituted impurities/guest atoms in a hexagonal carbon lattice. The up-to-date experimental growths show that carbon host atoms have been partially replaced by the guest atoms through the chemical vapor deposition method, such as Si [1], B [2], and N [2, 3]. A plenty of guest-atom-dressed graphene systems, which are very suitable for studying the unusual phenomena and the potential applications, are expected to be successfully synthesized in experimental laboratories. Obviously, these new 2D binary/ternary compounds can induce the non-equivalence of the original A and B sublattices; the energy-gap engineering and the distorted/destroyed Dirac cone will come to exist. According to the first-principles calculations on the Si-decorated graphene systems [4, 5], the π bonding extending on a hexagonal lattice presents a serious distortion or is absent under the different ionization potentials and the nonuniform hopping integrals. That is to say, there exist the drastic changes in the Dirac cone, the significant band gap, low-lying C-dominated & (C, Si)-co-dominated energy bands, and the atom-dependent wave functions. These will play critical roles in diversifying the magnetic quantization phenomena.

The diverse physical phenomena is very sensitive to the variations in the atomic components [6], the crystal symmetries [7, 8], the planar/buckled/rippled/folded structures [9–11], the stacking configurations [12–14], the number of layers [15, 16], the distinct dimensions [17, 18], the layer-dependent spin-orbital couplings [19, 20], the single- or multi-orbital hybridizations in chemical bonds [21], the gate voltage [22], and the uniform/spatially modulated magnetic field [19, 23]. The generalized tight-binding model, which can cover distinct geometric structures, all the significant atomic interactions, and the various external fields in the calculations simultaneously, is well developed to fully investigate the rich and unique magneto-electronic properties in any layered materials [24]. This model is able to directly combine with the dynamic and static Kubo formulas to explore the magneto-optical [Chaps. 4, 6 & 7] and quantum transport properties [Chap. 9], respectively. Furthermore, its combination with the modified random-phase approximation is useful in understanding the diversified magneto-Coulomb excitations [Chap. 10]. The Si-diversified magnetic quantization in monolayer graphene systems will be clearly illustrated by this method.

It is well known that a monolayer graphene exhibits a lot of the unusual fundamental properties, mainly owing to the honeycomb symmetry and the atom-scale thickness. The isotropic Dirac-cone structures, which is initiated from the K/K' valleys (corners of the 1st Brillouin zone), are magnetically quantized into the unique valence and conduction Landau levels, with the specific energy spectrum proportional to the square root of $B_z n^{c,v}$. This simple relation has been accurately confirmed by the high-resolution measurements by STS [25], optical spectroscopies [26], and transport instruments [27]. The magneto-optical absorption structures are thoroughly examined to satisfy a specific selection rule of $\Delta n = |n^v - n^c| = 1$, directly indicating the equivalence of A and B sublattices, This rule is responsible for the available scattering channels during the carrier transport; therefore, it leads to an unconventional half-integer Hall conductivity of $\sigma_{xy} = (m + 1/2)4e^2/h$ [27], where m is an integer and the factor of 4 represents the spin- and valley-induced Landau-level state degeneracy. Such unusual quantization is attributed to the quantum anomaly of $n^{c,v} = 0$ Landau levels arising from the gapless Dirac point.

To thoroughly investigate the diversified electronic and optical properties in Si-substituted graphene systems, the generalized tight-binding model directly links with the dynamic Kubo formula within the linear response. The complex combined effects, originating from the distinct ionization potentials, the nonuniform hopping integrals & bond lengths on a deformed honeycomb lattice, the various B_z-generated Peierls phases, and the vertical excitations due to electromagnetic waves, are accurately included in the giant magnetic Hamiltonian matrix. The exact diagonalization method is proposed to overcome the numerical issues and thus solve magneto-electronic properties and magneto-absorption spectra more efficiently [24]. How many kinds of Landau levels come to exist during the variation of Si-distribution configuration and concentration is the studying focus; that is, their relations are thoroughly researched.

Their main features, being characterized by probability distributions and oscillation modes, will be clearly illustrated by the different magneto-optical selection rules. Apparently, this work can create a new research category in the fundamental properties of 2D layered materials. The theoretical predictions require the further experimental verifications using STS [14, 25, 28, 29], magneto-optical spectroscopies [26, 30–33], and quantum transport instruments [27].

8.1 GUEST-ATOM-DIVERSIFIED ELECTRONIC PROPERTIES

From the theoretical point of view, the Si-doped graphene systems have a lot of well-behaved distribution configurations. The high-symmetry ones, which correspond to the enlarged rectangular unit cells with achiral or chiral edge structures, are very suitable for the numerical calculations. The substitution-related unit cell can be characterized by $(M_d\vec{a}_1, N_d\vec{a}_2)$, where $\vec{a}_1$ and $\vec{a}_2$ are primitive lattice vectors of monolayer graphene, and the coefficients of $[M_d, N_d]$ are similar to those defined in single-walled carbon nanotubes [34]. The doping concentration is defined by the ration between the guest-atom number and the total one. The specific enlarged unit-cell structures, closely related to $[M_d = 5, N_d = -1]$, are chosen for exploring the various magnetic quantizations and magneto-optical selection rules. For example, Figs. 8.1(a) and 8.1(b), have the high concentration of 2:16, respectively, corresponding to two Si guest atoms at the A_i-sublattice positions and the A_i- & B_i-sublattice ones. The nonuniform Si-C and C- bond lengths in the calculations are taken from the first-principles method [4, 5]. Furthermore, the equivalence or non-equivalence between A_i and B_i sublattices will be reflected in the essential properties.

The low-energy essential properties are mainly determined by the C-$2p_z$ and Si-$3p_z$ orbitals, with the site energies 1.3 eV and zero, respectively. The zero-field Hermitian Hamiltonian matrix covers the nonuniform site energies and nearest-neighboring hopping integrals, being expressed as

$$H = \sum_{\langle i \rangle} \epsilon_i c_i^\dagger c_i - \sum_{\langle i,j \rangle} t_{i,j} c_i^\dagger c_j. \tag{8.1}$$

i is the lattice site and $<i, j>$ only corresponds to the nearest neighbors. The hopping integrals and site energies are, respectively,

$$t_{i,j} = \begin{cases} t_{Si-C}, & \text{for the Si-C bonds,} \\ t_{C-C}, & \text{for the C-C bonds,} \end{cases} \tag{8.2}$$

and

$$\epsilon_i = \begin{cases} \epsilon_{Si}, & \text{for Si atoms,} \\ 0, & \text{for carbon atoms.} \end{cases} \tag{8.3}$$

Specifically, the 2:16 concentration, with two guest atoms at the Ai sublattices [Fig. 8.1(a)], presents 25 independent Hamiltonian matrix elements:

$$H_{A_1,B_1} = H_{B_1,A_1} = t_{Si-C}f_{-2},$$

$$H_{A_1,B_2} = H_{B_2,A_1} = t_{Si-C}f_{-3},$$

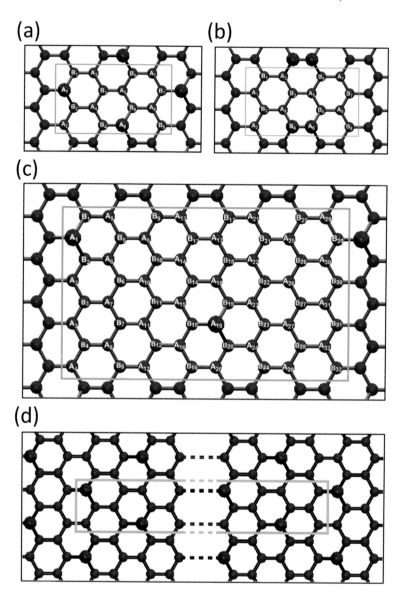

FIGURE 8.1 Distributions structures of Si-decorated graphene systems: under the 2:16 concentration with two guest atoms [the red balls] at (a) the A_i sublattices, (b) the A_i and B_i sublattices, (c) for the reduced 2:64 concentration at the Ai-sublattice distribution, (d) a supercell due to a uniform perpendicular magnetic field.

$$H_{A_1,B_7} = H_{B_7,A_1} = t_{Si-C}f_{-1},$$

$$H_{B_1,A_2} = H_{A_2,B_1} = t_{C-C}f_3,$$

$$H_{B_1,A_3} = H_{A_3,B_1} = t_{C-C}f_1,$$

$$H_{A_2,B_2} = H_{B_2,A_2} = t_{C-C}f_{-2},$$

$$H_{A_2,B_8} = H_{B_8,A_2} = t_{C-C}f_{-1},$$

$$H_{B_2,A_4} = H_{A_4,B_2} = t_{C-C}f_1,$$

$$H_{A_3,B_3} = H_{B_3,A_3} = t_{C-C}f_{-3},$$

$$H_{A_3,B_4} = H_{B_4,A_3} = t_{C-C}f_{-2},$$

$$H_{B_3,A_4} = H_{A_4,B_3} = t_{C-C}f_2,$$

$$H_{B_3,A_5} = H_{A_5,B_3} = t_{C-C}f_1,$$

$$H_{A_4,B_4} = H_{B_4,A_4} = t_{C-C}f_{-3},$$

$$H_{B_4,A_6} = H_{A_6,B_4} = t_{Si-C}f_1,$$

$$H_{A_5,B_5} = H_{B_5,A_5} = t_{C-C}f_{-2},$$

$$H_{A_5,B_6} = H_{B_6,A_5} = t_{C-C}f_{-3},$$

$$H_{B_5,A_6} = H_{A_6,B_5} = t_{Si-C}f_3,$$

$$H_{B_5,A_7} = H_{A_7,B_5} = t_{C-C}f_1,$$

$$H_{A_6,B_6} = H_{B_6,A_6} = t_{Si-C}f_{-2},$$

$$H_{B_6,A_8} = H_{A_8,B_6} = t_{C-C}f_1,$$

$$H_{A_7,B_7} = H_{B_7,A_7} = t_{C-C}f_{-3},$$

$$H_{A_7,B_8} = H_{B_8,A_7} = t_{C-C}f_{-2},$$

$$H_{B_7,A_8} = H_{A_8,B_7} = t_{C-C}f_2,$$

$$H_{A_8,B_8} = H_{B_8,A_8} = t_{C-C}f_{-3},$$

$$H_{A_1,A_1} = H_{A_6,A_6} = \epsilon_{Si-C}. \tag{8.4}$$

Here ϵ_{Si-C}=1.3 eV, t_{C-C}=2.7 eV, t_{Si-C}=1.3 eV, and $f_{\pm 1,2,3} = e^{i\vec{k}\vec{R}_{\pm 1,2,3}}$ where $\vec{R}_{\pm 1} = \pm(b,0)$, $\vec{R}_{\pm 2} = \pm(-\frac{b}{2}, -\frac{\sqrt{3}b}{2})$, and $\vec{R}_{\pm 3} = \pm(-\frac{b}{2}, \frac{\sqrt{3}b}{2})$.

Four kinds of typical Si-decorated graphene systems can clearly illustrate the diversified physical properties. They cover (I) the 2:16 concentration for the Si-[A$_1$, A$_6$]-sublattice distribution [red balls in Fig. 8.1(a)], (II) the 2:16 concentration under the -[A$_6$, B$_4$] configuration [Fig. 8.1(b)], (III) the 2:64 concentration associated with the [A$_1$, A$_{19}$] sublattices [Fig. 8.1(c)], and (IV) a pristine one with the equivalent A and B sublattices. The first and third kinds (the second kind) possess the nonequivalent (equivalent) A$_i$ and B$_i$ sublattices in a Si-induced enlarged unit cell, while both A and B sublattices are fully equivalent for graphene. For example, the first kind has a rectangular traditional cell, including two Si and fourteen C atoms, is consistent with the Landau gauge under $B_z\hat{z}$. There exist a slight buckling near the guest atoms (~ 0.93 Å deviation from graphene plane) and the distinct C–C and Si–C bond length [1.42 Å & 1.70 Å], according to the first-principles calculations [4,5]. This indicates the drastic modifications of the π bonding extending on a hexagonal lattice. However, the nonuniform site energies and nearest-neighboring hopping integrals, which are due to the major $2p_z$ orbitals of C host atoms and the minor $3p_z$ orbitals of Si guest atoms, are sufficient in understanding the low-lying energy bands. They are, respectively, assumed to be $\epsilon_{Si-C} = 1.3$ eV, γ_{C-C}=2.7 eV, and $\gamma_{Si-C} = 1.3$ eV, indicating the significant nonuniform environment. Such parameters have been examined for many distribution configurations and concentrations; that is, the calculated band structures match with those from the first-principles method.

The Si-decorated graphene exhibits the unusual low-energy electronic properties. For the first kind of Si-dressed graphene [the red curve in Fig. 8.2], the valence and conduction bands, nearest to the Fermi level, have the parabolic energy dispersions well separated by a direct energy gap of $E_g = 0.74$ eV. The electronic energy spectrum is anisotropic along the different **k**-directions, and it is asymmetric about E_F. The similar results are revealed in the third kind of lower-concentration system with a 0.26-eV band gap (the black curve). Energy gap appears when the guest atoms are only situated at the A$_i$ or B$_i$ sublattices. The nonuniform site energies and hopping integrals further induce the partial termination of the π bonding (the minor localized states), as observed in the zero-field and magnetic wave functions [Figs. 8.4(a), 8,4(b), and 8.4(d)]. On the other side, E_g vanishes under the Si-[A$_i$, B$_i$]-sublattice distribution configuration (the black solid curve). The equivalent guest-atom distribution induces the distorted π and thus the strongly modified Dirac cone structure with an obvious shift of Dirac point, the reduced Fermi velocity, and the anisotropic energy spectrum. Apparently, graphene exhibits a well-behaved Dirac cone (the dashed black curve) because of the purely hexagonal symmetry.

The main features of density of states [Fig. 8.3] mainly originate from those of energy bands [Fig. 8.2]. As for the Si-A$_i$-addressed graphene systems (the red and blue solid curves), a pair of valence and conduction shoulder structures crosses the Fermi level, clearly indicating the semiconducting properties with observable band gaps. Such van Hove singularities correspond to the highest occupied state and the lowest unoccupied one (the local maximum and minimum states). Furthermore, the higher-concentration case [2:16] has the larger density of states as a result of the smaller band curvature [Fig. 8.2]. On the other side, there exist the asymmetric and symmetric

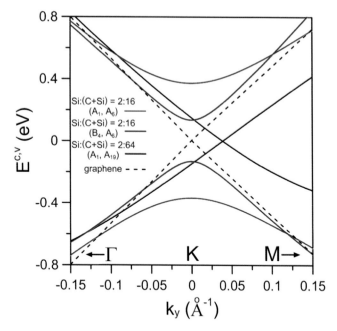

FIGURE 8.2 The low-energy band structures under four kinds of Si-guest-atom distributions in Fig. 8.1.

V-shape van Hove singularities, in which a zero density of states just at the Fermi level means the zero-gap semiconducting behavior. The former (the black solid curve) directly reflects the anisotropic and distorted Dirac-cone structure: the larger density of states (the smaller Fermi velocity) for valence states, compared with that of conduction ones. Specifically, a pristine graphene (the black dashed curve) has the smallest

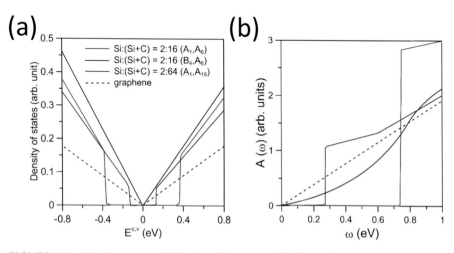

FIGURE 8.3 The low-energy (a) density of states and (b) absorption spectra due to band structures in Fig. 8.2.

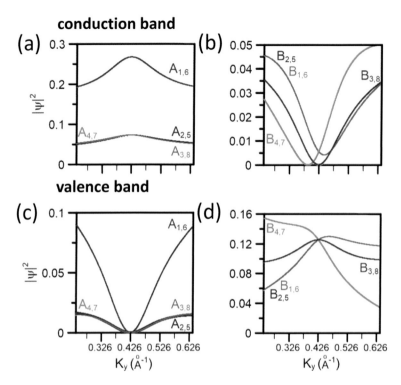

FIGURE 8.4 The low-energy zero-field subenvelope functions of the Si-dressed graphene in Fig. 8.1 on (a) & (c) 8-A_i and (b) & (d) 8-B_i sublattices for conduction and valence states, respectively.

density of states among all the Si-substituted cases. It should be noticed that the forms of van Hove singularities and the magnitudes of density of states are, respectively, closely related to the B_z-dependent magneto-electronic spectra and the neighboring Landau-level energy spacings [discussed in Figs. 8.5(a)–8.5(e)]. The theoretical predictions on the low-energy density of state could be verified from the high-resolution STS measurements [details in Sec. 2.1], which is very useful in understanding the Si-atom substitution effects, especially for the obvious changes in the band gap and energy dispersion.

The zero-field Bloch wave functions [Figs. 8.4(a)–8.4(d)], reflecting on the Landau-level ones [discussed later in Figs. 8.5(a)–8.5(e)], are worthy of a detailed investigation. Obviously, the A and B sublattices are equivalent in a pristine graphene [35]; that is, their tight-binding function has the same contributions to electronic states. Their wave-vector-dependent contributions are very sensitive to the Si-atom distribution configurations and concentrations. The Si-A_i-decorated graphene, with 2:16 concentration [Fig. 8.1(a)], could clearly illustrate the distinct dominance Si-guest and carbon-host atoms. For the low-lying conduction-band states, the dominating contributions are due to the Si-related A_1 & A_6 sublattices [the red curves in Fig. 8.4(a)]; furthermore, the minor ones arise from the fourteen C-created sublattices [the other curves in Figs. 8.4(a) and 8.4(b)]. Specifically, the B_3- & B_8-sublattice

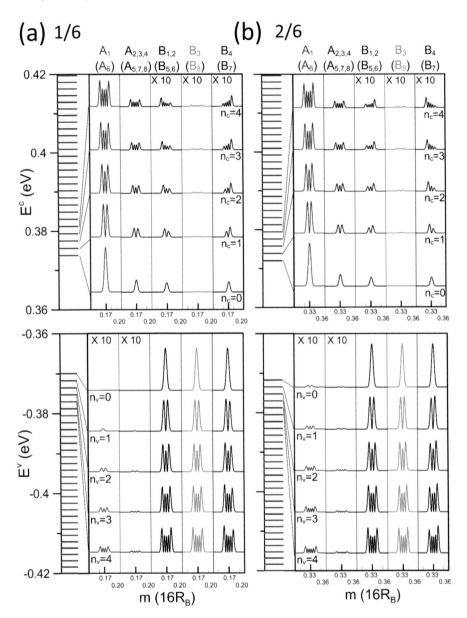

FIGURE 8.5 (a)(b)   At $B_z = 10$ T. The conduction and valence Landau-level energy spectra and probability distributions under 2:16 concentration with the Si-A_i-sublattice configuration near the (a) 1/6 and (b) 2/6 localization centers, (c) 2:16 concentration with the $[A_i, B_j]$-sublattice configuration, (d) 2:64 concentration with the Ai-sublattice distribution, and (e) pristine case.

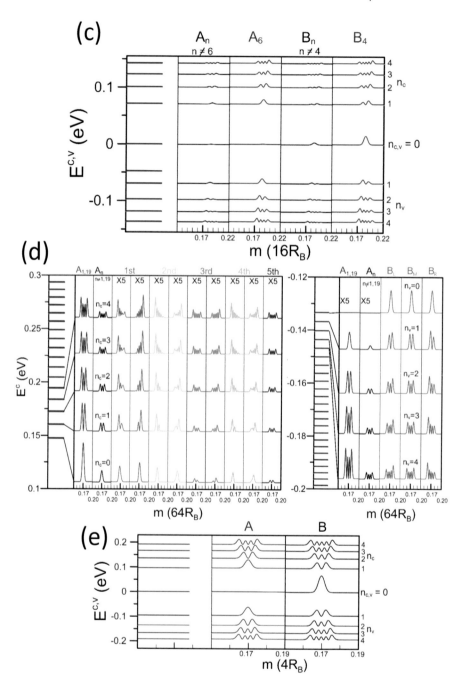

FIGURE 8.5 (c)-(e) At $B_z = 10$ T. The conduction and valence Landau-level energy spectra and probability distributions under 2:16 concentration with the Si-A_i-sublattice configuration near the (a) 1/6 and (b) 2/6 localization centers, (c) 2:16 concentration with the $[A_i, B_j]$-sublattice configuration, (d) 2:64 concentration with the A_i-sublattice distribution, and (e) pristine case.

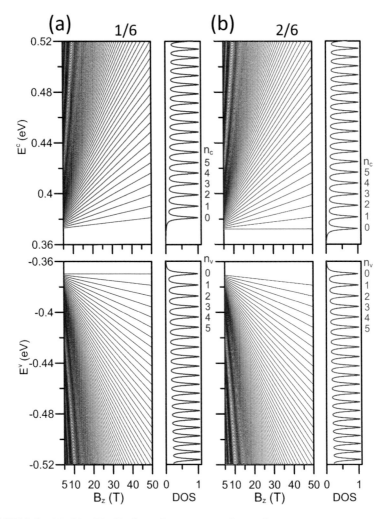

FIGURE 8.6 (a)–(b) The B_z-dependent magneto-electronic energy spectra corresponding to the different cases in Fig. 8.4. The density of states is also displayed for the first kind of Si-distribution configuration under the highest field strength.

envelope functions are vanishing near the K point, indicating the partial termination of the $2p_z$-orbital π bonding. This will lead to the localized magneto-electronic states, with the negligible B_z-dependence [the $n^c = 0$ Landau levels in Fig. 8.6(b)]. On the valence-state side, the dominant sublattices belong to all the B_i ($i = 1, 2, 3, 4, 5, 6, 7;$ 8) sublattices. The other A_i sublattices do not have any contributions at the K point, so the π bonding is fully terminated and the magneto-electronic energy spectrum is independent of field strength [the $n^v = 0$ Landau levels in Fig. 8.6(a)]. The above-mentioned results show that the equivalence of the A_i and B_i sublattices thoroughly disappear.

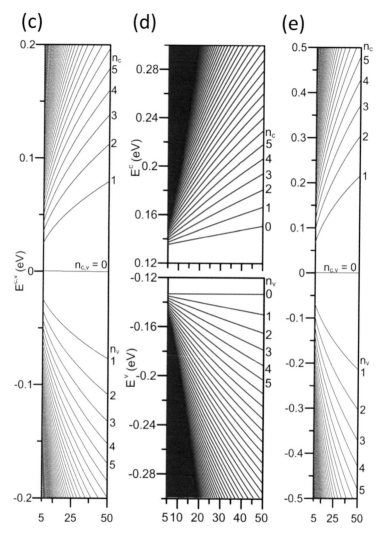

FIGURE 8.6 (c)–(e)   The B_z-dependent magneto-electronic energy spectra corresponding to the different cases in Fig. 8.4. The density of states is also displayed for the first kind of Si-distribution configuration under the highest field strength.

In addition to electronic properties, the Si-substitutions account for the dramatic changes of optical absorption spectra, as clearly illustrated in Fig. 8.3(b). A pristine graphene only exhibits the featureless structure (the dashed black curve). In which the spectral intensity is linearly proportional to frequency. The V-shaped density of states [Fig. 8.3(a)] and the constant dipole matrix element of $\sim v_F$ (the Fermi velocity) for the low-lying electronic states are the main mechanisms. On the other hand, the nonlinear relation between the absorption strength and frequency come to exist in Si-substituted graphene systems. This principally arises from the wave-vector-/energy-dependent electric dipoles, but not the linear joint density of states. For the highly modified Dirac-cone structure (the black curve), the (A_i, B_i)-substituted systems presents a lower-intensity spectrum, compared with that of an isotropic one. The dipole-moment transitions are reduced by the smaller hopping integrals associated with the longer Si-C bond lengths. Apparently, the threshold excitations, which reveal as the shoulder structures, appear in two semiconducting systems (the red and blue curves). Here, an optical gap is equal to band gap for a direct-gap semiconductor. The high-resolution measurements from optical spectroscopies [details in Sec. 2.2] can examine the spectral absorption structures and thus the nonuniform site energies and hopping integrals.

8.2 FOUR KINDS OF LANDAU LEVELS AND MAGNETO-OPTICAL SELECTION RULES

In a uniform perpendicular $B_z \hat{z}$, the dimension of the magnetic Hamiltonian matrix is determined by the guest-atom- and vector-potential-dependent periods, in which the latter might be much longer than the former, and their ratio is assumed to be an integer for more convenient calculations. The vector potential is chosen as $B_z x \hat{y}$, leading to an extra position-related Peierls phase in the tight-binding function/the nearest-neighbor hopping integral [details in Sec. 3.1]. Consequently, the magnetic unit cell becomes a super large rectangular one (a super cell), as illustrated in Fig. 8.1(d). For example, $R_B = 8,000$ at $B_z = 10$ T under the first kind of guest-atom distribution [Fig. 8.1(a)], corresponding to the total $16R_B$ atoms ($8R_B$ A and B atoms). This clearly implies that the Bloch magnetic wave functions can be expressed by the linear superposition of the $16R_B$ tight-binding functions in a super unit cell. The magnetic Hamiltonian matrix consists of the $50R_B$ non-vanishing elements, in which 24 components belong to the independent ones. When this giant Hermitian matrix is solved by the exact diagonalization method [24], the B_z-induced Landau-level energy spectrum and the magnetic subenvelope functions are solved more efficiently. Moreover, the reliable magneto-electronic properties are very useful in understanding the rich magneto-optical excitation spectra.

When the Si-decorated graphene systems are present in an electromagnetic wave, the occupied valence Landau-level states are vertically excited to the unoccupied conduction ones. In addition to $\Delta \mathbf{k} = 0$, the electric-dipole perturbations will make Landau levels generate a new magneto-optical selection rule $\Delta n = 0$, being never revealed in other layered condensed-matter systems. The broadening factor of $\Gamma = 1$ meV is used in the calculations of the dynamic linear Kubo formula [details in Eq. (3.3)]. The direction of the planar electric field hardly affects magneto-optical properties, so the electric polarization is assumed to be along $\hat{x}$. Whether the inter-Landau-level optical transitions could survive is evaluated from the electric dipole matrix elements under the gradient approximation. By the detailed and delicate analysis under the specific distribution configuration in Fig. 8.1(a), the vertical magneto-optical excitations are dominated by the A_2 subenvelope function of the initial occupied Landau-level states and the (B_1, B_2, B_8) subenvelope functions of the final unoccupied ones, mainly owing to the $\mathbf{k}$-dependent nonuniform hopping integrals. The similar results are revealed in other sublattice-dependent subenvelope functions. The spatial symmetric & anti-symmetric distribution modes due to the nearest-neighbor sublattices will lead to the specific magneto-optical selection rules, $\Delta n = 0$ and ± 1 [discussed in Figs. 8.6(a)–8.6(d)].

The magneto-electronic properties exhibit the rich and unique features. As to the Si-A_i-sublattice configuration under 2:16 concentration [Fig. 8.1(a)], the low-lying Landau-level energy spacings, as shown at $B_z = 10$ T in Figs. 8.4(a) and 8.4(b), are almost uniform and have an energy gap close to the zero-field value (~ 0.74 eV). In general, the quantum number of each highly degenerate Landau level is characterized by the zero points of the dominating oscillation mode. The magnetic Bloch wave function originates from the subenvelope functions of the 16 tight-binding functions on the 8-A_i and 8-B_i sublattices. Its spatial probability distribution of the $(k_x = 0, k_y = 0)$ state is localized at [1/6 & 4/6] and [2/6 & 5/6] of an enlarged unit cell [Fig. 8.1(c)]. Any (k_x, k_y) Landau-level states in the reduced first Brillouin are doubly degenerate except for the spin degree of freedom. Apparently, the decoration of Si guest atoms results in the destruction of the planar inversion symmetry and thus the non-degenerate 1/6 and 2/6 Landau-level states. According to the neighboring chemical environment, the original 16 sublattices could be classified into five subgroups: (A_1, A_6), (A_2, A_3, A_4, A_5, A_7, A_8), (B_1, B_2, B_5, B_6), (B_3, B_8), and (B_4, B_7), in which the probability distributions are identical for the sublattice envelope functions in each subgroup. The similar behaviors are revealed in the zero-field subenvelope functintion on 16 sublattices [Fig. 8.4]. The low-energy conduction Landau-level states are mostly contributed by the A_1 sublattice due to the Si-$3p_z$ tight-binding function, so that its zero-point number of the well-behaved probability distribution could serve as a good quantum number. $n^c = 0, 1, 2, ...$ come to exist in the normal ordering. Specifically, the contributions from the B_3 sublattice are very small, as seen in the zero-field wave functions. The oscillation modes are well described by n^c in the significant sublattices except for the weak $n^c \pm 1$ B_3-sublattice. On the other side, the valence Landau-level states mainly arise from all the B_i sublattices of the C-$2p_z$ tight-binding functions, where they have the similar oscillation modes in determining n^v. The contributions from the A_i sublattices are very weak, and the number of zero points is $n^v - 1$ and $n^v + 1$, respectively, corresponding to the 1/6 and 2/6 localization centers. The B_i-sublattice dominance and the specific mode differences between the B_i and A_i sublattices directly reflect the fact that the occupied valence states mainly originate from the C-$2p_z$ orbitals. In short, the sequence of n^c/n^v presents a good ordering, i.e., the Landau-level crossing or anti-crossing phenomena are thoroughly absent.

The spatial oscillation modes are very sensitive to the changes in the distribution configuration and concentration of Si-guest atoms, leading to four kinds of typical Landau levels. The diverse magneto-electronic states mainly arises from the Si-modulated A_i and B_i sublattices. For a very strong non-equivalence between A_i and B_i sublattices and enough high concentration [2:16 under the Si-A_1 configuration in Fig. 8.1(a)], only the significant sublattices exhibit similar oscillation modes for the low-lying conduction and valence Landau levels [the first kind in Figs. 8.5(a) and 8.5(b)]. However, the enhanced equivalence and the reduced concentration can create the composite/more complicated behaviors related to the heavily non-equivalent A_i & B_i sublattices and the fully equivalent ones (e.g., pristine graphene). The former [Fig. 8.1(b)], with two Si atoms in A_6 and B_4 sublattices, has the highly equivalent environment for A_i and B_i ones. The conduction and valence Landau states are described by the zero points of the B_i sublattices. All the sublattices make comparable contributions to the Landau-level wave functions, in which the difference of zero point number is ± 1 for A_i and B_i sublattices [the second kind in Fig. 8.5(c)]. Specifically, their spatial distributions are highly asymmetric and localization centers seriously deviate from 1/6 & 2/6, directly reflecting the seriously titled Dirac-cone [the black solid curve in Fig. 8.2]. Also, a seriously distorted distribution consists of the main $n^{c,v}$ mode and the side $n^{c,v} \pm 1$ ones. The localization centers are recovered to the normal positions under a further decrease of concentration with the Si-A_i distribution [e.g., 2:64 distribution configuration in Fig. 8.1(c)]. The dominating sublattices of the conduction and valence Landau levels are, respectively, (A_1, A_{19}) and B_i [the third kind in Figs. 8.5(d)]. For the former, the B_i sublattices have the n^c ($n^c = 1$) mode for the nearest and second-nearest neighbors [the others]. The latter might present the highly asymmetric distributions, indicating the significant side modes of $n^v \pm 1$. This property becomes more obvious as the valence Landau-level energy grows. Finally, a pristine graphene displays the well-behaved/oscillator-like Landau levels about the localization centers and the difference of ± 1 in the zero-point number due to the equivalent A and B sublattices [the fourth kind in Fig. 8.4(e)]. The above-mentioned concise relations between the A_i and B_i sublattices will determine the magneto-optical selection rules.

The B_z-dependent Landau-level energy spectra, as clearly displayed in Figs. 8.6(a)–8.6(e), exhibit the unusual features. The crossing or anti-crossing phenomena are absent for the low-lying Landau levels, illustrating the well separated Landau-level states and the specific-mode magnetic wave functions. For the first and third kinds of Landau levels [Figs. 8.6(a), 8.6(b), and 8.6(d)], the dispersion relation is almost linear, and the neighboring Landau-level energy spacing is approximately uniform. The higher is the density of stats [Fig. 8.3], the smaller is the Landau-level energy spacing. Specifically, the initial valence and conduction Landau levels, which are, respectively, related to the 1/6 and 2/6 localization centers, remain the fixed energies during the variation of field strength. They purely come from the localized electronic states, since the magnetic wave functions vanish in all the A_i/the (B_3, B_8) sublattices, as observed from Fig. 8.5(a)/Fig. 8.5(b). That is, the termination of the π bonding appears on a guest-host mixed hexagonal lattice. A uniform perpendicular magnetic field can create the splitting of the localized and extended electronic states; otherwise, they are hybridized each other and are revealed near the K and K' valleys. Such Landau-level states could be examined from the STS measurements on the van Hove singularities of the density of states, e.g., the delta-function-like prominent peaks across the Fermi level [the right-hand side figures of Figs. 8.5(a) and 8.5(b)]. On the other side, the second and fourth kinds of Landau levels show the $\sqrt{B_z}$-dependent energy spectra except for the constant energy of the degenerate $n^{c,v} = 0$ Landau levels. The anisotropic and distorted Dirac-cone structure [the black solid curve in Fig. 8.2] do not change the $\sqrt{B_z}$ dependence [Fig. 8.6(c)], while it creates the high asymmetry about the Fermi level. For a fixed energy range, there are more valence Landau levels, compared with the conduction ones. Apparently, the latter, as indicated in Fig. 8.6(e), has the largest energy spacing among four kinds of Landau levels because of the lowest density of states.

The main features of Landau levels are directly reflected in magneto-optical absorption spectra with a lot of delta-function-like peaks in Figs. 8.7(a)–8.7(d). For the Si-A$_1$-dressed graphene with 2:16 concentration [Fig. 8.7(a)], the spectral intensity gradually declines with the increasing frequency, while the energy spacing between two neighboring absorption peaks is almost uniform. Only the inter-Landau-level vertical excitations, which absorption frequency correspond to the identical quantum mode in the valence and conduction states, are revealed as the significant absorption peaks. For example, the threshold due to $0^v \rightarrow 0^c$ is 0.743 eV, which is very close to the energy gap; that is, an optical gap is equal to a band gap. The magneto-optical selection rule, $\Delta n = 0$, could be thoroughly examined from the electric-dipole momentum in Eq. (3.4) [details in the Sec. 3.2]. It is mainly determined by the specific Hamiltonian matrix elements covering the three nearest-neighboring hopping integrals. By the delicate analyses, the effective magneto-optical excitations strongly depend on the subenvelope functions of the B_i/A_i sublattices with the nearest-neighbor relations for the n^v/n^c Landau-level states. Furthermore, the significant sublattices present the same zero-point number. These results are responsible for a new selection rule, which has never been found until now.

Both $\Delta n = 0$ and 1 magneto-optical selection rules come to exist together under the reduced non-equivalence of A_i and B_i sublattices, as clearly shown in Figs. 8.7(b) and 8.7(c). As to the Si-[A$_6$, B$_4$]-decorated graphene of 2:16 concentration, two categories of inter-Landau-level excitation channels frequently appear during the variation of absorption frequency. The absorption peaks of $\Delta n = 1$ decrease quickly, while the opposite is true for those of $\Delta n = 0$. The former and the latter, respectively, come from the neighboring A_i and B_i sublattices with the mode difference of ± 1 and 0. The lower-frequency absorption peaks are dominated by $\Delta n = 1$, since the corresponding Landau levels, being similar to those of monolayer graphene, are magnetically quantized from the low-lying tilted Dirac cone. However, with the increasing energy, the enlarged derivation of localization center and the enhanced distortion of spatial probability/the strengthened side mode [Fig. 8.5(c)] create and enhance the available channels of $\Delta n = 0$. Apparently, these lead to the strong competition between these two kinds of magneto-selection rules.

On the other hand, the coexistent selection rules present another kind of behavior for the Si-A$_i$-dressed graphene with reduced concentration [Fig. 8.7(c)]. The $\Delta n = 0$ channels dominate the lower-frequency absorption spectrum, since the significant sublattices possess the same oscillation modes [Fig. 8.5(d)]. Their peak intensities slowly grow with the increasing frequency, clearly indicating the significant competition/cooperation between two categories of inter-sublattice transitions. $B_i \rightarrow A_i$ and $A_i \rightarrow B_i$ (except for the farthest ones) appear under the $n^v \rightarrow n^c$ inter-Landau-level excitations, in which the second category is negligible for a sufficiently high concentration in Fig. 8.7(a). Especially, the quick enlargement of the $\Delta n = 1$ absorption peaks is due to the strengthened side modes in all the B_i sublattices, the enhanced oscillations in the A_i sublattices of the valence Landau levels, and the $n^c \pm 1$ modes of certain B_i sublattices for conduction ones. Finally, it is well known that graphene only presents the $\Delta n = 1$ absorption intensity with a uniform optical spectrum [Fig. 8.7(d)], as a result of the full equivalence of A and B sublattices. Among all the Si-dressed graphene systems, the pristine one has the highest absorption intensity

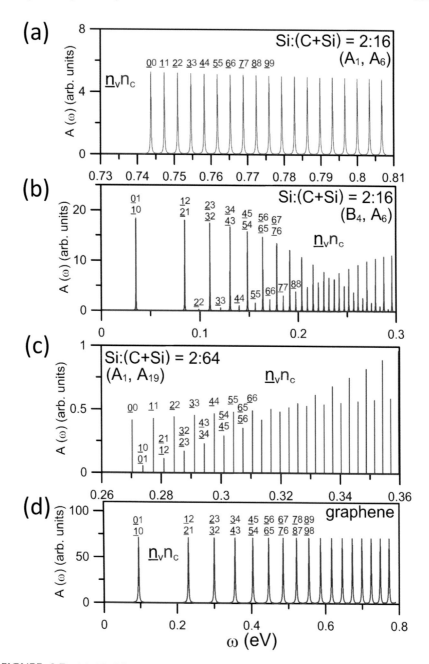

FIGURE 8.7 (a)–(d) Magneto-optical absorption spectra for four kinds of Si-dressed graphene systems in Fig. 7.4.

and energy spacing. These are, respectively, closely related to the strongest dipole matrix elements and the smallest density of states of the isotropic Dirac cone, as indicated from the well-behaved magnetic wave functions [Fig. 8.5(e)] and the symmetric V-shape van Hove singularity across the Fermi level [the dashed black curve in Fig. 8.3].

The diverse magneto-electronic properties and absorption spectra could be verified by STS and optical spectroscopies, respectively [details in Secs. 2.1 and 2.2]. The gate-voltage-modulated STS measurements are available in identifying the various features of magneto-electronic spectra (van Hove singularities in density of states), covering a lot of symmetric peaks due to the highly degenerate Landau-level states, band gap with/without the B_z-independent $n^{c,v} = 0$ Landau-level energies, the $n^{c,v} = 0$ Landau levels at the Fermi level dependent/independent on/of the magnetic-field strength, the symmetric or asymmetric electron-hole energy spectra, and the linear or square-root B_z-dependence. Furthermore, the energy-fixed STS experiments are very powerful in the direct examinations on four kinds of Landau-level states from the spatial probability distributions of the magnetic wave functions; that is, they would investigate the zero points, amplitudes, symmetric/anti-symmetric/asymmetric forms of the localized oscillation distributions. Moreover, the magneto-optical transmission/absorption/reflection spectroscopies [36–41] are very suitable for the full investigations on the significant inter-Landau-level vertical excitations, covering the threshold absorption frequencies equal or unequal to band gap in density of states, the available excitation channels being characterized by the specific magneto-optical selection rules of $\Delta n = 0$, ($\Delta n = 0$ & ± 1) and $\Delta n = 1$, and the B_z-enriched absorption frequencies. In short, the experimental examinations on four kinds of Landau levels and the distinct magneto-optical selection rules could provide the full information about the diversified essential properties, establish the emergent binary or ternary graphene compounds, and confirm the developed theoretical framework.

REFERENCES

1. Zhou W, Kapetanakis M D, Prange M P, Pantelides S T, Pennycook S J, and Idrobo J C 2012 direct determination of the chemical bonding of individual impurities in graphene *Phys. Rev. Lett.* **109 (20)** 206803.
2. Panchakarla L S, Subrahmanyam K S, Saha S K, Govindaraj A, Krishnamurthy H R, Waghmare U V, et al. 2009 Synthesis, structure, and properties of boron- and nitrogen-doped graphene *Adv. Mater.* **21 (46)** 4726.
3. Qu L, Liu Y, Baek J B, and Dai L 2010 Nitrogen-doped graphene as efficient metal-free electrocatalyst for oxygen reduction in fuel cells *ACS Nano* **4 (3)** 1321.
4. Zhang S J, Lin S S, Li X Q, Liu X Y, Wu H A, Xu W L, et al. 2015 Opening the band gap of graphene through silicon doping for the improved performance of graphene/GaAs heterojunction solar cells *Nanoscale* **8 (1)** 226.
5. Shahrokhi M and Leonard C 2017 Tuning the band gap and optical spectra of silicon-doped graphene: many-body effects and excitonic states *J. Alloys Compd.* **693** 1185.
6. John R and Merlin B 2017 Optical properties of graphene, silicene, germanene, and stanene from IR to Far UV—a first principles study *J. Phys. Chem. Solids* **110** 307.
7. Weinberg M, Staarmann C, Schler C, Simonet J, and Sengstock K 2016 Breaking inversion symmetry in a state-dependent honeycomb lattice: artificcial graphene with tunable band gap *2D Mater.* **3 (2)** 024005.

8. Do T N, Shih P H, Chang C P, Lin C Y, and Lin M F 2016 Rich magneto-absorption spectra of AAB-stacked trilayer graphene *Phys. Chem. Chem. Phys.* **18 (26)** 17597.
9. Shih P H, Chiu C W, Wu J Y, Do T N, and Lin M F 2018 Coulomb scattering rates of excited states in monolayer electron-doped germanene *Phys. Rev. B* **97 (19)** 195302.
10. Lee K J, Kim D, Jang B C, Kim D J, Park H, Jung D Y, et al. 2016 Multilayer graphene with a rippled structure as a spacer for improving plasmonic coupling *Adv. Funct. Mater.* **26 (28)** 5093.
11. Liu F, Song S, Xue D, and Zhang H 2012 Folded structured graphene paper for high performance electrode materials *Adv. Mater.* **24 (8)** 1089.
12. Lin C Y, Do T N, Huang Y K, and Lin M F 2017 *Optical Properties of Graphene in Magnetic and Electric Fields*. IOP Publishing IOP Concise Physics. San Raefel, CA, USA: Morgan & Claypool Publishers. 2053–2563, ISBN: 978-0-7503-1566-1.
13. Huang B L, Chuu C P, and Lin M F 2019 Asymmetry-enriched electronic and optical properties of bilayer graphene *Sci. Rep.* **9** 859.
14. Jhang S H, Craciun M F, Schmidmeier S, Tokumitsu S, Russo S, Yamamoto M, et al 2011 Stacking-order dependent transport properties of trilayer graphene *Phys. Rev. B* **84 (16)** 161408.
15. Pontes R B, Miwa R H, da Silva A J R, Fazzio A, and Padilha J E 2018 Layer-dependent band alignment of few layers of blue phosphorus and their van Der Waals heterostructures with graphene *Phys. Rev. B* **97 (23)** 235419.
16. Zhao S, Lu Y, and Yang X 2011 Layer-dependent nanoscale electrical properties of graphene studied by conductive scanning probe microscopy *Nanoscale Res. Lett.* **66 (1)** 498.
17. Mostofizadeh A, Li Y, Song B, and Huang Y 2018 Synthesis, properties, and applications of low-dimensional carbon-related nanomaterials *J. Nanomater.* **2011** 685081.
18. Meunier V, Souza Filho A G, Barros E B, and Dresselhaus M S 2016 Physical properties of low-dimensional sp^2-based carbon nanostructures *Rev. Mod. Phys.* **88 (2)** 025005.
19. Do T N, Shih P H, Gumbs G, Huang D, Chiu C W, and Lin M F 2018 Diverse magnetic quantization in bilayer silicene *Phys. Rev. B* **97 (12)** 125416.
20. Krasovskii E E 2011 Spin-orbit coupling at surfaces and 2D materials *Phys. Chem. Chem. Phys.* **13 (25)** 11929.
21. Wang S A 2011 Comparative first-principles study of orbital hybridization in two-dimensional C, Si, and Ge *Phys. Chem. Chem. Phys.* **13 (25)** 11929.
22. Novoselov K S, Geim A K, Morozov S V, Jiang D, Zhang Y, Dubonos S V, et al. 2004 Electric field effect in atomically thin carbon films *Science* **306 (5696)** 666.
23. Park S and Sim H S 2008 Magnetic edge states in graphene in nonuniform magnetic fields *Phys. Rev. B* **77 (7)** 075433.
24. Chen S C, Wu J Y, Lin C Y, and Lin M F 2017 *Theory of Magnetoelectric Properties of 2D Systems*. IOP Publishing IOP Concise Physics. San Raefel, CA, USA: Morgan & Claypool Publishers. 2053–2563, ISBN: 978-0-7503-1674-3.
25. Miller D L, Kubista K D, Rutter G M, Ruan M, and de Heer W A 2009 Observing the quantization of zero mass carriers in graphene *Science* **324 (5929)** 924.
26. Jiang Z, Henriksen E A, Tung L C, Wang Y J, Schwartz M E, Han M Y, et al. 2007 Infrared spectroscopy of Landau levels of graphene *Phys. Rev. Lett.* **98 (19)** 197403.
27. Novoselov K S, Geim A K, Morozov S V, Jiang D, Katsnelson M I, Grigorieva I V, et al. 2005 Two-dimensional gas of massless Dirac fermions in graphene *Nature* **438 (7065)** 197.
28. Rutter G M, Jung S, Klimov N N, Newell D B, Zhitenev N B, and Stroscio J A 2011 Microscopic polarization in bilayer graphene *Nat. Phys.* **7 (8)** 649.
29. Li G and Andrei E Y 2007 Observation of Landau levels of Dirac fermions in graphite *Nat. Phys.* **3 (9)** 623.

30. Orlita M, Faugeras C, Borysiuk J, Baranowski J M, Strupinski W, and Sprinkle M 2011 Magneto-optics of bilayer inclusions in multilayered epitaxial graphene on the carbon face of SiC *Phys. Rev. B* **83 (12)** 125302.
31. Plochocka P, Faugeras C, Orlita M, Sadowski M L, Martinez G, Potemski M, et al. 2008 High-energy limit of massless Dirac fermions in multilayer graphene using magneto-optical transmission spectroscopy *Phys. Rev. Lett.* **100 (8)** 087401.
32. Plochocka P, Solane P Y, Nicholas R J, Schneider J M, Piot B A, Maude D K, et al 2012 Origin of electron-hole asymmetry in graphite and graphene *Phys. Rev. B* **85 (24)** 245410.
33. Nicholas R J, Solane P Y, and Portugall O 2013 Ultrahigh magnetic field study of layer split bands in graphite *Phys. Rev. Lett.* **111 (9)** 096802.
34. Odom T W, Huang J L, Kim P, and Lieber C M 1998 Atomic structure and electronic properties of single-walled carbon nanotubes *Nature* **391** 62.
35. Borensztein Y, Prevot G, and Masson L 2014 Large differences in the optical properties of a single layer of Si on Ag (110) compared to silicene *Phys. Rev. B* **89** 245410.
36. Mak K F, Shan J, and Heinz T F 2010 Electronic structure of few-layer graphene: experimental demonstration of strong dependence on stacking sequence *Phys. Rev. Lett.* **104** 176404.
37. Li Z Q, Henriksen E A, Jiang Z, Hao Z, Martin M C, Kim P et al. 2009 Band structure asymmetry of bilayer graphene revealed by infrared spectroscopy *Phys. Rev. Lett.* **102** 037403.
38. Jiang Z, Henriksen E A, Tung L C, Wang Y J, Schwartz M E, Han M Y, et al. 2007 Infrared spectroscopy of Landau levels of graphene *Phys. Rev. Lett.* **98** 197403.
39. Plochocka P, Faugeras C, Orlita M, Sadowski M L, Martinez G, Potemski M, et al. 2008 High-energy limit of massless Dirac fermions in multilayer graphene using magneto-optical transmission spectroscopy *Phys. Rev. Lett.* **100** 087401.
40. Mak K F, Lui C H, Shan J, and Heinz T F 2009 Observation of an electric-field-induced band gap in bilayer graphene by infrared spectroscopy *Phys. Rev. Lett.* **102** 256405.
41. Mak K F, Sfeir M Y, Wu Y, Lui C H, Misewich J A, and Heinz T F 2008 Measurement of the optical conductivity of graphene *Phys. Rev. Lett.* **101** 196405.

9 Unusual Quantum Transport Properties

Thi-Nga Do,[c,d] *Chiun-Yan Lin,*[a] *Jhao-Ying Wu,*[b]
Po-Hsin Shih,[a] *Shih-Yang Lin,*[e]
Ching-Hong Ho,[b] *Ming-Fa Lin*[a,f,g]

[a] Department of Physics, National Cheng Kung University, Tainan 701, Taiwan
[b] Center of General Studies, National Kaohsiung University of Science and Technology, Kaohsiung 811, Taiwan
[c] Laboratory of Magnetism and Magnetic Materials, Advanced Institute of Materials Science, Ton Duc Thang University, Ho Chi Minh City, Vietnam
[d] Faculty of Applied Sciences, Ton Duc Thang University, Ho Chi Minh City, Vietnam
[e] Department of Physics, National Chung Cheng University, Chiayi 621, Taiwan
[f] Quantum Topology Center, National Cheng Kung University, Tainan 701, Taiwan
[g] Hierarchical Green-Energy Materials Research Center, National Cheng Kung University, Tainan, Taiwan

CONTENTS

The quantum Hall effects in sliding bilayer graphene and a AAB-stacked trilayer system are investigated using the Kubo formula and a generalized tight-binding model. The various stacking configurations can greatly diversify the magnetic quantization and thus create rich and unique transport properties. The quantum conductivities are very sensitive to the Fermi energy and magnetic-field strength. The diverse features cover the specific non-integer conductivities, the integer conductivities with distinct steps, the splitting-created reduction and complexity of quantum conductivity, a vanishing (AA and AB stackings) or non-zero conductivity (other stacking configurations) at the neutral point, and the well-like, staircase, composite, and abnormal plateau structures in the field dependencies. Such stacking-dependent characteristics mainly originate from the crossing, anti-crossing and splitting Landau-level energy spectra and three kinds of quantized modes.

Since the first discovery of few-layer graphene systems in 2004 by the mechanical exfoliation, its unconventional quantum Hall effects have stirred many theoretical and experimental researches [1–11]. Most importantly, the various stacking configurations, which dominate the low-energy essential physical properties, have been successfully synthesized and identified through the distinct physical [12–19] and chemical [20–28] methods. For example, the large-area graphene layers, with the high mobility and highly symmetric configurations, present the well-behaved AAA, ABA, and ABC stackings through the chemical vapor deposition [20–25]. Specifically, the unusual AAB stacking [15, 27, 29–32], an intermediate bilayer stacking under the interlayer shift/twist [33–36], is created by various methods. AAB stacking could be produced by the manipulation of STM tip [29], cleaving with tape [30], the liquid phase exfoliation of graphite [31], and the growth on substrates [27, 32]. Moreover, the sliding of graphene flakes on a graphene substrate could also be initiated by the STM tip to overcome the weak, but significant van der Waals interactions [33, 35]. In addition, the micrometer-size graphite flakes present a spontaneous sliding after being stirred by an STM tip [34, 36]. Similarly, the second layer of bilayer graphene will slide with respect to the first one by this method with the requirement of an initial activation out of the equilibrium state. Both four types of typical trilayer stackings and sliding bilayer ones should be very suitable in fully exploring the configuration-enriched quantum Hall effects.

Up to now, the generalized tight-binding model is well developed to thoroughly investigate the essential properties of layered graphene systems, especially for the magnetic quantization phenomena [details in Refs. (our books and review articles)]. For example, the relative shift of two graphene layers can create the continuous stacking configurations [37], such as those in the dramatic transformation between AA and AB stacking, thus leading to the diversified band structures and magneto-electronic properties. Under a uniform perpendicular magnetic field, the sliding bilayer systems possess two groups of valence and conduction Landau levels. Furthermore, they present three kinds of magneto-electronic states, namely, the well-defined, perturbed, and undefined Landau levels [37], respectively, with a specific oscillation mode, a main mode and several side ones, and certain dominating components. The rich and unusual phenomena are clearly revealed in the magnetic-field-dependent energy spectra, energy spectra, such as the initiated state energies of distinct Landau-level groups, the monotonous or non-monotonous B_z-dependences, and the frequent crossing/anti-crossing behaviors. These are attributed to the stacking-dominated interlayer atomic interactions of carbon $2p_z$ orbitals. Obviously, the trilayer AAA, ABA, ABC, and AAB stackings possess the unique and higher/lower stacking symmetries, in which three pairs of valence and conduction bands, respectively, cover the vertical Dirac-cone structures, the monolayer- and bilayer-like energy dispersions, the partially flat, sombrero-shaped, and linear bands; the oscillatory, sombrero-shaped and parabolic ones [38]. As a result, their magnetically quantized states exhibit the stacking-diversified phenomena in the Landau-level energies and sublattice-related subenvelope functions. Among these four systems, the interlayer hopping integrals are most complicated in AAB stacking, and so do the fundamental physical properties. Specifically, its state degeneracy for each Landau level is reduced to half in the absence of mirror symmetry about $z = 0$ plane. The above-mentioned results clearly show that the stacking-enriched magneto-electronic properties will greatly diversify the quantum Hall conductivities.

A lot of experimental [2, 3, 39–42] and theoretical [5, 43–47] researches are conducted on the quantum transport properties of graphene and graphene-related systems. Especially, the low-energy quantum Hall effect of few-layer graphenes have been successfully identified for monolayer, AB and ABC stacking systems. Magnetic transport measurements on monolayer graphene clearly verify the unconventional half-integer Hall conductivity $\sigma_{xy} = (m + 1/2)4e^2/h$ [2, 3], where m is an integer and the factor of 4 represents the spin and sublattice-dependent degeneracy. This unusual quantization is attributed to the quantum anomaly of the $n^{c,v} = 0$ Landau levels initiated from the Dirac point [5]. Concerning AB-stacked bilayer graphene, the Hall conductivity is confirmed to be $\sigma_{xy} = 4m'e^2/h$ (m' a non-zero integer) [39, 43]. Furthermore, there exists an unusual integer quantum Hall conductivity, a double step of $\sigma_{xy} = 8e^2/h$, at zero energy and low magnetic fields [44]. This result mainly originates from the $n^v = 0$ and $n^c = 1$ Landau levels of the first group across the Fermi level [40, 41]. The low-energy quantum-Hall plateaus of ABA-stacked trilayer graphene are clearly revealed as the sequence of $\pm 2e^2/h$, $\pm 4e^2/h$, $\pm 6e^2/h$, and $\pm 8e^2/h$ with a step height of $2e^2/h$, especially for the energy range of $\sim \pm 20$ meV [40]. This observation is consistent with the calculated Landau-level energy spectra [45]. The neighboring and next-neighboring interlayer hopping integrals, respectively, create the separated Dirac cone and parabolic bands, and the valley splitting of the latter. At very low energy, this further leads to six quantized Landau levels with spin degeneracy and thus the quantum-Hall step of $2e^2/h$. However, the higher-energy quantum Hall effect in ABA stacking could be regarded as the superposition of those in monolayer and AB bilayer [40, 41, 45]. The ABC-stacked trilayer graphene, on the other hand, presents the important differences in the main features of quantum Hall effect, compared to the ABA trilayer system. The quantum Hall conductivity is quantized in the unusual form of $\sigma_{xy} = 4(\pm|m'| \pm 1/2)e^2/h$ in the absence of the $\sigma_{xy} = \pm 2e^2/h$, $\pm 4e^2/h$, and $\pm 8e^2/h$ plateaus [41, 42]. Specifically, a $\sigma_{xy} = 12e^2/h$ step comes to exit near zero energy, being associated with the $n = 0$, 1 and 2 Landau levels of the first group due to the surface-localized flat bands [48]. In fact, the four-fold spin and valley degeneracy remains unchanged for the ABC trilayer due to the inversion symmetry, resulting in the plateau height of $4e^2/h$. It should be noticed that the effective-mass model can deal with the low-energy Hall effect in the well-stacked graphene systems [45]. However, this method might become cumbersome and ineffective in quantum transports closely related to the intragroup/intergroup anti-crossing Landau levels, e.g., those of the second groups of trilayer ABC stacking [48]. Up to now, there are no experimental examinations on the theoretical predictions of unusual quantum Hall effects in trilayer AA and AAB stackings and sliding bilayer systems. The above-mentioned interesting phenomena of electronic transport properties in graphene opens the door to exploring the configuration-enriched quantum Hall effects in other layered graphenes, such as the twisted bilayer graphenes [33–36], the chemical modified systems [49], and the defect-enriched ones [50–52].

The magnet-electronic specific heats could present another magnetic quantization phenomenon. The Landau levels in monolayer graphene are very suitable for a model investigation. It is well known that the hexagonal crystal structure accounts for the unique low-energy electronic properties, the isotropic and linear valence and conduction bands intersecting at the Dirac point and the square-dependent Landau-level energy spectrum of $E^{c,v} \propto \sqrt{n^{c,v} B_z}$. The Zeeman effect further splits the magneto-electronic states into the spin-up and spin-down ones. In general, the level spacings between two neighboring quantum numbers are much wider than the Zeeman splitting energy. This clearly indicates that the former is insensitive to temperature. Furthermore, the latter might be comparable to the thermal energy at low temperature. The Zeeman effect is expected to play an important role in the thermal properties. On the other hand, monolayer graphene could be rolled up to become a hollow cylinder (a single-walled carbon nanotube). Apparently, the periodic boundary condition is responsible for the creation of one-dimensional energy bands with linear and parabolic energy dispersions. However, it is almost impossible to induce the dispersionless Landau levels in nanotube surfaces under various high magnetic fields [53]. Dimensionality and magnetic quantization have a great influence on the van Have singularities of the density of states (DOS), so the magneto-electronic specific heats would be very different between these two systems.

From the sublattice-based theoretical framework, the linear static Kubo formula, which is directly linked to the generalized tight-binding model [details in Sec. 3.3], is developed to thoroughly explore the unique quantum Hall effect in few-layer graphenes with the high- and low-symmetry stacking configurations. Such an association would be very useful in identifying the magneto-electronic selection rules under the static scattering events; that is, the available transition channels in magneto-transport property could be examined in detail by the delicate calculations/analyses. The dependencies of quantum conductivity on the Fermi energy/the free carrier density and magnetic-field strength are fully investigated. The current work will show that the feature-rich Landau levels can create the extraordinary magneto-transport properties. The sliding bilayer systems, with three kinds of Landau levels, present the unusual quantum Hall effect, covering the integer and non-integer conductivities, the zero and finite conductivities at the neutral point, the well-like, staircase, and composite quantum structures, and the different step heights. Furthermore, the reduced conductivity, the complex plateaus and the abnormal structures are revealed in the AAB-stacked trilayer graphene, with the split and perturbed Landau levels. These results are deduced to be dominated by the crossing and anti-crossing energy spectra, and the spatial oscillation modes. There are significant differences among the trilayer AAB-, ABC-, ABA-, and AAA- stackings in the E_F- and B_z-dependent quantum Hall conductivities, e.g., the diversified quantum structures, plateau heights, and step sequences. The theoretical predictions on the diverse quantum conductivities could be experimentally verified, as done for well-stacked graphene systems [2–3, 39–42]. Another magnetic quantization phenomenon, the magneto-electronic specific heat, is also illustrated by the unusual Landau levels in monolayer graphene. Its behavior is expected to be greatly enriched by the strong dependences on the Zeeman splitting, temperature, magnetic-field strength, and doping density of free carriers.

9.1 SLIDING BILAYER GRAPHENES

The geometric symmetries, band structures, Landau-level energy spectra and wave functions, and quantum Hall conductivities are thoroughly studied for the sliding bilayer graphenes through the generalized tight-binding model and the static Kubo formula. The theoretical framework is very reliable in establishing the close and complex relations between the magneto-electronic and transport properties, since it could propose the concise physical pictures (the specific selection rules) from the delicate analysis in the available scattering events of the initial and final Landau states. Moreover, such bilayer systems, with the various stacking configurations, are very suitable in the typical studies of diverse quantum transport phenomena, since their dramatic transformations create three kinds of Landau levels with the distinct oscillations modes and the frequent anti-crossing & crossing behaviors. Hall conductivities are expected to be very sensitive to the changes in the sliding shift, the Fermi energy and the magnetic-field strength, e.g., their values at zero energy, the heights of various plateaus, the sequence of quantum structures, and the normal or unusual E_z- and B_z-dependences.

The low-energy Hamiltonians of the sliding bilayer graphenes mainly originate from the $2p_z$-orbital tight-binding functions in a primitive unit cell. When the upper layer, relative to the lower one [Fig. 9.1], present a shift along the specific armchair direction, there are four carbon atoms in the sublattices of (A^1, B^1, A^2, B^2). The stacking configurations can be transformed according to AA $(\delta = 0) \to$ AB $(\delta = b) \to$ AA' $(\delta = 1.5b)$. Obviously, the zero-field Hamiltonian matrix is a Hermitian 4×4 one, in which the dimension remains the same under the specific sliding of armchair direction. That is, the numerical calculations are similar for the various stacking configurations. The interlayer distance and the C–C bond length are, respectively, $d_0 = 3.37$ Å and $b = 1.42$ Å. The various Hamiltonian matrices are dominated by the intralayer and the interlayer atomic hopping integrals, in which the typical interaction parameters are chosen from the reliable empirical formula, as clearly illustrated in Eq. (5.2).

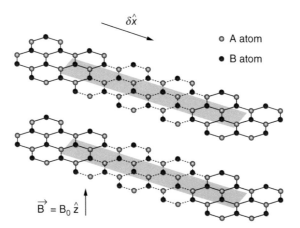

FIGURE 9.1 The geometric structure of the sliding bilayer graphene systems, with a relative shift between two layer along the armchair direction, under a uniform perpendicular magnetic field.

The similar numerical calculations could be done on the sliding bilayer graphenes along the zigzag direction [details in Ref. 54].

When layered graphenes exist in a uniform perpendicular magnetic field, the quantized magneto-electronic states are fully explored using the generalized tight-binding model even for low-symmetry stacking configurations. This model is based on the subenvelope functions of the distinct sublattices, in which the magnetic Hamiltonian is built from the bases of tight-binding functions coupled with a Peierls phase factor [details in Sec. 3.1]. A zero-field hexagonal unit cell, with $2N$ carbon atoms, is changed into an enlarged rectangular cell including $4NR_B$ carbon atoms [Fig. 9.1 for a sliding bilayer graphene], since the vector potential ($\mathbf{A} = [0, B_0x, 0]$) can induce the periodical Peierls phases, being characterized as the path integration of $\mathbf{A}$ between two lattice sites [the detailed calculations on the vector-dependent hopping integrals in Refs.]. R_B is the ratio of the flux quantum ($\phi_0 = hc/e$) versus the magnetic flux through a hexagon ($\phi = 3\sqrt{3}b_0^2B_0/2$), e.g., $R_B = 8 \times 10^3$ under $B_z = 40$ T.

The essential electronic properties of sliding bilayer graphenes are very sensitive to the changes in stacking configurations, as clearly shown in Figs. 9.2(a)–9.2(f). The well-behaved bilayer AA stacking of $\delta = 0$ exhibits has two pairs of linear conduction and valence bands under the strong overlap [Fig. 9.2(a)]. The upper and lower Dirac-cone structures are linearly intersecting at the same wave vectors for the Dirac points; that is, they present the vertical forms at two valleys of K and K'. Furthermore, they are, respectively, initiated from 0.32 eV and –0.36 eV, expected to behave so for the Landau-level states [discussed later in Fig. 9.4(a)]. On the other hand, the bilayer AB stacking, with $\delta = b$ in Fig. 9.4, presents two pairs of parabolic energy dispersions with a weak band overlap near the Fermi energy [the inset in Fig. 9.4(d)]. Apparently, the first and second pairs occur at ~ 0 eV and (0.32 eV & –0.36 eV), mainly owing to the vertical interlayer hopping integral [37]. The dramatic transformation between the Dirac cones and the parabolic bands, which is induced by the drastic changes of interlayer hopping integrals during the geometric sliding, leads to the drastic and significant hybridizations of the two neighboring valence and conduction bands (conduction/valence and conduction/valence bands) [37]. Most importantly, the intermediate stackings of $\delta = b/8$ [Fig. 9.2(b)], $\delta = 6b/8$ [Fig. 9.2(c)], and $\delta = 11b/8$ [Fig. 9.2(e)] clearly show the special band structures, where electronic states in the lower cone of the first pair and those of the upper cone of the second pair are seriously hybridized each other. An eye-shape stateless region comes to exist near the Fermi level along $\hat{k}_x$ or $\hat{k}_y$. The band structures are drastically changed even in the high-symmetry AA' stacking [$\delta = 1.5b$ in Fig. 9.2(f)]. Specifically, this system possesses two pairs of titled cone structures with the non-vertical Dirac points, as thoroughly examined from the projected constant-energy loops in Ref. 37. Such unusual energy bands will induce the variable localization centers for each Landau-level group and thus diversify the magneto-optical selection rules. Apparently, the relative sliding of graphene layers is a very efficient way in delicately modulating the essential physical properties.

The various electronic structures are magnetically quantized into the diverse Landau levels, which are characterized by the rich and unique features. Two groups of valence and conduction Landau levels, the first and the second ones [$n_1^{c,v}$ and $n_2^{c,v}$; blue and red curves in Figs. 9.3(a)–9.3(f)], are initially quantized from the zero-field

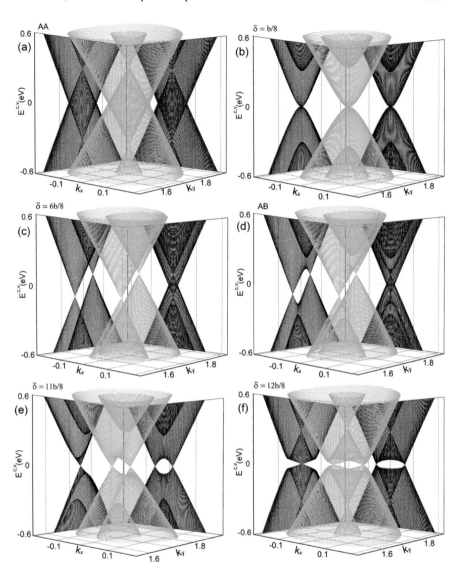

FIGURE 9.2 Band structures for the relative shifts of (a) $\delta = 0$, (b) $\delta = b/8$, (c) $\delta = 6b/8$, (d) $\delta = b$, (e) $\delta = 11b/12$, and (f) $\delta = 1.5b$. Also shown in those of insets are the valence and conduction bands near the Fermi level/the K valley.

band-edge states. In general, the Landau-level degeneracy are four-fold/eight-fold in the absence/presence of spin degree of freedom for each (k_x, k_y) state [details in Ref. 37]; furthermore; the magnetic subenvelope functions are localized about for 1/6, 2/6, 4/6, and 5/6 except for those near the $\delta = 1.5b$ stacking configurations. For the magnetic-field-dependent Landau-level energy spectra exhibit the crossing behaviors for all the well-behaved magneto-electronic states with the specific oscillation modes. On the other hand, the anti-crossing phenomenon will appear when two

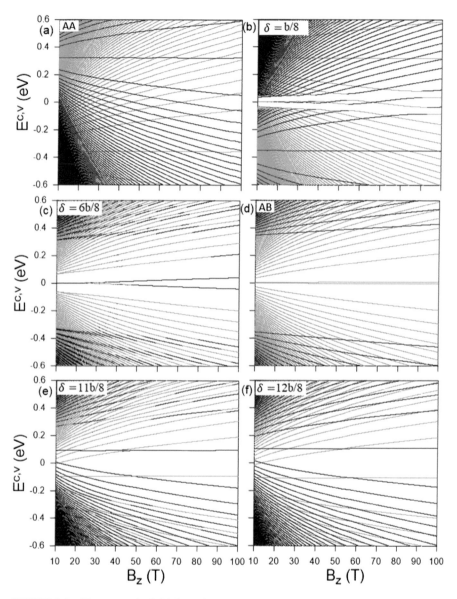

FIGURE 9.3 The magnetic-field-dependent energy spectra for the Landau levels under the various shifts: (a) $\delta = 0$, (b) $\delta = b/8$, (c) $\delta = 6b/8$, (d) $\delta = b$, (e) $\delta = 11b/12$, and (f) $\delta = 1.5b$.

Landau levels possess certain identical components. The anti-crossing energy spectra, arising from the perturbed and undefined Landau levels, are clearly revealed in the lower-symmetry stacking systems, such as the $\delta = b/8$, $\delta = 6b/8$, and $\delta = 11b/8$ ones [Figs. 9.3(b), 9.3(c), and 9.3(e)]. Especially, for the latter two stacking configurations, there exist a lot of undefined Landau levels at the higher-energy range, clearly indicating the very frequent intergroup anti-crossings. It should be noticed that, whether two or one groups of Landau level make contributions to Hall conductivity strongly depend on the range of energy overlap and their spatial probability distributions. The main features of the magnetic subenvelope functions on the four sublattices of (A^1, B^1, A^2, B^2) could be found in Ref. 37, and they are expected to play a critical role in quantum transport properties.

Diverse transport phenomena can be derived from equation (3.5) using the static Kubo formula. At zero temperature, monolayer graphene exhibits the unusual E_F-dependent plateau structures in σ_{xy}, with a uniform height of $4e^2/h$ and nonuniform widths. as clearly observed in Figs. 3.5(a) and 3.5(b). Here, the quantum Hall conductivities of a AA bilayer graphene could be regarded as the superposition of those from two directly overlapping Dirac-cone structures [Figs. 9.4(a) and 9.4(b)]. That is to say, only electronic scatterings between two intragroup Landau levels survive and the available ones are consistent with the specific selection rules of $\Delta n = \pm 1$. The step structures present a unit height of $4e^2/h$, in which the plateaus are situated at 0, $\pm 4e^2/h, \pm 8e^2/h$, and so on. A wide plateau, which covers zero Fermi energy, presents the insulating behavior (the vanishing conductivity) [Fig. 9.4(a) under $B_z = 40$ T]. This result is purely due to the absence of Landau-level states within the specific energy range; that is, there is a finite magneto-electronic gap identical to the width of the first plateau structures. Apparently, it sharply contrasts with the conducting behavior of monolayer graphene except for the neutral point E [Fig. 3.5(a)]. In monolayer system, the available interband transitions of the $\Delta n = n^c - n^v = 1$ and -1 selection rules, respectively, possess the positive and negative Hall conductivities. Moreover, the valence and conduction Landau levels are symmetric about $E_F = 0$, leading to the vanishing conductivity at zero energy. As for AA bilayer graphene, the total Hall conductivity mainly comes from the electronic transitions in two separated Landau-level groups which are almost symmetric about $E_F = 0$. At the neutrality point, the available transitions of the first group (blue lines) are dominated by the valence Landau levels, while those of the second group (red lines) are governed by the conduction ones. The former and the latter, respectively, satisfy the selection rules of $\Delta n_1^{v \to v} = -1$ and $\Delta n_2^{c \to c} = 1$; therefore, they cancel each other out. In general, a plateau of $4e^2/h$ height frequently appear when a certain Landau level becomes occupied or unoccupied during the variation of Fermi energy [Figs. 9.4(a) and 9.4(b)]. The higher-/deeper-energy plateaus are dominated by the conduction Landau levels/the valence one of the second/first group [Fig. 9.4(b)]. A fixed-height structure might be changed into a double step for the two merged Landau levels in distinct groups (the intergroup Landau-level crossings), such as $8e^2/h$ at $E_F \sim 0.026$ eV. It is worth noting that the quantized Landau levels at two Dirac points in an AA bilayer system [$n_1 = 0$ and $n_2 = 0$] are fully occupied/unoccupied, being different from the half-occupied $n = 0$ Landau level in a monolayer system. Apparently, the quantum Hall conductivity does not show the steps of $2e^2/h$ height near $E_F = 0$ as in monolayer graphene [55].

The transverse Hall conductivity in AB bilayer graphene is quantized as $\sigma_{xy} = 4me^2/h$ or $\sigma_{xy} = 4m'e^2/h$ during the variation of E_F [Figs. 9.5(a) and 9.5(b)], depending on the strength of magnetic field. The lower-E_F quantum Hall conductivity exhibits the usual step structures, in which the plateaus become higher, but their widths decline during the increase of Fermi energy. Especially, when there appears

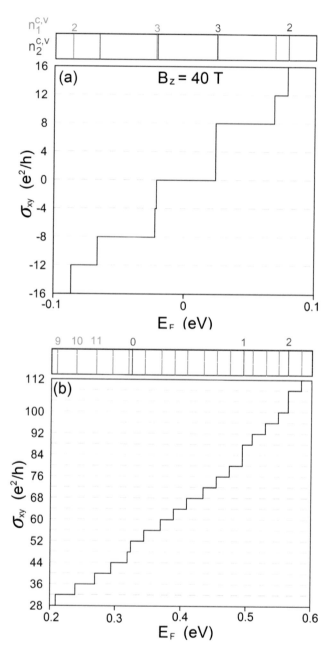

FIGURE 9.4 The quantum Hall conductivity of bilayer AA stacking at $B_z = 40$ T for the (a) lower and (b) higher Fermi energies.

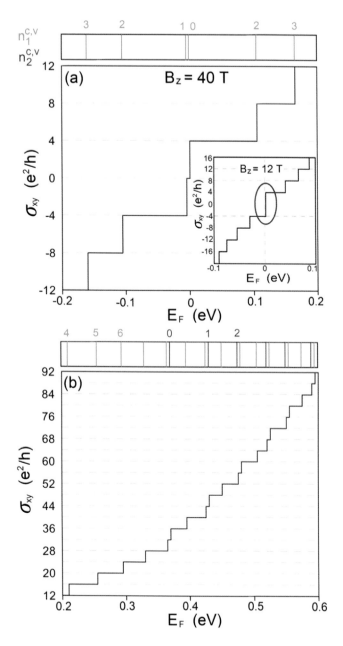

FIGURE 9.5 Similar plot as Fig. 9.4, but shown for bilayer AB stacking.

to be a very narrow plateau near zero energy [Fig. 9.5(a)]. Such plateau is observable only under the sufficiently large field strength. Its width corresponds to a small, but significant energy spacing between the $n_1 = 0$ and $n_1 = 1$ Landau levels of the first group. This energy spacing is monotonically dependent on the field strength, as shown in Fig. 9.3(d) [55]. It will be infinitesimal and even disappears below the critical field strength [$B_z < 20$ T, e.g., 12 T in the inset of Fig. 9.7(a)], creating a double step in the Hall conductivity. These results are in agreement with the previous theoretical predictions from the effective-mass approximation [39–41] and experimental measurements [43, 44]. At the higher/deeper Fermi energy [$E_F > 0.35$ eV/$E_F < -0.35$ eV in Fig. 9.5(b)], the quantum conductivity is dominated by the first and second groups of Landau levels (blue and red lines). There are four categories of transition channels during the static scattering events, covering the intragroup [$n_1^c \rightarrow n_1^c, n_2^c \rightarrow n_2^c$]/[$n_1^v \rightarrow n_1^v, n_2^v \rightarrow n_2^v$] and intergroup ones [$n_1^c \rightarrow n_2^c$, $n_2^c \rightarrow n_1^c n_1^c$]/[$n_1^v \rightarrow n_2^v, n_2^v \rightarrow n_2^1$] [55]. The number of steps is significantly enhanced [Fig. 9.5(b)]; furthermore, the change in plateau widths no longer follows a simple relation with the Fermi energy. Specifically, the available intergroup transitions, with the same quantized mode on the A^l (B^l) sublattice [Eq. (3.6)], only happen in two Landau levels with a large energy spacing. They only make negligible contributions, since the quantum Hall conductivity is inversely proportional to the square of energy difference between the initial and final states [Eq. (3.5)]. This is in great contrast with the frequency-dependent magneto-optical absorption spectra [38]. As a result, the quantum Hall conductivity of AB bilayer graphene could be considered as the superposition of those arising from two intragroup transition channels.

The magnetic-field-dependent Hall conductivity of AA bilayer graphene exhibits the wells, the monolayer-like or non-monotonic staircases, and the composite ones [Figs. 9.6(a)–9.6(d)], being sensitive to the B_z-induced Landau-level energy spectrum [Fig. 9.3(a)]. In particular, there are four types of step structures corresponding to the different Landau-level spectral regions. Within the low-energy range ($|E_F| < 0.1$ eV), the n_1^v valence Landau levels [blue curves in the inset of Fig. 9.6(a)], and the n_2^c conduction ones [red curves] cross one another and thus result in the rhombus-pattern area, directly corresponding to the well-like quantum conductivity [Fig. 9.6(a)]. The type-I electrical conductivity only presents two values of 0 and $4e^2/h$ in the well form during the variation of magnetic field. Since the intragroup transitions of the n_1^v Landau levels and those of the n_2^c ones have the opposite conductivities (the positive and negative contributions, respectively), the Hall conductivity is a constant value at the specific B_z-ranges in the absence of any crossing between the Landau levels and the Fermi level. At $E_F = 0$, the number of the unoccupied n_1^v Landau levels is identical to that of the occupied n_2^c ones; therefore, the Hall conductivity is vanishing. When E_F is slightly increased [$E_F = 0.01$ eV in Fig. 9.6(a)], the variation of B_z can create the asymmetric Landau-level distribution. This is responsible for the step heights of 0 and $4e^2/h$. For example, within the range of 22.5 T $\leq B_z \leq$ 24 T, the $n_1^v = 6$ Landau level becomes the occupied state from the initial magnetic field, while the $n_2^c = 6$ is getting into the unoccupied one at the final B_z [inset in Fig. 9.6(a)]. The occupation-number changes of two distinct Landau-level groups are also revealed in other B_z-ranges. Roughly speaking, the former and the latter, respectively, lead to the creation and destruction of a step with $4e^2/h$-height; that is, they make opposite contributions to quantum conductivity and thus create a sequence of well-like structures. With the increasing Fermi energy, the well-like and staircase structures (type-II) could coexist within the range of 0.1 eV $\leq E_F < 0.2$ eV. The n_1^v and n_2^c Landau levels have the highly non-equivalent distributions. For example, at $E_F = 0.1$ eV [Fig. 9.6(b)], the unoccupied $n_2^c = 11$, occupied $n_1^v = 3$ and unoccupied $n_2^c = 10$

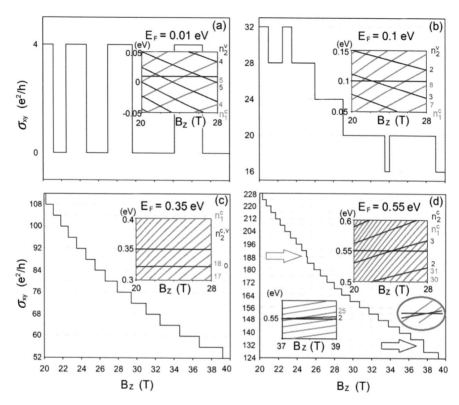

FIGURE 9.6 The magnetic-field-dependent Hall conductivities for bilayer AA stacking under (a) $E_F = 0.01$ eV, (b) $E_F = 0.1$ eV, (c) $E_F = 0.35$ eV, and (d) $E_F = 0.55$ eV. The insets show the field- and energy-covered Landau-level spectra.

Landau levels respectively, come to exist at $B_z=21$, 22.5 & 23.5 T's, so that the well-like structure occurs there [inset in Fig. 9.6(b)]. However, only the n_2^c Landau levels become unoccupied in the range of 23.5 T $\leq B_0 \leq$ 33.5 T, indicating the further decrease of quantum conductivity and thus the formation of staircase, as observed in monolayer graphene [Fig. 3.5(b)]. Specifically, as for the Fermi energy just above the initial Landau level of the first group, the $n_1^{c,v}=0$ one, [e.g., $E_F = 0.35$ eV in Fig. 9.6(c)], only the n_2^c Landau level displays a series of dramatic occupation dramatic transformation in the increase of field strength [inset in Fig. 9.6(c)]. Consequently, the B_z-dependent quantum conductivity behaves similarly to that of monolayer graphene [Fig. 3.5(b); Ref. 3]. The plateaus are reduced by a step of $4e^2/h$ and their widths gradually grow (type-III) whenever an extra n_2^c Landau level becomes unoccupied. The enhancement of the plateau width directly reflects the significant B_z-dependence between two neighboring unoccupied n_2^c Landau levels. At the sufficiently high energy ($E_F > 0.4$ eV), the n_1^c and n_2^c conduction Landau levels might overlap together. These two groups of Landau levels possess the similar dependencies on the magnetic field; therefore, the number of the step structures is greatly enhanced, e.g., σ_{xy} at $E_F = 0.55$ eV in 9.6(d). Particularly, with the growth of B_z within the range of

20 T → 40 T, the Hall conductivity shows more step structures, with few nonuniform heights (type-IV), as the n_1^c Landau levels appear [green and red arrows in Fig.
9.6(d)]. A very narrow plateau width near 25 T (green arrow) corresponds to the field-
strength difference in the unoccupied $n_1^c = 3$ and $n_2^c = 38$ Landau levels [green circle;
the right-hand-side inset in Fig. 9.6(d)]. Especially, a plateau of $132e^2/h$ is absent at
the crossing point of the unoccupied $n_1^c = 2$ and $n_2^c = 25$ Landau levels (the left-hand-
side inset), leading to a step of height $8e^2/h$ (red arrow). Apparently, the diverse Hall
plateau structures are closely related to the magneto-electronic energy spectra.

A bilayer AB stacking only presents the type-III and type-IV behaviors in the
magnetic-field dependence of Hall conductivity, as clearly displayed in Figs. 9.7(a)
and 9.7(b) under $E_F = 0.2$ and 0.5 eV, respectively. At the lower Fermi energy,
the step structures [Fig. 9.7(a)] are dominated by the electronic transitions between

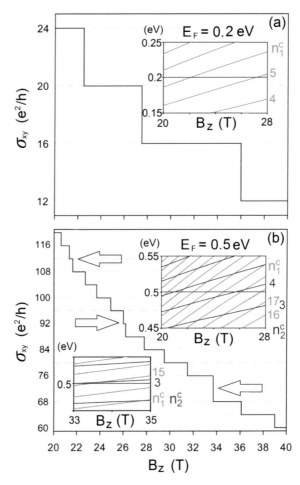

FIGURE 9.7 The B_z-induced electrical conductivities in bilayer AB stacking at (a) $E_F = 0.2$
eV and (b) $E_F = 0.5$ eV. Also displayed in the insets are the field- and energy-covered Landau-
level spectra.

the low-lying Landau levels of the first group (blue curves in the inset) [55]. More n_1^c states become unoccupied in the increment of B_z, leading to the monolayer-like quantum Hall effect [type-III]. However, for the higher energy range ($E_F > 0.35$ eV), the significant overlap of the two conduction Landau-level groups greatly enriches the Hall conductivity spectrum, e.g., $E_F = 0.5$ eV in Fig. 9.6(b). Although the step structures originate from the combination of all the available electronic transitions among two groups of Landau levels, only the intragroup ones are dominant (as discussed earlier). Consequently, a simple relation between quantum conductivity and field strength is absent. The small energy spacings between the n_1^c and n_2^c Landau levels create a number of narrow plateaus as indicated by the blue arrows. Furthermore, a double step of $8e^2/h$ appears at around 33.7 T (red arrow), mainly owing to the critical crossing of the $n_1^c = 15$ and $n_2^c = 3$ Landau levels [inset of Fig. 9.7(b)]. As could be predicted from the crossing behavior of the B_z-dependent Landau-level energy spectrum [Fig. 9.3(d)], the weaker the magnetic-field strength is, the more frequently the double-height steps will occur.

Obviously, the quantum Hall conductivities are largely diversified by the shift-dependent stacking configurations, e.g., the even and odd Hall conductivities, and the significant intragroup & intergroup Landau-level scatterings with the diverse selective rules. The lower-symmetry bilayer graphene of $\delta = b_0/8$ shows the unusual step structure, as clearly indicated in Fig. 9.8(a) under $B_z = 40$ T. The low-lying plateaus across zero energy, being illustrated by the colored arrows, are quantized as a unique sequence of $(-9, -4, -3, 4, 7)$ (in unit of e^2/h); therefore, they present the exclusive step heights of $5e^2/h$, e^2/h, $7e^2/h$, and $3e^2/h$. Such odd-integer steps are closely related to the Landau levels of $n_2^c = 2, 3, 4,$ and 5 (red lines) and $n_1^v = 2, 3, 4,$ and 5 (blue lines). Such magneto-electronic states are quantized from the hybridized electronic states near the eye-shape region of the band structure [Fig. 9.2(b)]. There exist extra intergroup scattering vents which satisfy the selection rules of $\Delta n = 0$ & ± 2, e.g., $n_2^c = 3 \rightarrow n_1^v = 3$, $n_2^c = 4 \rightarrow n_1^v = 2$, and others. Especially, only the transitions due to the $\Delta n = 2$ rule will contribute a positive quantum conductivity. As a result, the low-energy quantum Hall conductivity is dominated by the intragroup and intergroup Landau-level transition channels. When E_F crosses zero from a negative value, the $n_2^c = 3$ becomes occupied. The electronic transition from this Landau level to the $n_1^v = 3$ is available, and it generates the quantum conductivity of $-3e^2/h$. The zero-energy Hall conductivity originates from two intragroup transition channels of $[n_2^c = 3 \rightarrow n_2^c = 4 \ \& \ n_1^v = 4 \rightarrow n_1^v = 3]$ and the $n_2^c = 3 \rightarrow n_1^v = 3$ intergroup transition. The former two cancel each other out, being responsible for the $-3e^2/h$ conductivity (red arrow). This system is totally different from AA and AB bilayer graphenes with zero conductivities at the neutrality point. With the Fermi energy above the $n_1^v = 3$ Landau level, the intragroup transitions of $[n_1^v = 3 \rightarrow n_1^v = 2$ and $n_2^c = 3 \rightarrow n_2^c = 4]$ make the main contribution, and they create the quantum conductivity of $4e^2/h$ (blue arrow). These two special plateaus ($-3e^2/h$ & $4e^2/h$) result in a step height of $7e^2/h$. Also, the competitive or cooperative relations between the intragroup and intergroup transitions (two intragroup ones) is revealed at the lower and higher Fermi energies. The dominating transitions of $[n_1^v = 4 \rightarrow n_1^v = 4, n_2^c = 2 \rightarrow n_1^v = 4, n_2^c = 2 \rightarrow n_2^c = 3]$, $[n_1^v = 4 \rightarrow n_1^v = 3, n_2^c = 2 \rightarrow n_2^c = 3, n_1^v = 4 \rightarrow n_2^c = 4]$, and $[n_1^v = 3 \rightarrow n_1^v = 2, n_2^c = 4 \rightarrow n_2^c = 5, n_2^c = 4 \rightarrow n_1^v = 2]$, respectively,

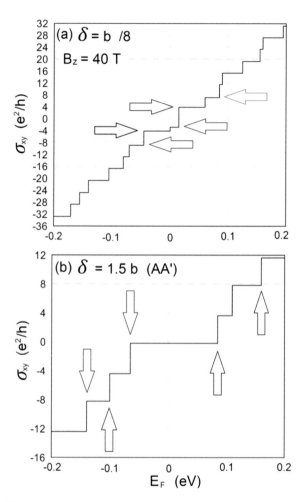

FIGURE 9.8 The dependence of quantum Hall conductivities on the Fermi energy for the sliding bilayer graphenes with the relative shifts of (a) $\delta = b/8$ and (b) $\delta = 12b/8$ at $B_z = 40$ T.

induce the $-9e^2/h$, $-4e^2/h$, and $7e^2/h$ quantum conductivities at $E_F = -0.046$, 0, and 0.06 eV (purple, black, and green arrows).

Specifically, the high-symmetry AA′ stacking presents the unusual non-integer plateaus [Fig. 9.8(b)], showing great contrast to the AA system [Fig. 9.4]. Two groups of well-behaved Landau levels, which are quantized from the tilted Dirac cones [Fig. 9.2(f)], only contribute the intragroup Landau-level scatterings of $\Delta n = \pm 1$ to Hall conductivities [55]. When an extra $n_1^{c,v}$ Landau level becomes occupied in the increase of E_F, the quantum conductivity [blue arrows in Fig. 9.8(b)] is enhanced by a height of $3.8e^2/h$. On the other hand, the occupation of one $n_2^{c,v}$ Landau levels can create a $4.2e^2/h$-height step [red arrows]. The non-integer Hall conductivities might be associated with the tilted Dirac cones, in which the localization centers of Landau levels gradually deviate from the normal ones during the increase of state energy [55]. Furthermore, the energy-dependent distribution centers could induce a

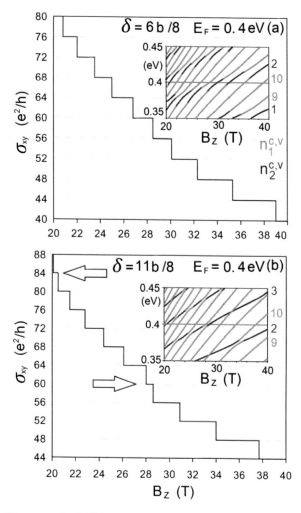

FIGURE 9.9 The magnetic-field-dependent Hall conductivities of the (a) $\delta = 6b/8$ and (b) $\delta = 11b/8$ sliding bilayer graphenes at $E_F = 0.4$ eV. The inset shows the corresponding Landau-level energy spectra.

small conductivity of $0.2e^2/h$ near the neutrality point, but a vanishing one appears in the AA stacking.

As for the undefined Landau levels in $\delta = 6b_0/8$ and $11b_0/8$ bilayer stackings [Figs. 9.9(a) and 9.9(b)], their Hall conductivities are worthy of a detailed examination. The former presents the $n_1^{c,v}$ conduction and valence Landau levels, with the perturbed modes, in the low-lying spectral region under the absence of the $n_2^{c,v}$ Landau levels [Fig. 9.2(c)]. At the higher-/deeper-energy range (e.g., $E_F = -0.4$ eV), where the $n_2^{c,v}$ Landau levels come to exist, two groups of magneto-electronic states continuously anti-cross together, as shown in the inset of Fig. 9.9(a). The localization modes of each Landau level cannot be well defined, since they vary with the field strength

[Fig. 9.2(c)]. Such highly degenerate magneto-electronic states, without a specific main mode, are classified as the undefined Landau levels [37]. It is noted that $n_2^{c,v}$ in Fig. 9.9(a) [also in Fig. 9.9(b)] only denotes the ordering of Landau levels, but not the real oscillation modes. The available transition channels cover the intragroup and intergroup ones in the absence of selection rule, however, they are dominated by the neighboring Landau levels near the Fermi energy. The quantum Hall conductivity monotonically declines with the increase of B_z, similar to that of AB bilayer graphene [Figs. 9.7(a) and 9.7(b)]. Since the undefined Landau levels do not possess any crossing behaviors, all the plateaus are quantized as single steps of height $4e^2/h$ without the double structures. A uniform height appears even for the irregular spatial distributions of the undefined Landau levels. When one n_1^c/n_2^c Landau level is getting unoccupied, the related transition channels can reduce an integer quantum conductivity of $4e^2/h$. Moreover, an inverse relation between plateau width and B_z disappears because of the frequently irregular anti-crossings. The similar quantum conductivity is revealed by the undefined Landau levels of the $\delta = 11b_0/8$ stacking. It might have some narrow plateaus [black arrows in Fig. 9.9(b)], e.g., the plateau at $B_z \sim 28$ T arising from the $n_1^c = 13$ and $n_2^c = 3$ Landau levels. Such plateaus directly reflect the very weak anti-crossings.

On the experimental side, the high-resolution transport measurements, as discussed earlier in Sec. 2.3, are available in verifying the theoretical predictions on the quantum Hall conductivities of the sliding bilayer graphene systems. Also, these delicate examinations are capable of identifying the diverse magnetic quantization phenomena due to the various stacking configurations and thus clarifying the specific one-to-one correspondence between the quantized Landau levels and the zero-field energy bands. The experimental focuses on the stacking-enriched quantum transport properties cover the existence/absence of the intragroup and intergroup inter-Landau-level scatterings, the well-behaved, extra & vanishing selection rules, the conducting or insulating behaviors at the neutrality point, the integer or non-integer Hall conductivities at zero temperature, the splitting-reduced quantum conductivities, the well-like, staircase & composite plateau structures, and the crossing-induced complicated B_z-dependence of their widths. In short, the experimental verifications are very useful in directly proving the close relations among the stacking symmetries, the complex hopping integrals, the magneto-electronic properties, and the quantum Hall effects.

9.2   AAA-, ABA-, ABC-, AND AAB-STACKED TRILAYER GRAPHENES

The well-stacked trilayer graphenes, the AAA, ABC, and ABC stackings [Figs. 9.10(a)–9.10(c)], possess six carbon atoms in a primitive unit cell. The various intralayer and interlayer hopping integrals, which are available in the tight-binding model, have been successfully established from a detailed comparison with the first-principles calculations on the low-lying valence and conduction bands on the high-symmetry points. The significant parameters for the trilayer AAA stackings, as clearly displayed in Fig. 9.10(a), cover $\alpha_0 = 2.569$ eV, $\alpha_1 = 0.361$ eV, $\alpha_2 = 0.013$ eV, and $\alpha_2 = 2.569$ eV, in which the first one and the others, respectively, belong to the intralayer and interlayer atomic interactions. α_1 and α_3 are the vertical ones due to the neighboring and next-neighboring planes, respectively. However, the non-vertical α_2 arise from the A and B atoms of the neighboring layers. Obviously,

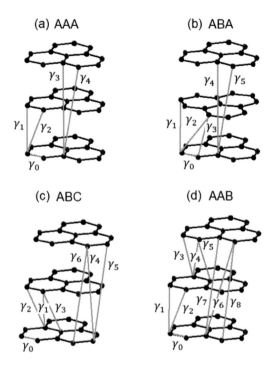

FIGURE 9.10   Trilayer stacking configurations for (a) AAA, (b) ABA, (c) ABC, and (d) AAB with the distinct interlayer hopping integrals due to the carbon-$2p_z$ orbitals.

the low-energy band structure of trilayer graphene, as shown in Fig. 9.11(a), is almost identical to the linear superposition of those from monolayer and bilayer systems. That is, there exist three pairs of vertical Dirac-cone structures, being initiated from the K and K′ valleys. The low-energy energy dispersions are linear and isotropic; furthermore, the Dirac points appear at 0 and $\pm\alpha_1$. Among all the trilayer stacking configurations, this system presents the largest free carrier density purely due to the interlayer hopping integrals, clearly indicating the highest geometric symmetry. The predicted energy dispersions near the Fermi level could be verified from the high-resolution angle-resolved photoemission spectroscopy (ARPES) measurements [56].

Apparently, there are more complicated hopping integrals in trilayer ABA stacking, as clearly shown in Fig. 9.10(b). By the delicate derivations from the SWMcC model [57,58], all the significant atomic interactions include one intralayer hopping integral (γ_0), three neighboring-layer ones (γ_1, γ_3, γ_4), two next-neighboring-layer ones (γ_2, γ_5), and the chemical environment difference (γ_6). Their magnitudes are as follows: $\gamma_0 = 3.12$ eV, $\gamma_1 = 0.38$ eV, $\gamma_2 = 0.021$ eV, $\gamma_3 = 0.28$ eV, $\gamma_4 = 0.12$ eV, $\gamma = 0.003$ eV, and $\gamma_6 = 0.0366$ eV [59]. This system presents a weak overlap between the valence and conduction bands [Fig. 9.11(b)], illustrating a semimetallic behavior with a very low free carrier density. Furthermore, two parabolic bands and one separated & distorted Dirac-cone structure, with the anisotropic energy dispersions, come to exist. That is, near the Fermi level, such electronic energy spectrum has been identified from the ARPES measurements [56]. In short, the low-lying band structure

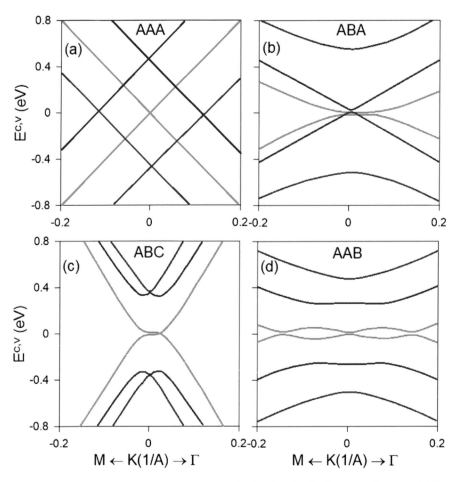

FIGURE 9.11   The low-lying valence and conduction bands of trilayer graphene (a) AAA, (b) ABA, (c) ABC, and (d) AAB stackings.

of ABA trilayer graphene could be regarded as the superposition of monolayer- and bilayer-like ones [59]. The electronic states, which are initiated from the K and K', will be quantized into the low-lying Landau levels.

As for the trilayer ABC stacking, each of the graphene sheets shifts a relative distance of C–C bond length with respect to the lower (upper) neighboring layer along the armchair direction of $\hat{x}$. The sublattice A of one layer is just located above the center of a hexagon in the adjacent lower layer, while the sublattice B lies above the lower-layer sublattice A. The intralayer and the interlayer hopping integrals, being the significant parameters in the tight-binding model, are illustrated as β_i's, in which $i = 0, 1, 2, 3, 4, \& 5$. β_0 accounts for the nearest-neighbor intralayer hopping integral, β_1, β_3, and β_4 correspond to the adjacent two layers, and β_2 and β_5 denote the next-neighboring interlayers. β_1 and β_2 arise from two vertical carbon atoms, and β_3, β_4, and β_5 couple the non-vertical sites. Moreover, their values are chosen as follows: $\beta_0 = 3.16eV$, $\beta_1 = 0.36$ eV, $\beta_2 = 0.01$ eV, $\beta_3 = 0.32$ eV,

$\beta_4 = 0.03$ eV, and $\beta_5 = 0.0065$ eV [48]. Apparently, the low-energy band structure of trilayer ABC stacking is characterized by three pairs of conduction and valence subbands in Fig. 9.11(c), with the partially flat, sombrero-shaped and linear energy dispersions. Among them, only the first one just crosses the Fermi level; therefore, a high DOS, being due to the surface localized states [48], appears there. The unusual or non-monotonous electronic energy spectrum clearly illustrates a giant difficulty of the magnetic quantization using the effective-mass approximations.

The trilayer AAB stacking is very suitable for studying the unique phenomena, compared with the well-stacked/high-symmetry AAA, ABA, and ABC. Figure 9.10(d) clearly shows that the A atoms in three layers possess the same (x,y) coordinates, while the B atoms on the third layer are projected into the hexagonal centers of the other two layers. To be consistent with the first-principles calculations [60], there exist 10 kinds of atom-atom interactions in creating the unusual energy dispersions, being responsible for the oscillatory, sombrero-shaped and parabolic energy bands [Fig. 9.11(d); Refs]. $\gamma_0 = -2.569$ eV denotes the nearest-neighbor intralayer atomic interaction; $\gamma_1 = -0.263$ eV, $\gamma_2 = 0.32$ eV, $\gamma_3 = -0.413$ eV, $\gamma_4 = -0.177$ eV, $\gamma_5 = -0.319$ eV, $\gamma_6 = -0.013$ eV, $\gamma_7 = -0.0177$ eV, and $\gamma_8 = -0.0319$ eV stand for the various interlayer atomic interactions among three graphene layers, as clearly indicated in Fig. 9.10(d). $\gamma_9 = -0.012$ eV accounts for the difference in the chemical environment of A and B atoms. Such critical interactions, which are built in the 6×6 Hamiltonian matrix, create the unusual energy dispersions with a very small indirect band gap of $\delta_i = 8$ meV [Fig. 9.11(d)]. The predicted valence-state energy spectra require the high-resolution verifications from the ARPES measurements. Apparently, they can greatly complicate the magnetic Hamiltonian and thus diversify the magneto-electronic properties. It should be noticed that the magnetic Hamiltonians possess the identical dimension for the AAB, ABC, ABA, and AAA stackings [the independent matrix elements in Ref. 38]. Generally speaking, the magneto-electronic states are initiated from the K and K' valleys.

The magneto-electronic properties of the trilayer graphene systems are further explored by solving the giant magnetic Hamiltonians with the real matrix elements for the specific $(k_x = 0, k_y = 0)$ state. The Landau-level localization centers are identical to those in bilayer systems [details in Sec. 9.1; Ref. 59]. However, the state degeneracy is very sensitive to the geometric symmetries (discussed later). Apparently, there exist important differences among four kinds of stacking configurations in the main features of Landau levels, as observed in the zero-field band structures [Fig. 9.11]. Concerning the AAA and ABA stackings [Figs. 9.12(a) and 9.12(b)], their magneto-electric properties are the approximate superposition of monolayer- and bilayer-like ones [59]. Three groups of valence and conduction Landau levels come to exist in the frequent crossings, and only a few anti-crossing phenomena appear in the latter. Specifically, the first, second, and third groups of the AAA (ABA) systems are initiated from $E^c \sim \alpha_1, \sim 0$, and $E^c \sim -\alpha_1$ ($E^{c,v} \sim 0, \sim 0$, and $E^{c,v} \sim \pm\alpha_1$) [59]. As to each (k_x, k_y) state in the reduced first Brillouin zone, the Landau levels possess the eight-fold degeneracy except for initial one and two Landau levels in the AAA and ABA stackings, respectively [Refs.]. Specifically, the $n_1^{c,v} = 0$, $n_2^{c,v} = 0$, and $n_2^{c,v} = 1$ Landau across the neutrality point are four-fold degenerate magneto-electronic states. mainly owing to the weak breaking of mirror symmetry about $z = 0$

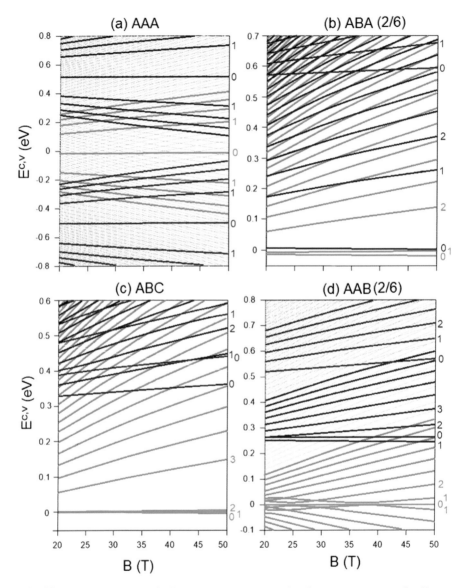

FIGURE 9.12 The magnetic-field-dependent Landau-level energy spectra of trilayer graphene systems under the (a) AAA, (b) ABA, (c) ABC, and (d) AAB stacking. The green, blue, and red curves, respectively, correspond to the first, second, and third groups of valence and conduction Landau levels.

due to the $(\gamma_2, \gamma_5, \gamma_6)$ interlayer hopping integrals and thus the state splitting. Generally speaking, the magnetic wave functions exhibit the well-behaved spatial distributions being similar to those of a harmonic oscillator [59].

Obviously, the ABC and AAB stackings possess the non-monotonic energy spectra in the magnetic-field-dependences, with the coexistent non-crossing, crossings, and anti-crossing behaviors [Figs. 9.12(c) and 9.12(d)]. Furthermore, the sublattice-based magnetic subenvelope functions exhibit the well-behaved or perturbed oscillation modes. A simple relation between the Landau-level energy and magnetic-field strength is absent; that is, it is rather difficult to get the analytic results for the low-lying magneto-electronic spectra from the effective-mass approximation except for ignoring certain significant interlayer hopping integrals. The first group of the ABC stacking, which is quantized from the partially flat energy dispersion [Fig. 9.11(c)], has three Landau levels very close to the Fermi within a rather narrow energy window. Such localized surface states will determine the Hall conductivity across zero energy [Fig. 9.14(a)]. Its second group presents the unusual intragroup anti-crossings, mainly owing to the main and sides in each Landau level and the sombrero-shaped energy dispersions. On the other side, the anti-crossing energy spectra frequently appear in the first group of the AAB stacking [Fig. 9.12(d)] as a result of the oscillatory energy bands [Fig. 9.11(d)]. Such group is expected to play the critical roles for the unusual Hall conductivities in the B_z-dependence at low energy. All the Landau levels in three groups have the four-fold degeneracy in the presence of spin degree of freedom. Their localizations about 1/6 and 2/6 [4/6 and 5/6] belong to the valley-split Landau-level states because of the breaking of mirror symmetry on the $z = 0$ plane. This AAB system might present the lower Hall conductivities, compared with the AAA, ABA, and ABC stackings.

The trilayer AAA stacking, with three groups of valence and conduction Landau levels, exhibits the linearly direct superposition of monolayer and bilayer graphene systems in the main features of quantum transport properties, as clearly indicated in the Fermi- and magnetic-field-dependent Hall conductivities [Figs. 9.13(a) and 9.13(b), respectively]. Only three intragroup channels are available during the scattering events, while six intergroup ones vanish there. Furthermore, the specific selection rule in each intragroup scattering is $\Delta n \pm 1$. Apparently, these results arise from the significant characteristics of the magneto-electronic states, covering the oscillator-like [well behaved] localization modes and the equivalent sublattices on the same & distinct layers. As a result, the usual height of plateau structures in the Fermi-energy dependence is $4e^2/h$. However, it becomes double only under the almost emergence of two Landau level from the different groups. Apparently, the plateau width is very sensitive to the variation of E_F because of the total contribution directly due to three groups of Landau levels. Moreover, four types of quantum structures are also revealed in the B_z-dependence of quantum conductivities, as observed in the bilayer AA stacking [Figs. 9.6(a) and 9.6(d)]. For a very small E_F (0.01 eV in Fig. 9.13(b), the Hall conductivity presents a series of well-like structures with the same height of $4e^2/h$ and the nonuniform widths, further illustrating the linear superposition from of conductive contributions arising from three groups of Landau levels.

The trilayer ABA stacking exhibits the unique plateau structures in quantum transport properties, as clearly shown in the Fermi-energy- and magnetic-field-dependent Hall conductivities [Figs. 9.14(a) and 9.14(b), respectively]. This system has three intragroup and two intergroup inter-Landau-level scattering channels [38], in which the latter come from the second and third groups [the blue and red curves in Fig. 9.12(b)]. Generally speaking, the available static transitions, excepted for very few anti-crossings, are characterized by the specific selection

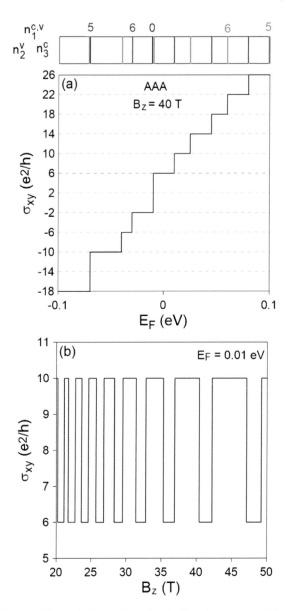

FIGURE 9.13   For the trilayer AAA stacking, (a) the Fermi-energy- and (b) magnetic-field-dependent Hall conductivities. The green, blue, and red lines, respectively, represent the first, second, and third groups of valence and conduction Landau levels.

of $\Delta n = \pm, 1$, as revealed in the AAA stacking [Figs. 9.13(a) and 9.13(b)]. At very low Fermi energies [$\sim E_F$, < 25 meV in Fig. 9.14(a)], there exists quantized Landau levels, with a four-fold degeneracy for each (k_x, k_y) state, and thus create the quantum-Hall step of $2e \cdot \& , 2/h$. Such magneto-electronic states consist of two $n_1^{c,v} = 0$, $n_2^{c,v} = 0$ and $n_2^{c,v} = 0$, being attributed to the significant ($\gamma_2 \gamma_5 \gamma_6$) hopping

The magneto-electronic properties of the ABC stacking quite differ from those in other ones [Figs. 9.12(a)–12(d)], and so do the quantum Hall conductivities. Their three groups of valence and conduction Landau levels can create nine categories of effective inter-Landau-level scatterings, covering three intragroup and six intergroup channels. In general, the specific $\Delta n = \pm 1$ and extra $[0, \pm 2]$ selection rules are, respectively, responsible for the non-anti-crossing and anti-crossing phenomena [38], e.g., the latter frequently appearing in the intragroup anti-crossings of the second group ($|E^{c,v}| \sim 0.3 - 0.4$ eV). However, the low-energy transport properties [Figs. 9.15(a) and 9.15(b)] only present the rich, but well-behaved plateau structures, since they are dominated by the intragroup inter-Landau-level scatterings due to the first group. Most importantly, three $n^{c,v} = 0$, 1, and 2 Landau levels, which are quantized from the surface localized states [38], cross zero energy and their energy spacings might be very narrow (or could be regarded as the merged states). Apparently, such magneto-electronic states create high conductivity at the neutrality point and one double height of $8e^2/h$ under $B_z = 40$ T, in which the plateau width declines in the increment of Fermi energy, as shown in Fig. 9.15(a). Furthermore, a normal staircase structure is revealed in the magnetic-field-dependent Hall conductivity at $E_F = 0.15$ eV [Fig. 9.15(b)]. In addition, the quantum conductivities are expected to become rather complicated for the higher Fermi energy ($|E_F| \sim 0.3 - 0.4$ eV), being attributed to the intragroup anti-crossing phenomena of the second group [Fig. 9.12(c)].

Apparently, the feature-rich Landau levels of the trilayer AAB stacking will induce the unusual quantum Hall effect. The Fermi-energy-dependent quantum Hall conductivity presents the step height of $2e^2/h$ instead of the usual one of $4e^2/h$ and the irregular widths of plateau structures [Fig. 9.16(a)], mainly owing to the destruction of the $z = 0$-plane mirror symmetry and the very complicated interlayer hopping integrals [Fig. 9.10(d)]. At low Fermi energies, there also exist certain double-height steps of $4e^2/h$, being closely related to the crossings of entangled Landau levels around the neutrality point or/and the linear superposition due to two localization centers under the identical Landau-level energies, as clearly displayed in Fig. 9.12(d). As a result of the splitting energy spectra, the Landau levels, which are localized at 1/6 and 2/6 localization centers, are very different from each other in the quantum conductivities. When the Fermi level approaches to zero energy, a plateau structure comes to exist under $B_z = 40$ T for the 1/6 center [a blue curve in Fig. 9.16(b)]; furthermore, two neighboring steps originate from the 2/6 center (a red curve). The absence/presence of the step riser depends on whether the anti-crossing Landau levels could survive there. For example, at the 1/6 localization center, the anti-crossing $n_1^{c,v} = 0$ and $n_1^c = 3$ Landau levels of the first group makes the former deviate from $E_F = 0$ [inset in Fig. 9.17(a)], leading to the absence of step riser at zero energy [marked by the green circle in Fig. 9.16(b)]. On the other hand, as to the 2/6 localization center, the $n_1^v = 1$ and $n_1^{c,v} = 0$ Landau levels continuously become occupied at $E_F = -3$ meV and 0, respectively. Therefore, the dominant transition channels of $n_1^v = 1 \rightarrow n_1^{c,v} = 0$ and $n_1^{c,v} = 0 \rightarrow n_1^c = 1$ create a step riser at $E_F = 0$ from a narrow-width step [a green circle in Fig. 9.16(b)]. Concerning the neutrality point, the competition between these two localization centers result in a zero conductivity [Fig. 9.16(a)]. In short, the splitting and anti-crossing energy spectra have generated the complex quantum structures in the half-reduced conductivities mixing with some

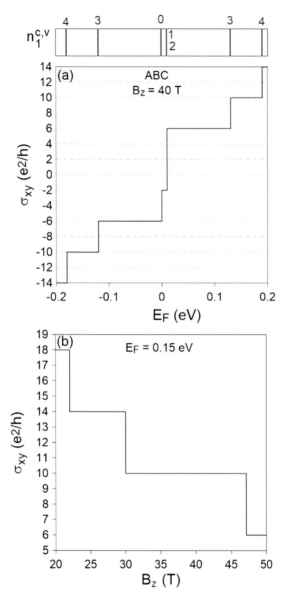

FIGURE 9.15 Similar plot as Fig. 9.14, but displayed quantum transport properties of the trilayer ABC stacking.

double heights [Figs. 9.16(a) and 9.16(b)]. To explore the intergroup-induced quantum transport properties, the Fermi energy need to be enhanced as $|E_F| > 0.3$ eV.

The magnetic-field-created Hall conductivity, being associated with the low-lying Landau levels, present the extraordinary quantum structures [Fig. 9.17(c)]. The extremely abnormal plateaus are absent in the sliding bilayer graphene systems [Figs. 9.6, 9.7, and 9.9]. Such structures mainly originate from the intragroup

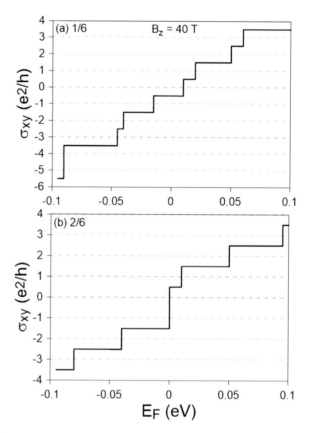

FIGURE 9.16 The Fermi-energy-dependent Hall conductivities of the trilayer AAB stacking for the (a) 1/6 and (b) 2/6 localization centers under $B_z = 40$ T.

anti-crossing Landau levels [related to the oscillatory energy bands in Fig. 9.12(d)]. As for the 1/6 localization center, the positive/negative quantum conductivity appears and then disappears within a certain range of B_z, as shown in Fig. 9.17(a). When the magnetic field reaches the critical value of $B_z = 24.75$ T, the $n_1^c = 4$ Landau level becomes occupied and the dominating $n_1^v = 0 \rightarrow n_1^c = 4$ transition is thus forbidden [inset in Fig. 9.17(a)]. The original scattering channel makes a contribution of $-2e^2/h$, so an upward step comes to exist there. And then, a higher plateau is changed into the original one at another critical field of $B_z = 25.5$ T, where the $n_1^c = 4$ Landau level is recovered to an unoccupied state. With a further increase of field strength, the opposite quantum structure occurs in the range of 37 T $\leq B_0 \leq$ 38.5 T. The reduced and enhanced quantum conductivities, respectively, correspond to the disappearance and recovery of the $n_1^v = 0 \rightarrow n_1^c = 3$ channel. It should be noticed that the extra selection rules, $\Delta = \pm, \pm 4, \Delta 0$, and ± 2, the available transition channels, mainly owing to the coexistent/comparable main and side modes in the perturbed Landau levels. On the other hand, the Landau levels localized at 2/6 center only present the well-like quantum structures, with the irregular plateau widths [Fig. 9.17(b)], i.e., the same enhancement and reduce of quantum conductivity appear continuously. Each structure

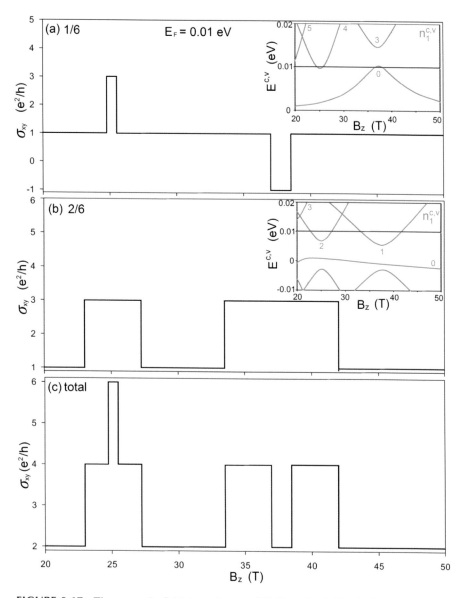

FIGURE 9.17 The magnetic-field dependences of Hall conductivities in the trilayer AAB stacking for the (a) 1/6 localization center, (b) 2/6 one, and (c) direct superposition at the Fermi energy of $E_F = 0.01$ eV. The insets in (a) and (b) show the B_z-dependent Landau-level energy spectra with the crossings and anti-crossings.

is dominated by the same Landau levels within two neighboring critical fields (the $n_1^v = 0 \to n_1^c = 2$ or $n_1^v = 0 \to n_1^c = 1$ transition in the inset). At higher/deeper Fermi energies, the Hall conductivities are expected to present more complicated plateau structures under the intragroup and intergroup scattering cases with the various selection rules.

There exist certain significant differences in transverse Hall conductivities among the AAA-, ABA-, ABC-, and AAB-stacked trilayer graphenes. Such systems exhibit the various inter-Landau-level scatterings, respectively, covering [3, 0], [3, 2], [3, 6], and [3, 6] (intragroup channels, intergroup ones). Furthermore, the scattering mechanisms are characterized by the following selection rules: $\Delta n = \pm 1$, $\Delta n = \pm 1$, ($\Delta n = \pm 1, 0$ & ± 2), and ($\Delta n = \pm 1, 0, \pm 2, \pm 3$ & ± 4). The extra selection rules in ABC stacking are also revealed in ABA systems with few anti-crossings. In general, the plateau height is $4e^2/h$, while it becomes double (reduces by a half) during the Landau-level crossings (under the breaking of mirror symmetry/inversion symmetry). This result is very sensitive to stacking configurations. The step height of $2e^2/h$ in a wide energy range, the localization-dependent plateau structures, and the irregular B_0-dependence are absent in the former three systems, since they have the higher-symmetry stacking configurations. The quantum conductivities of AAA and ABA stackings could be regarded as the linear superposition of those from the monolayer- and bilayer-like ones except for the latter under the low Fermi energy. Only the specific ABA system shows the insulating property within a very narrow range including zero Fermi energy and six plateau structures with a height of $2e^2/h$. Apparently, the well-like structures, staircases, or irregular quantum steps, which are observed in the magnetic-field dependences, strongly depend on the stacking-enriched magneto-electronic states. In addition, the unique transport phenomena of ABC and AAB stackings never appear in sliding bilayer systems. The above-mentioned important differences clearly illustrate the stacking-diversified quantum Hall effect; that is, the stacking-related hopping integrals dominate the magnetic quantization and thus the magneto-transport properties.

9.3 HEAT CAPACITY OF MONOLAYER GRAPHENE

The electronic-specific heat of monolayer graphene in response to a magnetic field is investigated within the Peierls tight-binding model. Generally speaking, the low-temperature thermal properties are dominated by the two lowest Landau levels, being induced by the Zeeman effect. Both temperature and magnetic field compete with each other in the specific heat, which reveals a composite form of $1/T^2$ and exponential function. In addition, a prominent peak at the critical temperature T_c (critical magnetic field B_{zc}) exists in the temperature (field) evolution, where T_c (B_{zc}) and B_z (T) are in a simple linear relation. In slightly doped cases, shoulder structures are further introduced in the specific heat, since the significant contributions originate from the second lowest unoccupied state and second highest occupied state. All those unusual features basically reflect the main characteristics of the magneto-electronic states near the chemical potential. There are certain important differences between graphene and carbon nanotubes in terms of the magneto-specific heat. Apparently, the rich thermal properties represent another kind of magnetic quantization.

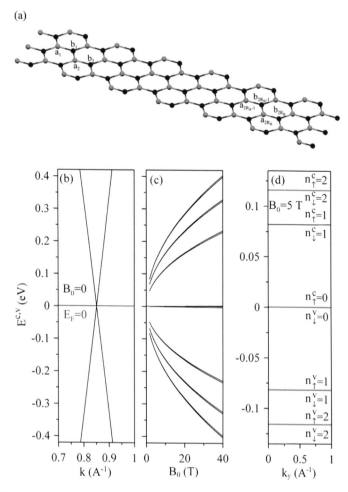

FIGURE 9.18 (a) The geometric structure of monolayer graphene in a uniform perpendicular magnetic field, (b) the low-lying zero-field Dirac cone, (c) the magneto-electronic energy spectrum, and (d) the Landau-level energies at $B_z = 5$ T. Inset in (d) shows a specific energy spacing between the spin-up and spin-down $n^{c,v} = 0$ landau levels.

The tight-binding model, which is built from the π bonding of carbon-$2p_z$ orbitals, is suitable in fully understanding the temperature-dependent electronic-specific heat. It is well known that monolayer graphene exhibits a linearly isotropic Dirac-cone structure initiated from the K & K′ valleys, as shown in Fig. 9.18(a). The linear energy dispersions correspond to $E^{c,v} = \pm v_F k$, where the Fermi velocity is $v_F = 3\gamma_0 b/2$. The valence band is almost symmetric to the conduction band about the zero chemical potential; therefore, μ hardly depends. Such a band structure creates a zero-gap semiconductor, with a vanishing DOS at zero energy. By the T-related Fermi-Dirac distribution function, the free carrier density [$\rho(T)$] of conduction electrons or valence holes, which is purely created by the thermal excitations, are characterized by the analytic formula: $\rho(T) = \pi k_B^2 T^2 / 6v_F^2$ (k_B the Boltzmann constant). Furthermore, such T-excited carriers have thermal energies of $\sim k_B T$ leading to the T^2-dependence of the electronic specific heat [Fig. 9.19(a); an analytic result in Ref. [61]]. Apparently, the significant electronic and thermal properties come from the honeycomb symmetry with the well-behaved π bonding. On the other hand, a normal metal, with $E_F \sim 1$ eV, exhibits the linearly proportional behavior, since only free carriers, being within a narrow energy range of $\sim k_B T$ near the chemical potential, make significant contributions to the thermal properties. This obvious difference between a zero-gap semiconducting graphene and metals further illustrates the critical roles of geometric symmetries and orbital hybridizations on the essential physical properties.

Obviously, the significant differences are presented for the 2D monolayer graphene and 3D graphites [62]. In general, the latter include the AAA, ABA, and ABC stacking configurations [63]. The first and third systems, respectively, possess the highest and the lowest free electron and hole density. Such free carriers purely arise from the important interlayer hopping integrals being closely related to the stacking symmetries. A simple hexagonal graphite exhibits the greatest specific heat and presents a linear temperature-dependence due to the nearly constant DOS [details in Ref. [64]]. A Bernal graphite deviates from the linear T-dependence as the temperature is above 100 K. Furthermore, a rhombohedral graphite shows the linear dependence only in the low temperature range. The low-lying band structures, which are dominated by the intralayer and interlayer atomic interactions, are responsible for the DOS across the Fermi level/the conduction carrier density and thus the thermal properties.

The highly degenerate Landau levels, being created by the magnetic quantization of the linear Dirac-cone structure, lead to the non-monotonic temperature dependence in the thermal properties. The magnetic-field-dependent energies are proportional to $\sqrt{B_z}$, as clearly displayed in Fig. 9.18(b). The $n^{c,v} = 0$ Landau levels, which are quantized from the Dirac points, just cross the undoped Fermi level. The higher unoccupied conduction Landau levels (the deeper occupied valence ones) are characterized by $n^c = 1$ [$n^v = 1$], 2, 3,... and so on. In general, there exists a sufficiently large energy spacing between $n^{c,v} = 0$ and $n^{c,v} = 1$, compared with the thermal energy at room temperature. For example, at $B_z = 5$ T [Fig. 9.18(c)] its magnitude is ~ 80 meV comparable to the specific temperature of 1,000 K. On the other hand, the Zeeman effect on the initial Landau levels is very effective in the low-temperature specific heat, since their splitting energy spacing is characterized by $E_z(\sigma B_z) \sim 0.29$ meV under $B_z = 5$ T. Apparently, the spin-up and spin-down $n^{c,v} = 0$ Landau levels play the critical roles on the low-temperature thermal properties under the undoped cases.

The low-T specific heat clearly shows the rich temperature dependence under the different magnetic fields, as clearly indicated in Fig. 9.19(b). In general, the T-dependent $C(T, B_z)$ could be classified into three temperature-regions based on the thermal excitations. The free conduction electrons and valence holes, being, respectively, created in the $n^c_\uparrow = 0$ and $n^v_\downarrow = 0$ Landau levels, are responsible for the unusual phenomena. (I) At extremely low temperatures, e.g., $T \ll 3.5$ K under $B_z = 5$ T, the electronic specific heat approaches to zero, mainly owing to the almost vanishing free carrier densities. (II) When the thermal energy starts from about 10% of the Zeeman splitting energy, $C(T, B_z)$ quickly grows and then reaches a maximum specific heat at the critical temperature (T_c indicated by an arrow). For example, under $B_z = 5$ T (the

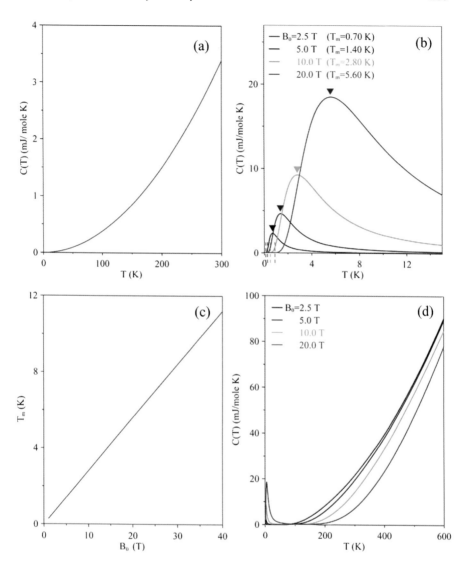

FIGURE 9.19 The temperature-dependent electronic specific heat under (a) a vanishing external field and the various magnetic strengths at the (b) lower and (d) higher temperature ranges; furthermore, (c) the critical temperatures during the variation of B_z. Also shown in (b) are the dashed lines indicating the threshold temperatures.

blue curve), the threshold and critical temperatures are $\sim 0.3 - 0.4$ K and 1.6 K, respectively. The specific heat, which is a complicated composite function of the $1/T^2$ and exponential forms, exhibits a prominent peak for any magnetic field. In addition, the thermally excited free carriers of the conduction and valence Landau levels equally contribute to the thermal properties as a result of the electron-hole symmetry. (III) With the further increase of temperature, the specific heat starts to rapidly decline and then is gradually vanishing. The rate of the temperature-enhanced free carrier density is lower than the linear T-form and thus the decreasing behavior in $C(T, B_z)$.

The threshold and critical temperatures are worthy of a closer examination. Apparently, the former is linearly proportional to the magnetic-field strength because of the characteristic Zeeman energy; that is, the stronger the magnetic field is, the higher the threshold temperature is [the dashed colored lines in Fig. 9.19(b)]. The latter could be evaluated from the zero derivative specific heat versus temperature, and an analytic formula is obtained through the detailed calculations [61]. A simple linear relation is clearly revealed in the magnetic-field-dependent critical temperatures [Fig. 9.19(c)]. Furthermore, the maximum specific heat quickly grows in the increment of B_z. This mainly arises from more thermal-excited free carriers accumulated at individual Landau levels in response to the larger field strength.

Also, the electronic specific heat shows another special feature for temperature above 100 K, as clearly indicated in Fig. 9.19(d) (the blue line). The thermal energy of $k_B T$ and the $n^{c,v} = 1$ Landau-level have the same order. Above this temperature, the half thermally excited electrons (holes) occupy the $n^c_\uparrow = 0$ [$n^v_\downarrow$] Landau levels. Such free carrier densities are close to being stable; therefore, these two magneto-electronic states have no impact on the specific heat. However, some electrons could occupy the $n^c = 1$ Landau levels, and the same density of holes would fill in the $n^v = 1$ ones. As a result, the specific heat ascends again as temperature is beyond 100 K, hole symmetry. With respect to other higher Landau levels [$n^{c,v} \geq 2$], the energy spacings between adjacent levels are getting narrower. In the higher temperature region, the thermally excited carriers rapidly occupy more Landau levels, which in turn enhance $C(T, B_z)$. Most importantly, the factors contributing to $C(T, B_z)$ at high temperatures are rather complicated, but could be delicately decomposed to the distinct contributions of individual Landau levels. It should be noticed that the presence of Zeeman splitting does not play any critical role in the high-temperature specific heat.

The magnetic field dependence of the specific heat at various temperatures is illustrated in Fig. 9.20, exhibiting only one obvious peak under a critical strength (B_{zc}). Apparently, this prominent structure mainly originates from the Zeeman splitting, as discussed earlier in Fig. 9.19(b). For rather weak magnetic fields, the Zeeman splitting energy is greatly reduced, so the thermally excited carriers are almost equally distributed at the $n^c_\uparrow = 0$ and $n^v_\downarrow = 0$ Landau levels [$f(n^c_\uparrow = 0) = f(n^v_\downarrow = 0) \sim 0.5$. The very slow growing rate in thermal carriers leads to the almost vanishing specific heat. However, the stronger field strengths enlarge the Zeeman splitting and thus enhance the density of thermally excited carriers. This is responsible for the rising behavior of the specific heat with the increasing B_z. The opposite is true as the magnetic field exceeds B_{zc}. That is to say, the increasing rate of excited carriers will be diminished in the further increment of B_z. The critical value of field strength under various temperatures could be determined by taking the first derivative of $C(T, B_z)$ versus B_z in Eqs. (3.8)–(3.10). A simple linear relation is revealed between B_{zc} and T [details in Ref. [61]], as observed for T_c and B_z [Fig. 9.19(c)]. In addition to a pronounced shift of peak position (B_{zc}'s), the specific heat at higher temperatures creates a more prominent peak, mainly owing to the great enhancement of Landau states.

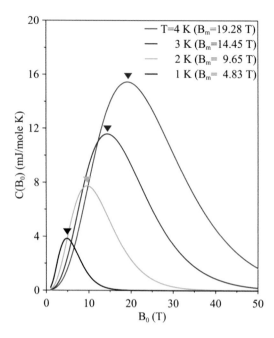

FIGURE 9.20 The magnetic-field-dependent specific heat under the specific temperatures.

Under the specific electron/hole doping, the temperature-dependent specific heat presents an extra shoulder structure, as clearly shown in Fig. 9.21 under the various dopings and magnetic fields. The distinct doping levels [n_d^c's], which is characterized by the shift of chemical potential, are only investigated for the full occupation of Landau levels. For example, the chemical potential of $n_d^c = 1.5$ corresponds to the middle of $n^c = 1$ and $n^c = 2$ Landau-level energies at $T = 0$, which is sensitive to the change of temperature (inset). Based on particle conservation, $\mu(T)$ hardly depends on the temperature at low T, and then declines as temperature grows. Apparently, the specific heat reveals two special structures, one peak and one shoulder, respectively, at lower and higher temperatures (the various arrows for distinct field strengths and doping levels). The former comes from the Zeeman splitting only (thoroughly discussed earlier); therefore, its position and height are independent of doping levels under a fixed B_z. Generally speaking, the extra shoulder structure is closely related to the thermally excited electrons (holes) occupying the lowest unoccupied (the second highest occupied) spin-up and spin-down Landau levels. As for the $n_d^c = 1$ [2] doping case, the initial peak is due to the $n_{d\downarrow}^c = 1$ [2] and $n_{d\uparrow}^c = 1$ [2] Landau states, while the shoulder structure is associated with $n_d^c = 2$ [3] and $n_d^c = 0$ [1] ones. Concerning the latter, both the spin-up and spin-down Landau levels make the identical contribution at higher temperature, and the T-induced conduction 1 electrons being transferred from $n_d^c = 0$ [1] to 2 [3]. With more electron dopings, the energy spacings arising from the highest Landau level is greatly reduced, shifting this shoulder structure toward lower temperature. Also, $C(T, B_z)$s of doped graphene systems are strongly affected

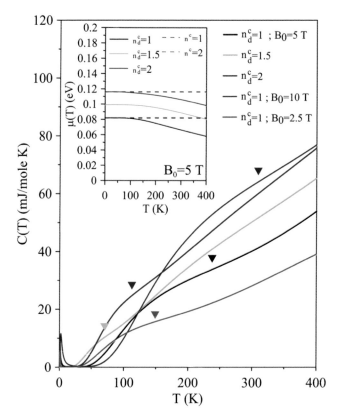

FIGURE 9.21 The wide-range temperature dependence of specific heat for the different magnetic fields and doping levels. Also, inset shows the temperature-dependent chemical potentials for the various electron dopings.

by magnetic fields. At a fixed doping level, the stronger fields lead to more prominent structures, including the intensities of peak and shoulder, respectively, at low and high temperatures. Furthermore, their positions are moved toward higher temperatures as a result of the wider level spacings.

The different fillings might lead to the drastic changes in the temperature-dependent specific heat, e.g., the absence/presence of a specific peak at lower temperatures. The full-filled case of $n_d^c = i = 1/2$, which the zero-T chemical potential is located exactly at the middle of $n_\uparrow^c = i$ and $n_\uparrow^c = i + 1$ is discussed in detail, as shown by the green curve in Fig. 9.21. Based on particle conservation, the chemical potential of $\mu(T)$ hardly changes in the low-temperature range, while it declines as T further grows [inset in Fig. 9.21]. The specific heat of $n_d^c = 1.5$ only reveals a shoulder structure at about 70 K (a green arrow), which results from the excited electrons (holes) occupying the lowest unoccupied (the highest occupied) spin-up and spin-down levels simultaneously. That is, such special thermal structure is associated with $n^c = 1$ and $n^c = 2$ Landau levels. However, it is almost independent of spin-up and spin-down levels, nearly making the same contributions. Compared with the fully filled cases, the half occupations do not present any peak at low temperature. Hence, their specific heats hardly survive in this temperature range, directly reflecting the absence of a very narrow energy spacing due to the Zeeman splitting across the chemical potential. Furthermore, the shoulder structure of the half-filled case comes to exist at a much lower temperature than the fully filled cases mainly owing to the smaller level spacing of the dominant Landau levels.

The magneto-electronic specific heats are quite different between 2D monolayer graphene and 1D single-walled carbon nanotubes, when a uniform magnetic field is, respectively, perpendicular and parallel to the honeycomb lattice and the tubular axis. The latter possesses the unique magneto-electronic properties which agree with the Aharonov-Bohm effect in the absence of Landau levels [65]. As a result, such 1D cylindrical systems are predicted to exhibit four kinds of temperature dependence [details in Ref. [66]]. The first kind dependence has a composite form of power and exponential functions, being induced by the 1D parabolic valence and conduction subbands with a small gap near the chemical potential. The second kind dependence is divergent in the inverse of $\sqrt{T}$. This reflects the most important characteristic of parabolic valance and conduction subbands intersecting just at $\mu\,(T)$. The third one reveals a peak structure and a finite value at zero temperature because of the overlapping parabolic subbands. However, no analytic relation is suitable for this behavior. For the last one, it is linearly proportional to temperature, mainly owing to the subbands linearity. The above-mentioned thermal properties are absent in few-graphene systems directly reflect the significant differences between the highly degenerate dispersionless (0D) Landau levels and the 1D parabolic/linear energy subbands in the van Hove singularities in the DOS. In addition, a net magnetic flux in carbon nanotube only creates a separate coupling with each discrete angular momentum; furthermore, a magnetic field perpendicular to a cylindrical surface cannot create the 0D Landau levels/the effective magnetic quantization [66].

The high-resolution thermal calorimeters, as thoroughly discussed in Sec. 3.2, are available in testing the theoretical predictions on the n-specific/specific heat of monolayer graphene and thus the Landau-level energy spectra/Dirac cone. For the linearly isotropic energy bands in monolayer graphene, the thermal measurements on the T^2-dependent specific heat provide another verification tool. The low-temperature magneto-specific heat, with an unusual peak structure, relates to a rather small Zeeman splitting energy only; furthermore, there exists a linear dependence between T_c and B_z. Since the Zeeman effects are proportional to the magnetic-field strength, so the experimental examinations on this relation are very useful in correcting the factor of g/m^* [Eq. (3.10)]. Under the doping cases, the specific heat shows more rich temperature-dependences. The thermal measurements could also be utilized to examine the doping levels (free carrier densities) by identifying the significant temperature corresponding to a shoulder structure. The higher/deeper the doping level is, the lower temperature the position of the shoulder will be. According to the previously experimental results, the specific heat has been measured for carbon-related systems, such as the specific heats of graphite [64] and carbon nanotubes [66]. The highest resolution of calorimeter measurement is down to 0.1 K [67]. It is expected that such resolution is sufficient to verify the specific-heat peak of graphene at lower temperatures purely induced by the Zeeman effect; therefore, it is surely eligible for detecting the shoulder structure at higher temperatures.

REFERENCES

1. Novoselov K S, Geim A K, Morozov S V, Jiang D, Zhang Y, Dubonos S V, et al. 2004 Electric field effect in atomically thin carbon films *Science* **306** 666.
2. Novoselov K S, Geim A K, Morozov S V, Jiang D, Katsnelson M I, Grigorieva I V, et al. 2004 Two-dimensional gas of massless Dirac fermions in graphene *Nature* **438** 197.
3. Zhang Y, Tan Y W, Stormer H L, and Kim P 2005 Experimental observation of the quantum Hall effect and Berry's phase in graphene *Nature* **438** 201.
4. Zheng Y and Ando T 2002 Hall conductivity of a two-dimensional graphite system *Phys. Rev. B* **65** 245420.
5. Gusynin V P and Sharapov S G 2005 Unconventional integer quantum Hall effect in graphene *Phys. Rev. Lett.* **95** 146801.
6. Pertes N M R, Guinea F, and Castro Neto A H 2006 Electronic properties of disordered two-dimensional carbon *Phys. Rev. B* **73** 125411.
7. Nomura K and MacDonald A H 2006 Quantum Hall ferromagnetism in graphene *Phys. Rev. Lett.* **96** 256602.
8. Alicea J and Fisher M P A 2006 Graphene integer quantum Hall effect in the ferromagnetic and paramagnetic regimes *Phys. Rev. B* **74** 075422.
9. Goerbig M O, Moesser R, and Doucot B 2006 Electron interactions in graphene in a strong magnetic field *Phys. Rev. B* **74** 161407.
10. Gusynin V P, Miransky V A, Sharapov S G, and Shovkovy I A 2006 Excitonic gap, phase transition, and quantum Hall effect in graphene *Phys. Rev. B* **74** 195429.
11. Ezawa M 2007 Supersymmetry and unconventional quantum Hall effect in monolayer, bilayer and trilayer graphene *Physica E* **40** 269.
12. Webb M J, Palmgren P, Pal P, Karis O, and Grennberg H 2011 A simple method to produce almost perfect graphene on highly oriented pyrolytic graphite *Carbon* **49** 3242.
13. Liu Z, Zheng Q S, and Liu J Z 2010 Stripe/kink microstructures formed in mechanical peeling of highly orientated pyrolytic graphite *Appl. Phys. Lett.* **96** 201909.
14. Zhao Y R and Pieter K 1993 Electronic effects in scanning tunneling microscopy: Moiré pattern on a graphite surface *Phys. Rev. B* **48 (23)** 17427.
15. Campanera J M, Savini G, Suarez-Martinez I, and Heggie M I 2007 Density functional calculations on the intricacies of Moiré patterns on graphite *Phys. Rev. B* **75** 235449.
16. Wu Y, Wang B, Ma Y, Huang Y, Li N, Zhang F, et al. 2010 Efficient and large-scale synthesis of few-layered graphene using an arc-discharge method and conductivity studies of the resulting films *Nano Res.* **3 (9)** 611.
17. Wu Z S, Ren W, Gao L, Zhao J, Chen Z, Liu B, et al. 2009 Synthesis of graphene sheets with high electrical conductivity and good thermal stability by hydrogen arc discharge exfoliation *ACS Nano* **3** 411.
18. Lee J K, Lee S C, Ahn J P, Kim S C, Wilson J I B, and John P 2008 The growth of graphite on (111) diamond *J. Chem. Phys.* **129 (23)** 234709.
19. Borysiuk J, Soltys J, and Piechota J 2011 Stacking sequence dependence of graphene layers on SiC (000-1)—Experimental and theoretical investigation *J. Appl. Phys.* **109 (9)** 093523.
20. Ellis C T, Stier A V, Kim M H, Tischler J G, Glaser E R, Myers-Ward R L, et al. 2013 Magneto-optical fingerprints of distinct graphene multilayers using the giant infrared Kerr effect *Sci. Rep.* **3** 03143.

21. Hwang J, Shields V B, Thomas C I, Shivaraman S, Hao D, Kim M, et al. 2010 Epitaxial growth of graphitic carbon on C-face SiC and sapphire by chemical vapor deposition (CVD) *J. Cryst. Growth*. **312** 3219.

22. Juang Z Y, Wu C Y, Lu A Y, Su C Y, Leou K C, Chen F R, et al. 2010 Graphene synthesis by chemical vapor deposition and transfer by a roll-to-roll process *Carbon* **48** 3169.

23. Lenski D R and Fuhrer M S 2011 Raman and optical characterization of multilayer turbostratic graphene grown via chemical vapor deposition *J. Appl. Phys.* **110** 013720.

24. Jayasena B and Subbiah S 2011 A novel mechanical cleavage method for synthesizing few-layer graphenes *Nanoscale Res. Lett.* **6** 95.

25. Lui C H, Malard L M, Kim S, Lantz G, Laverge F E, Saito R, et al. 2012 Observation of layer-breathing mode vibrations in few-layer graphene through combination Raman scattering *Nano Lett.* **12** 5539.

26. Mak K F, Sfeir M Y, Misewich J A, and Heinz T F 2010 The evolution of electronic structure in few-layer graphene revealed by optical spectroscopy *PNAS* **107** 14999.

27. Biedermann L B, Bolen M L, Capano M A, Zemlyanov D, and Reifenberger R G 2009 Insights into few-layer epitaxial graphene growth on substrates from STM studies *Phys. Rev. B* **79** 125411.

28. Zhang L, Zhang Y, Camacho J, Khodas M, and Zaliznyak I 2011 The experimental observation of quantum Hall effect of l=3 chiral quasiparticles in trilayer graphene *Nat. Phys.* **7** 953.

29. Xu P, Yang Y, Qi D, Barber S D, Schoelz J K, Ackerman M L, et al. 2012 Electronic transition from graphite to graphene via controlled movement of the top layer with scanning tunneling microscopy *Phys. Rev. B* **86** 085428.

30. Pong W T, Bendall J and Durkan C 2007 Observation and investigation of graphite superlattice boundaries by scanning tunneling microscopy *Surface Sci.* **601** 498.

31. Asieh S K, Crampin S, and Ilie A 2013 Stacking-dependent superstructures at stepped armchair interfaces of bilayer/trilayer graphene *Appl. Phys. Lett.* **102** 163111.

32. Que Y, Xiao W, Chen H, Wang D, Du S and Gao H J 2015 Stacking-dependent electronic property of trilayer graphene epitaxially grown on Ru (0001) *Appl. Phys. Lett.* **107** 263101.

33. Emtsev K V, Bostwick A, Horn K, Jobst J, Kellogg G L, Ley L, et al. 2009 Towards wafer-size graphene layers by atmospheric pressure graphitization of silicon carbide *Nature Mater.* **8** 203.

34. Lee D S, Riedl C, Krauss B, von Klitzing K, Starke U, and Smet J H 2009 Raman Spectra of epitaxial graphene on SiC and of epitaxial graphene transferred to SiO_2 *Nano Lett.* **8** 4320.

35. Lee D S, Riedl C, Beringer T, Castro Neto A H, von Klitzing K, Starke U, et al. 2011 Quantum Hall effect in twisted bilayer graphene *Phys. Rev. Lett.* **107** 216602.

36. Fallahazad B, Hao Y, Lee K, Kim S, Ruoff R S, and Tutuc E 2012 Quantum Hall effect in Bernal stacked and twisted bilayer graphene grown on Cu by chemical vapor deposition *Phys. Rev. B* **85** 201408(R).

37. Huang Y K, Chen S C, Ho Y H, Lin C Y, and Lin M F 2014 Feature-rich magnetic quantization in sliding bilayer graphenes *Sci. Rep.* **4** 7509.

38. Lin C Y, Do T N, Huang Y K, and Lin M F 2017 *Optical Properties of Graphene in Magnetic and Electric Fields*. IOP Publishing IOP Concise Physics. San Raefel, CA, USA: Morgan & Claypool Publishers. 2053–2563, ISBN: 978-0-7503-1566-1.

39. Novoselov K S, McCann E, Morozov S V, Fal'ko V I, Katsnelson M I, Zeitler U, et al. 2006 Unconventional quantum Hall effect and Berry's phase of 2p in bilayer graphene *Nat. Phys.* **2** 177.

40. Taychatanapat T, Watanabe K, Taniguchi T, and Jarillo-Herrero P 2008 Quantum Hall effect and Landau-level crossing of Dirac fermions in trilayer graphene *Nat. Phys.* **7** 621.

41. Bao W, Jing L, Velasco Jr J, Lee Y, Liu G, Tran D, et al. 2011 Stacking-dependent band gap and quantum transport in trilayer graphene *Nat. Phys.* **7** 948.

42. Zhang L, Zhang Y, Camacho J, Khodas M, and Zaliznyak I 2011 The experimental observation of quantum Hall effect of l=3 chiral quasiparticles in trilayer graphene *Nat. Phys.* **7** 953.

43. Nakamura M, Hirasawa L, and Imura K I 2008 Quantum Hall effect in bilayer and multilayer graphene with finite Fermi energy *Phys. Rev. B* **78** 033403.

44. McCann E and Falko V I 2006 Landau-level degeneracy and quantum Hall effect in a graphite bilayer *Phys. Rev. Lett.* **96** 086805.

45. Yuan S, Roldan R, and Katsnelson M I 2011 Landau level spectrum of ABA- and ABC-stacked trilayer graphene *Phys. Rev. B* **84** 125455.

46. Zhang X Q, Li H, and Liew K M 2007 The structures and electrical transport properties of germanium nanowires encapsulated in carbon nanotubes *J. Appl. Phys.* **102** 73709.

47. Zhong X L, Amorim R G, Scheicher R H, Pandey R, and Karna S P 2012 Electronic structure and quantum transport properties of trilayers formed from graphene and boron nitride *Nanoscale* **4** 5490.

48. Lin Y P, Lin C Y, Ho Y H, Do T N, and Lin M F 2015 Magneto-optical properties of ABC-stacked trilayer graphene *Phys. Chem. Chem. Phys.* **17 (24)** 15921.

49. Nethravathi C and Rajamathi M 2008 Chemically modified graphene sheets produced by the solvothermal reduction of colloidal dispersions of graphite oxide *Carbon* **46 (14)** 1994.

50. Zhou W, Kapetanakis M D, Prange M P, Pantelides S T, Pennycook S J, and Idrobo J C 2012 Direct determination of the chemical bonding of individual impurities in graphene *Phys. Rev. Lett.* **109 (20)** 206803.

51. Panchakarla L S, Subrahmanyam K S, Saha S K, Govindaraj A, Krishnamurthy H R, Waghmare U V, et al. 2009 Synthesis, structure, and properties of boron-and nitrogen-doped graphene *Adv. Mater.* **21 (46)** 4726.

52. Qu L, Liu Y, Baek J B, and Dai L 2010 Nitrogen-doped graphene as efficient metal-free electrocatalyst for oxygen reduction in fuel cells *ACS Nano* **4 (3)** 1321.

53. SYLin (heat capacity).

54. Tran N T T, Lin S Y, Glukhova O E, and Lin M F 2015 Configuration-induced rich electronic properties of bilayer graphene *J. Phys. Chem. C* **119 (19)** 10623.

55. Do T N, Shih P H, Chang C P, Lin C Y, and Lin M F 2016 Rich magneto-absorption spectra of AAB-stacked trilayer graphene *Phys. Chem. Chem. Phys.* **18 (26)** 17597.

56. Vogt P, De Padova P, Quaresima C, Avila J, Frantzeskakis E, Asensio M C, et al. 2012 Silicene: compelling experimental evidence for graphenelike two-dimensional silicon *Phys. Rev. Lett.* **108** 155501.

57. Charlier J C, Michenaud J P, and Gonze X 1992 First-principles study of the electronic properties of simple hexagonal graphite *Phys. Rev. B* **46** 4531.

58. Nguyen-Manh D, Saha-Dasgupta T, and Andersen O K 2003 Tight-binding model for carbon from the third-generation LMTO method: a study of transferability *Bull. Mater. Sci.* **26** 7.

59. Lin C Y, Wu J Y, Ou Y J, Chiu Y H, and Lin M F 2015 Magneto-electronic properties of multilayer graphenes *Phys. Chem. Chem. Phys.* **17** 26008.

60. Do T N, Lin C Y, Lin Y P, Shih P H, and Lin M F 2015 Configuration-enriched magneto-electronic spectra of AAB-stacked trilayer graphene *Carbon* **94** 619.

61. Lin S Y, Ho Y H, Huang Y C, and Lin M F 2012 Magneto-electronic specific heat of graphene *J. Phys. Soc. Jpn.* **81** 084602.

62. Partoens B and Peeters F M 2006 From graphene to graphite: electronic structure around the K point *Phys. Rev. B* **74** 075404.

63. Palser A H 1999 Interlayer interactions in graphite and carbon nanotubes. *Phys. Chem. Chem. Phys.* **1** 4459.
64. Lin S Y, Ho Y H, Shyu F L, and Lin M F 2013 Electronic thermal property of graphite *J. Phys. Soc. Jpn.* **82** 074603.
65. Ajiki H and Ando T 1996 Aharonov—Bohm effect on magnetic properties of carbon nanotubes *Physica B: Condensed Matter* **216** 358.
66. Lin M F and Shung K W K 1995 Magnetoconductance of carbon nanotubes *Phys. Rev. B* **51** 7592.
67. Popov V N 2002 Low-temperature specific heat of nanotube systems *Phys. Rev. B* **66** 153408.

10 Rich Magneto-Coulomb Excitations in Germanene

Jhao-Ying Wu,[b] *Chiun-Yan Lin,*[a] *Thi-Nga Do,*[c,d] *Po-Hsin Shih,*[a] *Shih-Yang Lin,*[e] *Ching-Hong Ho,*[b] *Ming-Fa Lin*[a,f,g]

[a] Department of Physics, National Cheng Kung University, Tainan 701, Taiwan
[b] Center of General Studies, National Kaohsiung University of Science and Technology, Kaohsiung 811, Taiwan
[c] Laboratory of Magnetism and Magnetic Materials, Advanced Institute of Materials Science, Ton Duc Thang University, Ho Chi Minh City, Vietnam
[d] Faculty of Applied Sciences, Ton Duc Thang University, Ho Chi Minh City, Vietnam
[e] Department of Physics, National Chung Cheng University, Chiayi 621, Taiwan
[f] Quantum Topology Center, National Cheng Kung University, Tainan 701, Taiwan
[g] Hierarchical Green-Energy Materials Research Center, National Cheng Kung University, Tainan, Taiwan

CONTENTS

The electron-electron Coulomb excitations are one of the main-stream topics in condensed-matter physics; furthermore, they could present the rich and unique magnetic quantization. Up to now, a plenty of theoretical and experimental investigations are conducted on the graphene-related systems [1–5]. Most importantly, the single-particle and collective excitations have been identified to be greatly diversified by the distinct dimensions [6–13], geometric symmetries [14–22], number of layers [23–30], stacking configurations [31–35], temperatures [32, 36, 37], free conduction electrons/valence holes due to dopings, gate voltages, and magnetic fields have been examined (predicted) to exhibit the diverse phenomena. For example, a 2D plasmon mode, with a specific $\sqrt{q}$ frequency dispersion relation, could survive in a finite-T pristine graphene [31, 37–39], an extrinsic one [36, 40, 41], bilayer AA stacking and trilayer ABC stacking [23,24,34,42,43]. The $(\mathbf{q}, \omega)$-phase diagrams are very sensitive to temperature, doping level, band overlap, and surface states, being in sharp contrast with a 2D electron gas [40,44]. This clearly illustrates the importance of band-structure effects. Moreover, the magneto-electronic bare and screened excitation spectra are investigated in detail for monolayer graphene [45–47], and bilayer AA & AB stackings [42, 48–50]. The strong competition/cooperation between the longitudinal Coulomb interactions and the transverse cyclotron forces is responsible for the unusual phenomena, covering the isotropic single- and many-particle behaviors under low-frequency excitations, many discrete inter-Landau-level excitation channels in the absence of specific selection rules, the prominent symmetric peaks in response functions, and the layer-, stacking-, magnetic-field-, & doping-enriched phase diagrams.

Up to date, only few theoretical researches are conducted on the many-body electronic-electron Coulomb interactions of layered germanene & silicene systems [9, 10, 51, 52]. The previous prediction clearly shows that the feature-rich electronic excitations of monolayer germanene purely lie in the important spin-orbit coupling and the buckled geometry. The collective and electron-hole excitations strongly depend on the magnitude and direction of transferred momentum, Fermi energy, and gate voltage. The former are classified into four kinds of plasmon modes, according to the unique frequency- and momentum-dependent excitation spectra. They behave as 2D acoustic modes at a long wavelength limit. However, under the larger momenta, they might change into another kind of undamped plasmons, become the seriously suppressed modes in the heavy intraband e-h excitations, keep the same undamped ones, or decline and then vanish in the strong interband e-h excitations. Moreover, germanene, silicene and graphene are very different from one another in the main features of the diverse plasmon modes. All the above-mentioned excitation spectra are further examined to be very efficient deexcitation channels of the excited quasiparticle states [53–56]. For example, as to the electron-doped germanene systems, the low-lying valence holes could decay through the undamped acoustic plasmon; therefore, such states present very fast inelastic Coulomb scatterings, non-monotonous energy dependence, and anisotropic behavior. However, the low-energy conduction electrons and holes behave as a 2D electron gas. The higher-energy conduction states and the deeper-energy valence ones exhibit the similar behaviors in the available deexcitation channels and thus the dependence of decay rate on the wave vector.

The magneto-Coulomb excitations of monolayer will be thoroughly explored by the close association of the generalized tight-binding model and the modified random-phase approximation [details in Sec. 3.4]. The bare and screened response functions are delicately evaluated and analyzed according to Eqs. (3.21)–(3.23). Their dependences on the transferred momenta and frequencies, the Fermi energies/the densities of free conduction electrons, the Coulomb on-site energies due to various gate voltages, and the magnetic-fields are investigated in detail. Furthermore, the concise physical pictures, being closely related to the inter-Landau level/band-edge-state transitions, are proposed to account for the diverse excitation phenomena in the real &

imaginary parts of dielectric functions, the energy loss spectra, and the (momentum, frequency)-phase diagrams. The dramatic transformation of electronic excitations, which is due to the magnetic quantization, will be clearly presented by the distinct electron-hole damping regions and the main features of plasmon/magnetoplasmon modes. Also, the calculated results are very useful in fully understanding the important differences among germanene, silicene, and graphene.

10.1 DOPING- AND GATE-VOLTAGE-ENRICHED SINGLE-PARTICLE AND COLLECTIVE EXCITATIONS

The generalized tight-binding model, being related to the Ge-$4p_z$ orbitals, is reliable for the field-modified band structures and magneto-electronic states in monolayer germanene [the calculation details in Sec. 3.4]. Apparently, this buckled system exhibits the feature-rich electronic structures which strongly depend on the spin-orbit coupling, wave vector, and electric-field-strength. The unoccupied conduction band is almost symmetric to the occupied valence one about the zero chemical potential under the intrinsic case. Electronic states in the presence of spin-orbital interactions are doubly degenerate for the spin degree of freedom [the black solid curve in Fig. 10.1]. That is, the wave functions possess the spin-up- and spin-down-dominated spin configurations simultaneously. These two different spin configurations can make the same contributions to Coulomb excitations. Most importantly, the significant spin-orbital coupling has generated the separation of Dirac points and thus created an energy spacing of $E_D = 93$ meV [the double of λ_{soc} in Sec. 3.4] near the K point. The low-lying valence and conduction bands present parabolic energy dispersions initiated from the K/K′ valley and then gradually become linear ones in the increment of wave vector. In addition, at higher energy ($|E^{c,v}| \geq 1$) eV, such energy bands recover into parabolic ones and have a saddle van Hove singularity at the M point with very high density of states. It is noticed that only the low-energy bands of $|E^{c,v}| \leq 0.2$ eV shows the isotropic electronic spectra [57].

The low-lying electronic properties are easily modulated by a gate voltage (a uniform perpendicular electric field). The Coulomb potential energy difference between the A and B sublattices can create an obvious destruction of mirror symmetry about the $z = 0$ plane. Furthermore, it leads to the spin-split electronic states, i.e., one pair of doubly-degenerate energy bands dramatically changes into the spin-down- and spin-up-dominated ones. These conduction and valence bands are, respectively, represented by $E_1^{c,v}$ and $E_2^{c,v}$ [Fig. 10.1(a)]. As for the electronic states near the K valley, the pair of energy bands, with the spin-down-dominated configuration, is relatively close to zero energy. However, the opposite is true for those near the K′ valley. When E_z grows, the first pair gradually approaches to zero energy and E_D is getting smaller. E_D is vanishing at the critical potential of $V_z = \lambda_{soc}$ (the green curves), where the intersecting linear bands comes to exist, or the electronic structure has a pair of linear Dirac-cone structure. With a further increment of V_z (the dashed orange curves), the energy spacing of parabolic valence and conduction bands is recovered and enlarged. On the other hand, the second pair of spin-up-dominated energy bands is away from zero energy. As a result, two splitting spin-dependent configurations are expected to greatly diversify the single-particle and collective excitations.

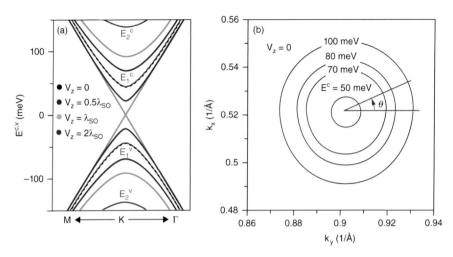

FIGURE 10.1 (a) Band structures of monolayer germanene under the various electric fields, and (b) the constant-energy loops at $V_z = 0$.

Very interestingly, the electric-field-enriched electronic states are directly quantized into the unusual Landau levels under the spin and valley-splittings, as clearly indicated in Fig. 10.2 for the various strengths at $B_z = 8$ T. Apparently, the magneto-electronic energy spectra exhibits the strong and non-monotonic E_z-dependence, in which the crossing and anti-crossing phenomena appear simultaneously. By the delicate analyses (details in book Chen), the former and the latter are, respectively, associated with the $[n^c_{1K} \ \& \ n^c_{1K} + 1]/[n^v_{1K'} \ \& \ n^v_{1K'} + 1]$ Landau levels and the $[n^c_{1K'} \ \& \ n^c_{1K'} + 1]/[n^v_{1K} \ \& \ n^v_{1K} + 1]$ ones. Such phenomena would be enlarged at higher magnetic fields. Compared with silicene, such are easily observed in germanene because of the smaller intralayer hopping integral, the larger spin-orbital coupling, and the larger effect due to electric field. It should be noticed that the critical electric field, corresponding to zero energy gap, is insensitive to the magnetic-field strength, but not the significant spin-orbital interactions.

Electronic excitations in the absence of magnetic fields are first discussed. The electron-hole excitation phenomena are characterized by the imaginary part of the dielectric function ($Im[\epsilon]$), being sensitive to the direction and magnitude of momentum transfer, Fermi energy, and gate voltage. The non-vanishing $Im(\epsilon)$ corresponds to the interband and intraband electron-hole excitations consistent with the Fermi-Dirac distribution and the conservation of energy and momentum. For the intrinsic case, it exhibits a prominent divergent peak in the square root form at the threshold energy of $\omega_{th} = 2\sqrt{\lambda^2_{soc} + v^2_F q^2/4}$ (v_F the Fermi velocity), as shown by the black dashed curve in Fig. 10.3(a) under $E_F = 0$, $\theta = 0°$, and $T = 0$ K. This structure principally arises from the interband excitations which are associated with the valence or the conduction band-edge states (the separated Dirac points). From the Kramers-Kronig relations between the imaginary and the real parts of the dielectric function, the former obviously shows a similar divergent peak at the left-hand side [Fig. 10.3(b)].

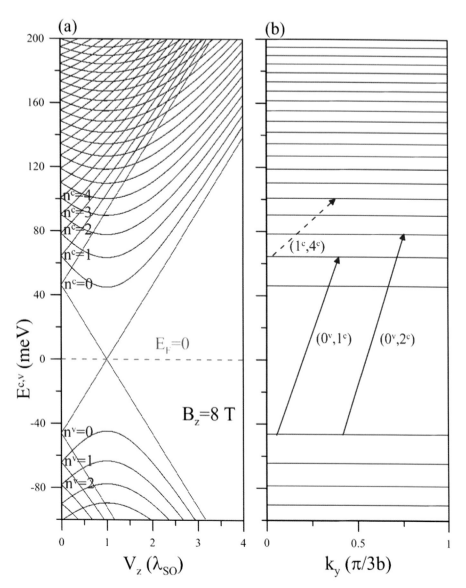

FIGURE 10.2 The electric-field-dependent Landau-level energy spectrum of monolayer germanene under a fixed magnetic field of $B_z = 8$ T, and (b) the available interband and intraband excitations by the solid and dashed blue arrows, respectively.

However, $Re(\epsilon)$ is always positive at the low-frequency range, clearly illustrating that collective excitations hardly survive.

Very interestingly, the intraband excitations are created by the carrier doping; therefore, these strong competitions with the interband ones lead to the diversified many-body phenomena. The latter are gradually suppressed and then replaced by the former in the increment of the Fermi energy. For example, both of them could coexist

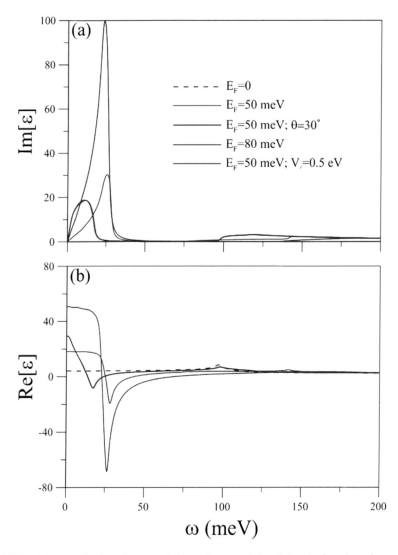

FIGURE 10.3 (a) The imaginary and (b) real parts of the dielectric functions for mono-layer germanene under the distinct Fermi energies, directions of the transferred momenta, and gate-voltages.

at $E_F = 0.05$ eV, in which the intraband and interband channels, respectively, present a prominent square root divergent peak and a shoulder structure in $Im(\epsilon)$ [red curve in Fig. 10.3]. Apparently, these two special structures, which mainly originate from the electronic excitations of the Fermi-momentum states, are strongly affected by an excitation gap due to the energy spacing of Dirac points. The prominent asymmetric peak comes to exist at $\omega_{ex} \sim v_F q$, directly reflecting the linear energy dispersion of conduction band. This specific excitation energy is just the upper boundary of the intraband single-particle excitations [Fig. 10.5(a)–(d) discussed in detail later].

Furthermore, the corresponding $Re(\epsilon)$ exhibits the square root and logarithmically divergent at lower and higher frequencies, as clearly indicated in Fig. 10.3(b). The two zero points in $Re(\epsilon)$, if at where $Im(\epsilon)$ is sufficient small, are associated with collective excitations. The intraband excitations become the dominating channels in the low-frequency single-particle spectrum at the larger Fermi energy (e.g., $E_F = 0.08\,\text{eV}$ by the blue curve). $Im(\epsilon)$ and $Re(\epsilon)$ have the pronounced peak structures, in which the second zero point of the latter appears under the very weak Landau dampings. In short, the intensity of intraband excitation peaks is getting stronger with the increasing E_F. However, the zero point of $Re(\epsilon)$ is revealed at the higher frequency, accompanied by the greatly decreased $Im(\epsilon)$. That is, the further increment of E_F will reduce the electron-hole Landau dampings and enhance the frequency/strength of collective excitations.

Apparently, the direction of transferred momenta, which characterizes the isotropic or anisotropic characteristics of single-particle and collective excitations, is closely related to the constant-energy contours. They are vertically flipped for the K and K′ valleys, as shown in Fig. 10.1(b). The energy variation, being measured from them along the direction of $\theta = 30°$, is, respectively, smaller and bigger, compared with the $\theta = 0°$ case. With the increasing θ, the widened frequency range of the electron-hole damping is expected to strongly affect the plasmon modes. The gate voltage can split valence and conduction bands and thus diversify the channels of single-particle excitations. There exist eight categories of excitation channels, in which two and six ones, respectively, belong to the intraband and the interband electron-hole excitations. Among of them, the four interband excitation channels, $E_2^v \rightarrow E_1^v, E_2^c \rightarrow E_1^v, E_1^c \rightarrow E_2^c$, and $E_2^c \rightarrow E_1^c$, are negligible within the low-frequency range, mainly owing to the almost vanishing Coulomb matrix elements induced by the spin configurations [Eq. (3.23)]. Consequently, the V_z-dominated electron-hole excitations mainly arise from the intraband excitations of $[E_2^c \rightarrow E_2^c\ \&\ E_1^c \rightarrow E_1^c]$ and the interband excitations of $[E_1^v \rightarrow E_1^c\ \&\ E_2^v \rightarrow E_2^c]$. They, respectively, cause $Im(\epsilon)$ to display one square root divergent peaks and two shoulder structures at lower and higher frequencies, as clearly shown in Fig. 10.3(a) at $V_z = 0.5\lambda_{SOC}$, $E_F = 0.05$ eV, $\theta = 0°$ and $q = 0.015\ \text{Å}^{-1}$ by the yellow curve. It should be noticed that the second peak due to the $E_1^c \rightarrow E_1^c$ channel has a characteristic excitation frequency almost independent of V_z. By the detailed calculations, moreover, the threshold excitation frequencies of both interband channels, which correspond to the energy difference of the valence Dirac point and Fermi-momentum state, are seriously suppressed by $\omega_{th} = 2\sqrt{[\lambda_{soc} \mp V_z]^2 + v_F^2 q^2/4}$ [details in [9,52,58]]. They will determine more complicated boundaries created by the interband single-particle excitations [discussed later in Fig. 10.5(d)]. The initial excitation frequency of the first shoulder becomes lower as V_z grows from zero to λ_{soc} and then higher in the further increase from λ_{soc} to $2\lambda_{soc}$. Differently, that of the second one is monotonically increasing with the enhancement of V_z. However, the zero-point frequency in $Re(\epsilon)$, related to the $E_1^c \rightarrow E_1^c$ electron-hole peak, hardly depends on V_z. This will induce the serious suppression of the interband Landau dampings on plasmon modes.

The energy loss function, defined as $Im(\epsilon)$, measures the response ability after Coulomb screenings and is useful to characterize the collective excitations. Without free carriers, only a weak shoulder structure, which arises from the interband

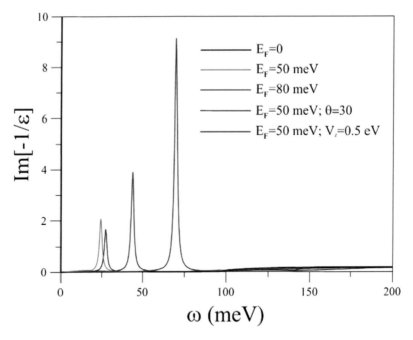

FIGURE 10.4 The energy loss spectra of monolayer germanene under the various cases as shown in Fig. 10.3.

single-particle excitations, is revealed in the screened response function at $E_F = 0$, $q = 0.015 \text{ Å}^{-1}$, and $\theta = 0°$ [Fig. 10.4 by the black dashed curve], leading to the disappearance of plasmon peak. That is, the plasmon mode is thoroughly damped through the interband electron-hole excitations. When the free carrier density is sufficiently high, a narrow sharp plasmon peak, being accompanied by a shoulder structure, comes to exist, e.g., $Im(-1/\epsilon)$ under $E_F = 0.05$ eV (red curve). The significant interband electron-hole dampings are greatly reduced in the further increase of conduction electron density, e.g., a prominent plasmon peak in the absence of shoulder structure at $E_F = 0.08$ eV (blue curve). This result directly reflects the fact that the interband excitations are progressively replaced by the intraband ones. On the other hand, the intensity of the plasmon peak grows with θ, e.g., the energy loss spectrum at $E_F = 0.05$ eV and $\theta = 30°$ [green curve in Fig. 10.4]. As for the gate-voltage effects, they have resulted in more shoulder structures due to the spin-split electron-hole excitation channels [Fig. 10.3(a)], and change the main features of the pronounced collective excitations [such as the momentum dependence in Fig. 10.5(d)].

Free carriers and gate voltages, as clearly indicated in Figs. 10.5(a)–10.5(d) for the (q, ω)-dependent phase diagrams, can create the unique single-particle excitations and plasmon modes. The boundaries of intraband and interband electron-hole transitions (the white dashed and solid curves) should be determined by the Fermi-momentum states and Dirac points. They might reveal an excitation gap [Figs. 10.5(a), 10.5(c), and 10.5(d)], mainly owing to the separation of valence and conduction Dirac points. By the delicate analyses on the momentum dependences

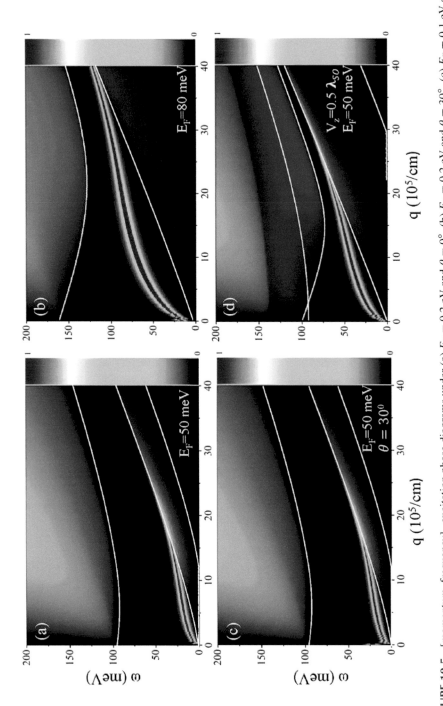

FIGURE 10.5 [momentum, frequency]–excitation phase diagrams under (a) $E_F = 0.2$ eV and $\theta = 0°$, (b) $E_F = 0.2$ eV and $\theta = 30°$, (c) $E_F = 0.1$ eV and $\theta = 0°$; (d) $E_F = 0.2$ eV, $\theta = 0°$, and $V_z = 2\lambda_{SOC}$.

and Landau dampings, the collective excitations are classified into four kinds of plasmons with the relatively strong intensities. First, the free-carrier-induced plasmon
is undamped at small qs under $E_F = 0.2$ eV and $\theta = 0°$ [Fig. 10.5(a)]. With the
increasing momentum, this plasmon will enter the interband excitation region and
experience the gradually enhanced damping. Specially, the partially undamped plasmon exhibits a sharp plasmon peak beyond a sufficiently large q; that is, this mode
could survive within the frequency gap between the interband and intraband excitation boundaries. Secondly, at large $\theta (\sim 30°)$ [Fig. 10.5(b)], the larger-q undamped
plasmon is replaced by the seriously damped plasmon in the intraband excitation region (the red curve). The weak plasmon intensity is due to the replacement of the
excitation gap by the extended and overlapped excitation regions. Thirdly, for a sufficiently low Fermi energy [e.g., $E_F = 0.1$ eV in Fig. 10.5(c)], the plasmon remains
as an undamped mode even at large qs. Such plasmon comes to exist in the enlarged
excitation gap associated with the comparable E_F and E_D. Moreover, a suitable gate
voltage could make this plasmon to disappear under the strong interband Landau
dampings, e.g., $V_z = 2\lambda_{soc}$ in Fig. 10.5(d). In addition, V_z leads to more complicated
single-particle excitation boundaries because of the split energy bands [Fig. 10.1(a)].
The above-mentioned plasmon frequencies present the $\sqrt{q}$-dependence at long wavelength limit, a feature of acoustic plasmon modes. Apparently, the similar plasmon
modes have been identified in a 2D electron gas [40, 44].

Free carriers and gate voltages, as clearly indicated in Figs. 10.5(a)–10.5(d) for
the [q, ω]-dependent phase diagrams, can create the unique single-particle excitations and plasmon modes. The boundaries of intraband and interband electron-hole
transitions (the white dashed and solid curves) should be determined by the Fermi-
momentum states and Dirac points. They might reveal an excitation gap, mainly owing to the separation of valence and conduction Dirac points. By the delicate analyses
on the momentum dependences and Landau dampings, the collective excitations are
classified into four kinds of features with the relatively strong intensities. First, the
free-carrier-induced plasmon is undamped at small qs under $E_F = 0.2$ eV and $\theta = 0°$
[Fig. 10.5(a)]. Specially, the partially undamped plasmon exhibits a sharp plasmon
peak beyond a sufficiently large q; that is, this mode could survive within the frequency gap between the interband and intraband excitation boundaries. Secondly,
the larger-q undamped plasmon is replaced by the seriously damped plasmon in the
intraband excitation region [Figs. 10.5(a) and 10.5(c)]. The weak plasmon intensity
is due to the replacement of excitation gap by the extended and overlapped excitation regions. Thirdly, for a sufficiently high Fermi energy [e.g., $E_F = 0.08$ eV in
Fig. 10.5(b)], the plasmon remains as an undamped mode even at large qs. Such plasmon comes to exist in the enlarged excitation gap associated with the comparable E_F
and E_D. In addition, V_z leads to more complicated single-particle excitation boundaries because of the split energy bands [Fig. 10.5(d)]. The above-mentioned plasmon
frequencies present the $\sqrt{q}$-dependence at long wavelength limit, a feature of acoustic plasmon modes. Apparently, the similar plasmon modes have been identified in a
2D electron gas [40, 44].

The above-mentioned intraband and interband excitations could serve as the very
efficiently Coulomb inelastic scattering channels for the quasiparticle states [53–56].
After the external perturbations of the electromagnetic waves or electron beams, the

doped germanene systems might create the excited conduction electrons above the Fermi energy and the hole quasiparticles in valence and conduction bands below E_F. The screened exchange energy, which characterized by Matsubara's Green's functions, is very suitable in fully understanding the lifetimes of three kinds of excited states [55]. By the delicate calculations, the Coulomb descattering rates of the excited states are predicted to strongly depend on the direction and magnitude of wave vector, valence and conduction bands, and Fermi energies. Furthermore, they clearly show that the intraband single-particle excitations, the interband ones, and the distinct plasmon modes play critical roles in determining the deexcitation behaviors. The unusual Coulomb decay rates are revealed as the oscillatory energy dependence, the strong anisotropy, the nonequivalent valence and conduction Dirac points, and the 2D-EGS behaviors for the low-energy conduction electrons and holes. The predicted Coulomb decay rates could be directly examined from the high-resolution measurements of angle-resolved photoemission spectroscopy on the energy widths of the quasiparticle spectrum at very low temperatures [53,59,60].

10.2 MAGNETOPLASMONS WITH SIGNIFICANT LANDAU DAMPINGS

The magneto-electronic excitation behaviors are fully explored by the close association of the modified random-phase approximation and the generalized tight-binding model [details in Sec. 3.4]. The hooping integrals, spin-orbit couplings, magnetic and electric fields, as well as the electron-electron interactions are responsible for the rich and unique dynamic Coulomb screenings/scattering. Most importantly, magnetic quantization will induce a lot of interband plasmon modes with discrete frequency dispersion relations restricted to highly-degenerate energy states. An intraband plasmon, with a higher intensity and continuous dispersion relation, appears only under the enough high dopings of free carriers. This mode is dramatically transformed into an interband plasmon when the magnetic field is greatly enhanced, leading to abrupt changes in the plasma frequency and its intensity. Specifically, an electric field could separate the spin and valley polarizations and create extra plasmon modes, a unique feature arising from the buckled structure and the existence of noteworthy spin-orbit couplings. Furthermore, the bare and screened response functions present the diverse single-particle and collective excitations.

An intrinsic monolayer germanene, which possesses the highest occupied Landau levels of $n^F = n^v = 0$, only presents the transitions from valence to conduction ones at zero temperature through the Coulomb interactions. By the conservations of momentum and energy, the available single-particle channels are characterized by the excitation frequency $\omega_{ex} = E(\mathbf{k} + \mathbf{q}; n^c) - E(\mathbf{k}; n^c)$ [Fig. 10.2(b)]. As a result, a pair of quantum numbers, $(n^{c,v}, n^{c,v})$, is very suitable in identifying an inter-Landau-level excitation channel and thus a specific magnetoplasmon mode. The similar notation is further utilized for the extrinsic cases (the various dopings). Generally speeaking, there exist the intraband and interband channels (the solid and dashed blue curves). Most importantly, the transition order, being defined as $\Delta n = n^v - n'^c$ or $|n^c - n'^c|$, is very useful for categorizing the single-particle excitation and collective excitations. Also noticed that both interband channels, (n^v, n'^c) and (n'^v, n^c), have the similar excitation behaviors. They have the identical transition frequency and intensity, while Fermi-Dirac distribution functions might be different from each other.

The magneto-bare response functions [Figs. 10.6(a)–10.6(h)], being determined by $Im(\epsilon)$, are very sensitive to the changes in the Fermi energies and gate voltages. For an intrinsic system, there are a lot of delta-function-like peaks in $Im(\epsilon)$ [the solid curve in Fig. 10.6(a)], directly reflecting the very strong van Hove singulariuties. Each prominent excitation structure represents a major inter-Landau-level transition channel. The peak intensity in Eq. (3.23) is proportional to the square of Coulomb matrix element $(< n^c; \mathbf{k} + \mathbf{q} |e^{i\mathbf{q} \cdot \mathbf{r}}|n^v; \mathbf{k} >)$ associated with the wave function overlap between the initial and final magneto-electronic states. Based on the characteristics of the well-behaved Hermite polynomials [61], the single-particle excitation channel, with a lower transition order of Δn, presents the larger Coulomb-matrix element under the smaller transferred momenta. The opposite $(0^v, 0^c)$ transition is true for that with a higher Δn. For example, the three lowest frequency peaks come from $(0^v, 1^c)$, $(1^v, 2^c)$, and $(2^v, 3^c)/(1^v, 0^c)$, $(2^v, 1^c)$, and $(3^v, 2^c)$ from the low to high excitation frequencies, as clearly illustrated by the black curve in Fig. 10.6(a). They belong to the interband channels of $\Delta n = 1$, in which the response intensity declines as ω_{ex} grows. It should be noticed that the threshold excitation channel does not correspond to the $(0^v, 0^c)$ transition because of the vanishing Coulomb matrix elements.

The bare excitation spectra are drastically changed by the doping effects, as shown in Figs. 10.6(c), 10.6(e), and 10.6(g). Apparently, the intraband channels come to exist at lower frequencies and become the dominating ones. For example, if the Fermi energy corresponds to the full occupation of the $n^c = 0$ Landau levels [$n_F = n^c = 0$; $E_F = 50$ meV in Fig. 10.6(c)], the response intensity of the original $(0^v, 1^c)$ interband channel is reduced to half because of the disappearance of the $(1^v, 0^c)$ excitation channel. Apparently, the doping-induced $(0^c, 1^c)$ intraband one presents a very high spectra weight within the low-frequency range. When the Fermi energy crosses the $n^c = 1$ Landau levels [$n_F = 1$; $E_F = 70$ meV in Fig. 10.6(e)], the two channels of $(0^v, 1^c)$ and $(0^c, 1^c)$ are Pauli blocked. The very strong single-particle response, which is due to the $(1^c, 2^c)$ transition, becomes the threshold characteristics. With the further increase of E_F, the initial intraband excitations will dominate the bare response function, e.g., $Im(\epsilon)$ at $E_F = 80$ meV in Fig. 10.6(d). In addition to doping effects, the magneto-Coulomb excitations are greatly enriched under the action of gate voltage. $Im(\epsilon)$ might present more complicated single-particle spectra in the presence of gate voltage [the red curves in Figs. 10.6(a), 10.6(c), 10.6(e), and 10.6(g)], e.g., the blue and (blue and red) shifts, respectively, in the intraband and interband transitions, and the split and reduced peaks of the latter. Apparently, these results are closely related to the spin- and valley-polarized Landau levels. The above-mentioned results clearly show that the strong competitions/cooperations between interband and intraband transitions are easily modulated by carrier dopings and electric fields.

Obviously, the real part of the dielectric function is closely to its imaginary partner through the well-known Kramers-Kronig relations, as clearly shown in Figs. 10.6(b) & 10.6(a), 10.6(d) & 10.6(c), 10.6(f) & 10.6(e), and 10.6(h) & 10.6(g). A pair of opposite asymmetric peaks in $Re(\epsilon)$ correspond to a symmetric peak, in which the drastic changes of the former might vanish at some frequencies. If a zero point in $Re(\epsilon)$ (a low value due to the broadening phenomena) occurs at a small $Im(\epsilon)$, then

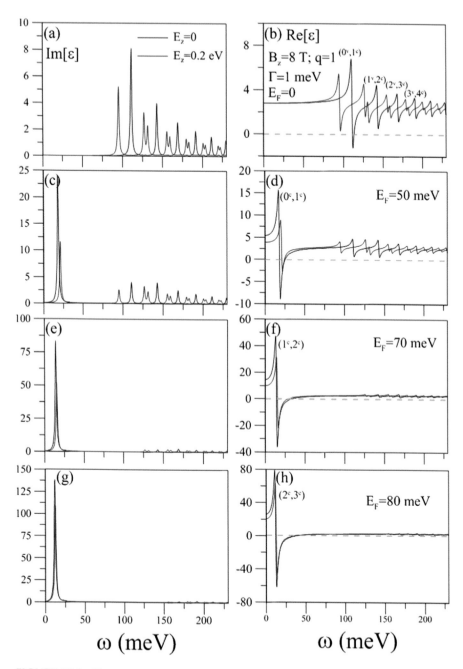

FIGURE 10.6 The magneto-electronic excitations are characterized by the [a, c, e, g] imaginary and real [b, d, f, h] parts of dielectric functions at at $V_z = 0$ & 0.2 eV, $B_z = 8$ T, [in unit of $10^{-3}/\text{Å}$] under the various doping cases: $E_F = 0$, 50, 70 V and 80 meV's, respectively.

this creates the weak Landau damping for the specific magnetoplasmon mode. For example, there are zero points in $Re(\epsilon)$ for the $(1^v, 0^c)$, $(0^c, 1^c)$, $(1^c, 2^c)$, and $(2^c, 3^c)$ excitation channels, respectively, under the various doping cases of $E_F = 0$, 50 meV, 70 meV, and 80 meV [Figs. 10.6(b), 10.6(d), 10.6(f), and 10.6(h)].

The magneto-electronic energy loss functions, which could be measured by inelastic light scatterings [details in Sec. 2.4], are very useful in understanding the close relations between the single-particle and collective excitations. Each spectral peak in $Im(-1/\epsilon (q, \omega; B_z))$, as clearly shown in Fig. 10.7(a)–10.7(d), is regarded as the strong screened response of a magnetoplasmon mode with the different degree of Landau damping. For a pristine system, the most intense interband magnetoplasmon mode is situated located between the single-particle excitation energies of the $(0^v, 1^c)$ and $(1^v, 2^c)$ excitation channels [the black curve in Fig. 10.7(a)]. The observable magnetoplasmon peaks, which possess the universal/dimensionless intensities higher than 1/2, are also created by the initial some inter-Landau-level transitions. The higher the quantum number of Landau level is, the weaker the magnetoplasmon strength is. That is to say, the Landau dampings become stronger in the increase of excitation frequency.

The frequencies $(\omega_p s)$ and intensities of magnetplamson modes strongly rely on the gate voltages and doping levels, V_z makes an undoped system display the neighboring double-peak structures, as indicated by the red curve in Fig. 10.7(a). All the peak intensities decline obviously; furthermore, the frequency (frequencies) of intraband magnetoplasmon (interband modes) present the red shift (the red and blue shifts simultaneously). The spitting effects are the main mechanisms. The interband collective excitations, being shown in Figs. 10.7(b)–10.7(e), are greatly reduced as the Fermi energy grows. For example, at $E_F = 50$ meV, the $(0^v, 1^c)$ peak intensity is less than half of that under the intrinsic case [Figs. 10.7(b) and 10.7(a)], mainly owing to the absence of $(1^v, 0^c)$ channel. It would be very difficult to observe the interband magnetoplamson modes at the higher Fermi energies, e.g., the energy loss spectra at $E_F \geq 50$ meV in Figs. 10.7(c)–10.7(e). On the other hand, the threshold intraband excitation channel creates a very prominent magnetoplasmon peak at the lower frequency for the various Fermi energies. This weakly damped mode might have a small blue-shift frequency under a rather strong bare response [Figs. 10.6(d), 10.6(f), and 10.6(h)].

The (momentum, frequency/magnetic field)-phase diagrams of magneto-electronic excitations could fully provide the informations on the main features of magneto-plasmon modes and Landau dampings, especially for their q- and B_z-dependences in Figs. 10.8–10.10. Generally speaking, the peak frequencies in energy loss secptra present the unusual dispersion relations with the transferred momentum, as clearly indicated by a lot of broadening purple curves in Fig. 10.8(a)–10.8(d). A pristine germanene system in Fig. 10.8(a) only exhibits the interband magnetoplasmon modes, in which the frequeencies of collective charge oscillations quickly grow, locally maximize and then decline in the increment of q. Apparently, the first critical and second critical momenta (q_{c1} and q_{c2}) respectively, correspond to the dramatic transformation of magnetoplasmon frequency and its disappearance. However, a simple relation between q_{c1}/q_{c2} and ω_p is absent. The non-monotonic dispersions directly characterize the strong competitions of the longitudinal Coulomb intercations and

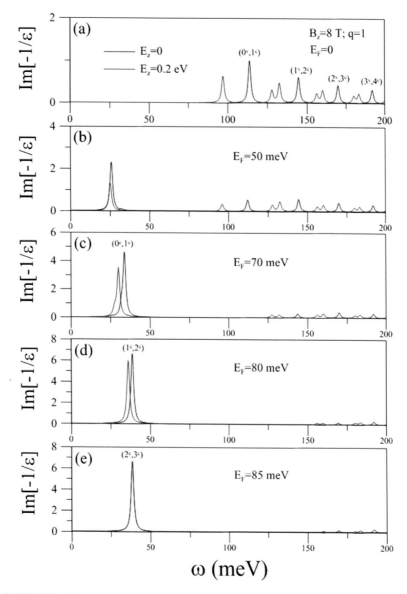

FIGURE 10.7 The magneto-energy loss spectra at $V_z = 0$ & 0.2 eV, B_z and $q = 1$ under the distinct dopings: (a) $E_F = 0$, (b) 50 meV, (c) 70 meV, (d) 80 meV, and (e) 85 meV.

the transverse cyclotron forces, since the plasma waves propagate along the opposite directions under $q < q_{c1}$ and $q > q_{c1}$. Also noticed that the magnetoplasmon modes could survive at smaller transferred momenta, e.g., $q_{c2} \sim 20$–30 10^5/cm. Moreover, the excitation energy spectra become more complicated under the application of gate voltage [e.g., $V_z = 0.2\lambda_{soc}$ in Fig. 10.8(b)]. The splitting of the interband magneto-plasmon modes appears except for the initial $(0^v, 1^c)$ one, in which their frequencies,

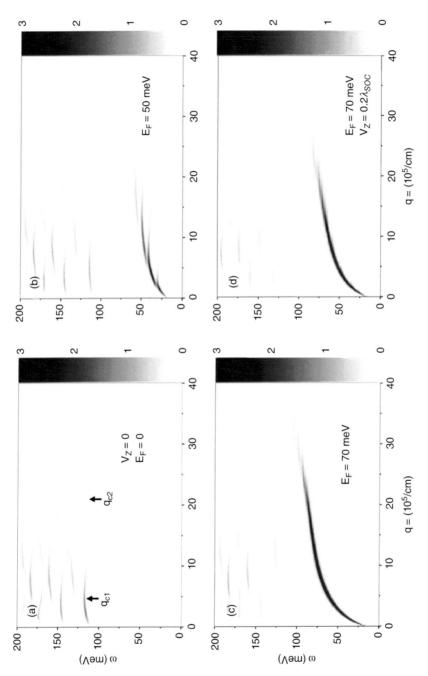

FIGURE 10.8 The [momentum, frequency]-phase diagrams of magneto-electronic excitations at $B_z = 8$ T under the various cases: (a) [$E_F = 0$, $V_z = 0$], (b) [$E_F = 0$, $V_z = 0.2$ eV], (c) [$E_F = 70$ meV, $V_z = 0$], and (d) [$E_F = 70$ meV, $V_z = 0$].

intensities, and critical momenta are greatly reduced. It would be getting difficult to verify the magneto-electronic collective carrier oscillations by the experimental measurements of light inelastic scatterings [21, 63].

The momentum-dependent magneto-excitation spectra present the dramatic transformations after the carrier dopings [Figs. 10.8(b)–10.8(c)]. Obviously, a very intraband magnetoplasmon mode, with a rather strong peak in energy loss function [Figs. 10.7(b)–10.7(d)], comes to exist at small qs. By the numerical fitting, its frequency is roughly proportional to the transferred momentum, i.e., $\omega_p \propto \sqrt{q}$ at long wavelength limit, being similar to the dispersion relation of 2D electron gas [40, 44]. This clearly illustrates the dominance of the longitudinal Coulomb forces on the propagating plasma waves with the very slow variations of the whole charge oscillations. For example, the first critical momenta are, respectively, $q_{c1} = 2$ and 20 for $E_F = 50$ meV and 70 meV. Furthermore, the multi-intraband inter-Landau-level magetoplamsons will appear beyond them and then disappear at the larger transferred momenta. Through the dynamic screening effects, the doping carrier densities have strong effects on the original interband magnetoplasmon modes, including their destruction, the shifts of collective excitation frequencies, and the changes of plasmon peaks.

The magnetic-field dependence of phase diagram is diversified by the doping levels; that is, it exhibits the drastic changes during the variation of Fermi energy, as shown in Figs. 10.9 and 10.10. For an undoped germanene, both frequencies and intensities of magnetoplasmon modes, being purely associated with the interband Landau-level transitions, grow with the increasing field strength monotonously under the fixed transferred momenta, such as, Figs. 10.9(a), 10.9(b), and 10.9(c) at $q = 1$, 5, and 10, respectively. However, a simple relation, e.g., $\omega_p \propto B_z$ or $\sqrt{B_z}$, is absent, mainly owing to the B_z- and SOC-dependent energy gap and the screened effects closely related to $Re(\epsilon)$. As a result, it is relatively to observe the interband magnetplasmons at smaller transferred momenta and higher magnetic-field strengths through inelastic light scatterings [62, 63]. On the other hand, the intraband magnetoplasmons, which arise from the free electrons in conduction Landau levels, display the discontinuous B_z-dependences, e.g., the magnet-electronic phase diagrams at $E_F = 70$ meV and various transferred momenta in Figs. 10.10(a) ($q = 1$), 10.10(b) ($q = 5$), and 10.10(c) ($q = 10$). The Fermi energy remains unchanged in the increment of B_z, while the conduction Landau-level energy quickly increases. Consequently, the $n^c = 1$ magneto-electronic states change into unoccupied ones. Apparently, this is responsible for the abnormal relations, as well as the reduced magnetoplasmon frequencies. Furthermore, a very strong intraband magnetoplasmon, as revealed in a 2D electron gas, might be separated into one prominent mode and several discrete intraband branches at $q = 5$ and 10.

As to Coulomb and magneto-electronic excitations, there are certain important similarities and differences among germanene, silicene, and graphene, mainly owing to the significant/negligible spin-orbit interactions and the distinct/vanishing buckled structures. Under the doping cases, all of monolayer systems can exhibit the similar 2D-EGS-like plasmon modes at long wavelength limit, in which the effects due to the SOC-induced energy spacing are fully suppressed. But at larger transferred momenta, the excitation gap can create an undamped plasmon mode in germanene. The first kind of plasmon is absent in silicene and graphene as a result of the rather small

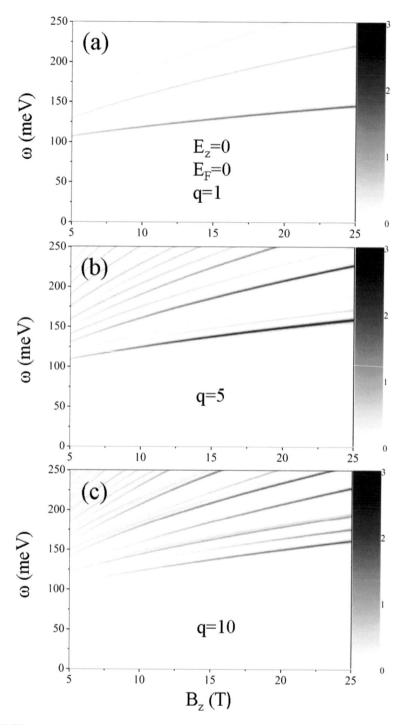

FIGURE 10.9 The [momentum, magnetic field]-phase diagrams at the vanishing Fermi energy and gate voltage under the various transferred momenta: (a) $q = 1$, (b) $q = 5$ and (c) $q = 10$ [in unit of $10^5/cm$].

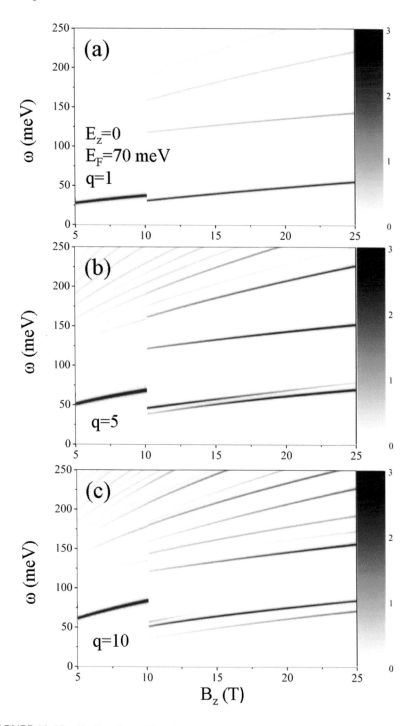

FIGURE 10.10 Similar plot as Fig. 10.9, but displayed for $E_F = 70$ meV.

E_Ds. To observe the angle-dependent plasmon modes, the required Fermi energies are, respectively, about 1.0 eV, 0.4 eV, and 0.1 eV for graphene, silicene, and germanene. The third kind of plasmon is revealed in germanene and silicene, with the requirement of extremely low Fermi energy (~ 0.01 eV) for the latter. Also, silicene can exhibit the fourth kind of plasmon at the lower gate voltage, e.g., V_z is comparable to the effective spin-orbital couplings. Apparently, germanene possesses four kinds of plasmon modes only under the small variation in the Fermi energy. On the other hand, the main features of the effective inter-Landau-level transitions and magneto-plasmon modes are similar for three systems, covering the prominent peaks in some initial bar response functions; the unusual momentum and field-strength dependences in the frequency, intensity, and number of magnetoplamson modes.

On the experimental side, the reflection electron energy loss spectroscopy and the inelastic hard-X-ray scatterings are, respectively, available in verifying the predicted Coulomb and magneto-electronic excitations, especially for the momentum-, doping-, gate-voltage-, and magnetic-field-dependences [details in Sec. 2.4]. The high-resolution measurements will cover four kinds of plasmon modes, the various electron-hole excitations and boundaries, the critical transferred momenta, and the intraband & interband magnetoplasmon modes only with very narrow regions for Landau dampings. Such examinations could provide the full informations about the composite effects due to the critical factors and thus the significant spin-orbital couplings and buckled structure. Also, they are useful in distinguishing the important differences among germanene, silicene, and graphene [9, 10, 32].

REFERENCES

1. Vogt P, Padova P D, Quaresima C, Avila J, Frantzeskakis E, Asensio M C, et al. 2012 Silicene: Compelling experimental evidence for graphenelike two-dimensional silicon *Phys. Rev. Lett.* **108** 155501.
2. Chen L, Liu C C, Feng B, He X, Cheng P, Ding Z, et al. 2012 Evidence for dirac fermions in a honeycomb lattice based on silicon *Phys. Rev. Lett.* **109** 056804.
3. Feng B, Ding Z, Meng S, Yao Y, He X, Cheng P, et al. 2012 Evidence of silicene in honeycomb structures of silicon on Ag(111) *Nano Lett.* **12** 3507.
4. Wang W and Uhrberg R I G 2017 Investigation of the atomic and electronic structures of highly ordered two-dimensional germanium on Au(111) *Phys. Rev. Mater.* **1** 074002.
5. Sadeddine S, Enriquez H, Bendounan A, Das P K, Vobornik I, Kara A, et al. 2017 Compelling experimental evidence of a Dirac cone in the electronic structure of a 2D silicon layer *Sci. Rep.* **7** 44400.
6. Zeppenfeld K 1969 Wavelength dependence and spatial dispersion of the dielectric constant in graphite by electron spectroscopy *Opt. Com.* **1** 119.
7. Papageorgiou N, Portail M, and Layet J M 2000 Dispersion of the interband π electronic excitation of highly oriented pyrolytic graphite measured by high resolution electron energy loss spectroscopy *Sur. Sci.* **454** 462.
8. Marinopoulos A G, Reining L, Olevano V, Rubio A, Pichler T, Liu X, et al. 2002 Anisotropy and interplane interactions in the dielectric response of graphite *Phys. Rev. Lett.* **89** 076402.
9. Wu J Y, Lin C Y, Gumbs G, and Lin M F 2015 Temperature-induced plasmon excitations for intrinsic silicene and effect of perpendicular electric field *RSC Advances* **5** 51912.
10. Shih P S, Chiu Y H, Wu J Y, Shyu F L, and Lin M F 2017 Coulomb excitations of monolayer germanene *Sci. Rep.* **7** 40600.

11. Suenaga K and Koshino M 2010 Atom-by-atom spectroscopy at graphene edge *Nature* **468** 1088.

12. Knupfera M, Pichlera T, Goldena M S, Finka J, Rinzlerb A, and Smalley R E 1999 Electron energy-loss spectroscopy studies of single wall carbon nanotubes *Carbon* **37** 733.

13. Reed B W and Sarikaya M 2001 Electronic properties of carbon nanotubes by transmission electron energy-loss spectroscopy *Phys. Rev. B* **64** 195404.

14. Lin M F and Chuu D S 1997 Impurity screening in carbon nanotubes *Phys. Rev. B* **56** 4996.

15. Ho Y H, Ho G W, Chen S C, Ho J H, and Lin M F 2007 Low-frequency excitation spectra in double-walled armchair carbon nanotubes *Phys. Rev. B* **76** 115422.

16. Low T, Roldn R, Wang H, Xia F, Avouris P, Moreno L M, et al. 2014 Plasmons and screening in monolayer and multilayer black phosphorus *Phys. Rev. Lett.* **113** 106802.

17. Ghosh B, Kumar P, Thakur A, Chauhan Y S, Bhowmick S, and Agarwal A 2017 Anisotropic plasmons, excitons, and electron energy loss spectroscopy of phosphorene *Phys. Rev. B* **96** 035422.

18. Jin F, Roldan R, Katsnelson M I, and Yuan S 2015 Screening and plasmons in pure and disordered single- and bilayer black phosphorus *Phys. Rev. B* **92** 115440.

19. Kato H, Suenaga K, Mikawa M, Okumura M, Miwa N, Yashiro A, et al. 2000 Syntheses and EELS characterization of water-soluble multi-hydroxyl Gd@C82 fullerenols *Chem. Phys. Lett.* **324** 255.

20. Oku T, Hirano T, Kuno M, Kusunose T, Niihara K, and Suganuma K 2000 Synthesis, atomic structures and properties of carbon and boron nitride fullerene materials *Mater. Sci. Eng. B* **74** 206.

21. Stockli T, Bonard J -M, Chatelain A, Wang Z L, and Stadelmann P 2000 Plasmon excitations in graphitic carbon spheres measured by EELS *Phys. Rev. B* **61** 5751.

22. Tomita S 2001 Structure and electronic properties of carbon onions *J. Chem. Phys.* **114** 7477.

23. Lin M F, Chuang Y C, and Wu J Y 2012 Electrically tunable plasma excitations in AA-stacked multilayer graphene *Phys. Rev. B* **86** 125434.

24. Chuang Y C, Wu J Y, and Lin M F 2013 Electric-field-induced plasmon in AA-stacked bilayer graphene *Ann. Phys.* **339** 298.

25. Wachsmuth P, Hambach R, Kinyanjui M K, Guzzo M, Benner G, and Kaiser U 2013 High-energy collective electronic excitations in free-standing single-layer graphene *Phys. Rev. B* **88** 075433.

26. Politanoa A, Radovic I, Borkab D, Miskovic Z L, Yu H K, Farias D, et al. 2017 Dispersion and damping of the interband π plasmon in graphene grown on Cu(111) foils *Carbon* **114** 70.

27. Fei Z, Iwinski E G, Ni G X, Zhang L M, Bao W, Rodin A S, et al. 2015 Tunneling plasmonics in bilayer graphene *Nano Lett.* **15** 4973.

28. Eberlein T, Bangert U, Nair R R, Jones R, Gass M, Bleloch A L, et al. 2008 Plasmon spectroscopy of free-standing graphene films *Phys. Rev. B* **77** 233406.

29. Chang C P, Wang J, Lu C L, Huang Y C, Lin M F, and Chen R B 2008 Optical properties of simple hexagonal and rhombohedral few-layer graphenes in an electric field *J. Appl. Phys.* **103** 103109.

30. Gamayun O V 2011 Dynamical screening in bilayer graphene *Phys. Rev. B* **84** 085112.

31. Shyu F L and Lin M F 2000 Plasmons and optical properties of semimetal graphite *J. Phys. Soc. Jpn. Lett.* **69** 3781.

32. Shyu F L and Lin M F 2001 Low-frequency π-electronic excitations of simple hexagonal graphite *J. Phys. Soc. Jpn.* **70** 897.

33. Chiu C W, Shyu F L, Lin M F, Gumbs G, and Roslyak O 2012 Anisotropy of π-plasmon dispersion relation of AA-stacked graphite *J. Phys. Soc. Jpn.* **81** 104703.
34. Lin C Y, Lee M H, and Lin M F 2018 Coulomb excitations in ABC-stacked trilayer graphene *Phys. Rev. B* **98** 041408.
35. Lin C Y, Lin M C, Wu J Y, and Lin M F 2019 Unusual electronic excitations in ABA trilayer graphene *arXiv*:1803.10715.
36. Das Sarma S and Li Q 2013 Intrinsic plasmons in two-dimensional Dirac materials *Phys. Rev. B* **87** 235418.
37. Gumbs G, Balassis A, and Silkin V M 2017 Combined effect of doping and temperature on the anisotropy of low-energy plasmons in monolayer graphene *Phys. Rev. B* **96** 045423.
38. Patel D K, Ashraf S S Z, and Sharma A C 2015 Finite temperature dynamicalpolarizat ion and plasmons in gapped graphene *Phys. Status Solidi B* **252** 1817.
39. Iurov A, Gumbs G, Huang D, and Balakrishnan G 2017 Thermal plasmons controlled by different thermal-convolution paths in tunable extrinsic Dirac structures *Phys. Rev. B* **96** 245403.
40. Shung K W -K 1986 Dielectric function and plasmon structure of stage-1 intercalated graphite *Phys. Rev. B* **34** 979.
41. Scholz A, Stauber T, and Schliemann J 2012 Dielectric function, screening, and plasmons of graphene in the presence of spin-orbit interactions *Phys. Rev. B* **86** 195424.
42. Ho J H, Lu C L, Hwang C C, Chang C P, and Lin M F 2006 Coulomb excitations in AA- and AB-stacked bilayer graphites *Phys. Rev. B* **74** 085406.
43. Chuang Y C, Wu J Y, and Lin M F 2012 Analytical calculations on low-frequency excitations in AA-stacked bilayer graphene *J. Phys. Soc. Jpn.* **81** 124713.
44. Jackson J D 1975 *Classical Electrodynamics*. New York, Wiley.
45. Wu J Y, Chen S C, Roslyak O, Gumbs G, and Lin M F 2011 Plasma excitations in graphene: Their spectral intensity and temperature dependence in magnetic field *ACS Nano* **5** 1026.
46. Wu J Y, Gumbs G, and Lin M F 2014 Combined effect of stacking and magnetic field on plasmon excitations in bilayer graphene *Phys. Rev. B* **89** 165407.
47. Wu J Y, Chen S C, Gumbs G, and Lin M F 2016 Feature-rich electronic excitations of silicene in external fields *Phys. Rev. B* **94** 205427.
48. Das Sarma S and Hwang E H 1998 Plasmons in coupled bilayer structures *Phys. Rev. Lett.* **81** 4216.
49. Ho J H, Chang C P, and Lin M F 2006 Electronic excitations of the multilayered graphite *Phys. Lett. A* **352** 446.
50. Chuang Y C, Wu J Y, and Lin M F 2013 Electric field dependence of excitation spectra in AB-stacked bilayer graphene *Sci. Rep.* **3** 1368.
51. Wu J Y, Chen S C, and Lin M F 2014 Temperature-dependent Coulomb excitations in silicene *New J. Phys.* **16** 125002.
52. Tabert C J and Nicol E J 2014 Dynamical polarization function, plasmons, and screening in silicene and other buckled honeycomb lattices *Phys. Rev. B* **89** 195410.
53. Bostwick A, Ohta T, Seyller T, Horn K, and Rotenberg E 2007 Quasiparticle dynamics in graphene *Nat. Phys.* **3** 36.
54. Leem C S, Kim C, Park S R, Kim M K, Choi H J, and Kim C 2009 Highresolution angle-resolved photoemission studies of quasiparticle dynamics in graphite *Phys. Rev. B* **79** 125438.
55. Shih P H, Chiu C W, Wu J Y, Do T N and Lin M F 2018 Coulomb scattering rates of excited states in monolayer electron-doped germanene *Phys. Rev. B* **97** 195302.
56. Chiu C W, Ho Y H, Chen S C, Lee C H, Lue C S, and Lin M F 2006 Electronic decay rates in semiconducting carbon nanotubes *Physica E* **34** 658.

57. Ho J H, Lai Y H, Chiu Y H, and Lin M F 2008 Landau levels in graphene *Phys. E* **40** 1722.
58. Ezawa M 2012 A topological insulator and helical zero mode in silicene under an inhomogeneous electric field *New J. Phys.* **14** 033003.
59. Valla T, Camacho J, Pan Z -H, Fedorov A V, Walters A C, Howard C A, et al. 2009 Anisotropic electron-phonon coupling and dynamical nesting on the graphene sheets in superconducting CaC6 using angle-resolved photoemission spectroscopy *Phys. Rev. Lett.* **102** 107007.
60. Fedorov A V, Verbitskiy N I, Haberer D, Struzzi C, Petaccia L, Usachov D, et al. 2014 Observation of a universal donor-dependent vibrational mode in graphene *Nat. Commun.* **5** 3257.
61. Lai Y H, Ho J H, Chang C P, and Lin M F 2008 Magnetoelectronic properties of bilayer Bernal graphene *Phys. Rev. B* **77** 085426.
62. Richards D 2000 Inelastic light scattering from inter-Landau level excitations in a two-dimensional electron gas *Phys. Rev. B* **61** 7517.
63. Eriksson M A, Pinczuk A, Dennis B S, Simon S H, Pfeiffer L N, and West K W 1999 Collective excitations in the dilute 2D electron system *Phys. Rev. Lett.* **82** 2163.

11 Topological Characterization of Landau Levels for 2D Massless Dirac Fermions in 3D Layered Systems

Ching-Hong Ho,[b] Jhao-Ying Wu,[b] Chiun-Yan Lin,[a] Thi-Nga Do,[c,d] Po-Hsin Shih,[a] Shih-Yang Lin,[e] Ming-Fa Lin[a,f,g]

[a] Department of Physics, National Cheng Kung University, Tainan 701, Taiwan
[b] Center of General Studies, National Kaohsiung University of Science and Technology, Kaohsiung 811, Taiwan
[c] Laboratory of Magnetism and Magnetic Materials, Advanced Institute of Materials Science, Ton Duc Thang University, Ho Chi Minh City, Vietnam
[d] Faculty of Applied Sciences, Ton Duc Thang University, Ho Chi Minh City, Vietnam
[e] Department of Physics, National Chung Cheng University, Chiayi 621, Taiwan
[f] Quantum Topology Center, National Cheng Kung University, Tainan 701, Taiwan
[g] Hierarchical Green-Energy Materials Research Center, National Cheng Kung University, Tainan, Taiwan

CONTENTS

11.1   SOME TOPOLOGICAL ASPECTS

Topology originates as a branch of mathematics, which is concerned about classifying the spatial properties of objects with the number of holes, termed *genus*, in the object surfaces [1]. If an object has nonzero genus, any closed path around a hole in the surface cannot be smoothly shrunk to a point. In case of that, the surface is *topologically nontrivial*; otherwise, one has a *topologically trivial* surface with zero genus. The central idea is the *topological equivalence* between objects that have the same genus and can be mapped to each other by continuous deformation with the genus kept invariant. There is the homotopy group, which contains all the topologically equivalent objects of a certain class. The concept of topology can be naturally generalized to abstract things, even to those in the physical world. For condensed matter physics, the discovery of the integer quantum Hall effect (QHE) [2] and the following topological interpretation [3–5] ought to be cited as the pioneering works. Nowadays, a new branch of physics has been built with extensive research on the topological matter [6–8].

In the paradigmatic grand unification theory, a given thermodynamic equilibrium state of a quantum field or matter in Universe is characterized by the symmetry group (denoted $\mathcal{H}$) of, say, the Hamiltonian [9], which leads to the relevant equation for the considered system. Fundamentally, $\mathcal{H}$ is a subgroup of the largest symmetry group $\mathcal{G}$ of physical laws. Through cooling downward the ground state, $\mathcal{H}$ gets more and more reduced by successive spontaneous symmetry breaking processes. This fact puts the theoretical basis for classifying the states of physical systems in connection to phase transitions as described in, for example, the Landau's theory [10]. That is, as symmetry is reduced, particles in a system tend to organize themselves so that the state can transit into more ordered phases, where the phase transitions occur due to spontaneous symmetry breaking in company with changes of certain local order parameter. In this approach, it is possible that there can appear topological defects [11], such as solitons or domain walls, in real space of an ordered medium, like holes in the surface of an object. The determination of such phases is generally given by the nontrivial elements belonging to the homotopy group $\pi_n(\mathcal{H}/\mathcal{G})$, which is defined on S^n, the n-sphere. Also, phase transitions of topological defects can occur with changing $\pi_n(\mathcal{H}/\mathcal{G})$, as a result of spontaneous symmetry breaking.

So far, one might conclude that topology is a consequence of symmetry in the cold Universe he lives in. This is, however, not in the trend of current topological research on condensed matter. There is, indeed, an alternative approach that starts with the ground states at zero temperature, where topology is presumed to show up *a priori* while symmetry is taken to be emergent [12]. In contrast to the reciprocal relation between local order and symmetry in the approach of the grand unification theory, the topological phases appearing here have distinguished kinds of order, which cannot be characterized by symmetry alone. Such *topological order* is characterized and classified by suitable quantum numbers, namely, topological invariants [8, 13]. Like the genus in an object's surface, a nonzero topological invariant marks a nontrivial topological phase while a zero topological invariant marks trivial. The relevant transition between different topological phases, named topological *quantum phase transition* (QPT), does not involve spontaneous symmetry breaking, whereas it occurs when the topological invariant changes at the critical point [8, 14]. In the integer

QHE, the topological QPT between Hall plateaus takes place with changing the first Chern number (TKNN), which is the topological invariant as given by the Hall conductivity [3].

With respect to long- or short-range entanglement of wave functions in the systems considered, there could be nontrivial topological phases being robust against arbitrary or symmetry-constrained local perturbations, respectively. The phases are then divided into two respective categories: the topologically ordered phases [8, 15, 16] and the symmetry protected topological (SPT) phases, respectively [8, 17–24]. For example, the fractional QHE [25] and high T_c superconductors [26] are nontrivial topological phases in the first category [8]. On the other hand, there are abundant nontrivial topological phases theoretically or experimentally found in noninteracting fermionic lattice, which are largely categorized as SPT phases [8]. An SPT phase is nontrivial as the disorder or parameter tuning respects the relevant symmetries and is only perturbative. In the single-particle band structure, the topological invariant of an SPT phase is determined with respect to the Berry phase describing the global phase evolution of the wave functions [27]. If the bulk band is fully gapped for an insulating phase, the Berry flux over the entire Brillouin zone (BZ) is well defined [28]. Therefore, the topological invariant can change only when the energy gap is closed. One can think of two spatially adjacent insulators with different topological invariants in their bulk. Then, there must be a gapless state localized at the boundary between them. Since the vacuum can be taken as a trivial insulator, an SPT insulating phase of a topological insulator (TI) must be uniquely characterized by a gapless state on its surface or edge, which faces the vacuum. That is the so-called *bulk-surface correspondence*, a relationship notionally invoked by the topological QPT. For an SPT phase, the surface states are protected by the bulk symmetry and take a form depending on the symmetry [29–33].

Also, the notion of topological QPT may lead to the emergence of topological semimetals (TSMs) in nature. Recall the thought underlined in the bulk-surface correspondence; now again, think of two spatially adjacent systems, with one of them being a TI while the other, a TSM. This two-system picture can be identified as a three-system picture, where the TSM is intervened between the TI and the vacuum (a trivial insulator). Thus, the possibility of a TSM mediating between two insulators is expected. In general, a TSM has band crossing between bands with the same symmetry, as opposed to the avoided level crossing stated in the von Neumann-Wigner's theorem for trivial semimetals [34]. Here a node can be identified where the band crossing (almost) is generated at the Fermi level. To described the nodes, a condition, say, for a two-band model, is given by $d_f - d_n + c_s = 2 \times 2 - 1 = 3$ with d_f the degrees of freedom, d_n the nodes' dimension, and c_s the number of symmetry constraints. It is noted that d_f depends on how the nodes are constrained. If the maximum $c_s = 3$ is required, the one acquire $d_n = d_f$ so that the nodes are fixed at the symmetry points or lines. These so-called *essential* band crossings are completely protected by symmetries, which are crucially nonsymmorphic group symmetries, possibly together with nonspatial symmetries [35–37], and are not considered here. On the other hand, in case that $c_s < 3$, the band crossing is *accidental*. The significance of *accidental* nodes was issued in the early days for time reversal invariant crystals [38]. Nodes at *accidental* band crossings acquire degrees of freedom to adapt to perturbative parameter tuning, which preserves the required symmetries, to be moved over generic points in the BZ. Their stability depends on if they are resided in the interior of a finite and connected region in the parameter space, and they are unstable, or vanishingly improbable, at the margin of the region. That is, the nodes are only perturbatively stable and they can be removed when the deformations are too large. Return to the topological QPT. Now, for the phase transition from an SPT insulating phase to a trivial insulating phase, at the critical point the band gap should be closed somewhere in the BZ via *accidental* band crossing, which is marginal [39,40]. Inspired by the two-system picture thought above, one would expect the possibility of a TI-to-TSM transition, which amounts to broadening the margin between TI-to-trivial insulator so as to have a region in the parameter space for *accidental* band crossing to reside inside. These stable nodes give rise to a TSM [41,42].

In comparison to the topologically ordered phases, which are robust against arbitrary perturbation, the SPT phases are, in a sense, less robust since they need the protection of certain symmetry. The perturbation an SPT phase can survive is limited to preserve the symmetry specific to the very phase. Hence, it is of fundamental importance to characterize and classify the SPT phases of condensed matter. For this purpose, one should know that the severest perturbation comes from disorder. Since early days, physicist have known that in a noninteracting electronic system in spatial dimension $d \leq 2$, the ground-state wave function (at zero temperature) can be localized by disorder in the thermodynamic limit [43]. This so-called *Anderson localization* stands as the basis for the QPT. The advent of the scaling theory introduced the universal classification of the localization behavior for noninteracting fermionic systems in accordance to symmetry and spatial dimensionality [44,45]. This classification was solidly expressed later by using the random matrix theory [46,47]. After including the new class of anti-localization found in the time reversal invariant TIs and topological superconductors (TSCs), a *tenfold way* classification was established for noninteracting insulating systems, where ten universal symmetry classes are given, with characteristics depending on the spatial dimensionality in a period-8 periodic table [17,48]. The ten symmetry classes comprise the combination of three nonspatial symmetries with regard to the presence or absence of electron spin $SU(2)$ rotation symmetry. Spatial symmetries are excluded from the *tenfold way* classification because they are prone to disorder. The three nonspatial symmetries are time-reversal symmetry (TRS), particle-hole symmetry (PHS), and sublattice symmetry (SLS). It is noted that in the context of the random matrix theory, SLS is alternatively called chiral symmetry (CS) owing to that the two sublattices lead to a spinor. Both TRS and PHS are antiunitary while SLS is unitary. Of them, anyone can be derived from the product of the other two. Such a way of classification can be proved mathematically using the K theory [18].

The situation of SPT semimetallic phases is more involved because of their nodal band structures, where the nodes are pinned to or close to the Fermi energy $E_F = 0$. A node leads to singularity in the BZ and, therefore, one cannot derive a topological invariant by integration over the whole BZ. Thus, another *tenfold way* classification has also been devoted to semimetallic phases with nodes of dimension d_n at *accidental* band crossings [20–23,50], where the codimension $p = d - d_n$ is taken into account. This classification is defined for a submanifold that is gapped around the node so that the topological invariant is acquired in the submanifold. Such nodes are stable individually, to the extent of symmetry-preserving disorder or perturbation of parameter tuning, when they are characterized by nonzero topological invariants. In contrast to the case of SPT insulating phases, the protection of nodes in the BZ often is achieved by the nonspatial symmetries together or in combination with certain spatial symmetries [49,50]. In the continuum limit in the vicinity of a single node between two bands the quasiparticles show up as two-component fermions, which are, however, forbidden in a lattice by the Nielsen-Ninomiya's no-go theorem [51–53]. The theorem in the framework of topological condensed matter states that in a local-action, real and noninteracting fermionic lattice, the number of species of existing fermions must be double with opposite topological charges, say, chiralities. That is, there must be as many fermions of positive chirality as of negative chirality and consequently the components of the fermionic wave function are doubled. One thus obtains four-component Dirac fermions, instead of two-component fermions. This *fermion doubling* problem arises when a continuous field are discretized into a lattice. It has a topological origin. Due to the opposite topological charges of the nodes existing in pair, the Berry phase obtained by integration along the border of the BZ must vanish. Hence, the fate of node annihilation cannot be got rid of as the nodes at accidental band crossing are moved toward each other under symmetry-preserving disorder or perturbation. Around linearly crossing nodes, i.e., Dirac points (DPs), the Dirac fermions hosted are massless. Massless Dirac particles are ubiquitous in condensed matter, for they obey certain loosely identified Dirac equations [54] as compared to the genuine Dirac equation in the relativistic quantum mechanics [55]. In this context the Lorentz invariance has dropped out from the mimic Dirac equations. Moreover, the components of wave functions need not comply with the requirement of the genuine Dirac equation (2^d for $d = 2, 3$); rather, it depends on the number of spinors adapted to the system under consideration. Those spins are acquired *a priori* in contrast to the electron spin, the latter being intrinsically derived from the genuine Dirac equation. In the forbidden case of two-component Weyl fermions around a single nodes, the Hamiltonian is naively casted as $H(\mathbf{p}) = v\sigma \cdot \mathbf{p}$, with v the Fermi velocity and σ the Pauli matrices acting on the spinor. Obviously, the eigenstates of $H(\mathbf{p})$ are chiral since they are simultaneous eigenstates of the helicity operator $h = \sigma \cdot \mathbf{p}/|\mathbf{p}|$ [18, 19]. This means that $H(\mathbf{p})$ has a CS effectively derived in d dimensional space, differing from the CS defined in $(d + 1)$ dimensional Minkowski spacetime.

The first demonstration of 2D massless Dirac fermions in condensed matter was addressed on the 2D honeycomb lattice [56], where spinless condition or spin degeneracy ($SU(2)$ symmetry) was assumed while the two sublattices come into play as a pseudospin. With only the nearest neighbor tight binding hoppings (minimal model), there would exist two DPs in the valleys around which 2D massless Dirac fermions of chirality $+1$ and chirality -1 are separately hosted. This leads to another pseudospinor [57]. Hence, four-component 2D massless Dirac fermions have been recognized to be promising in an analog to the $(2 + 1)$-dimensional gauge field. The possibility has been proposed in the 2D limit of graphite, which was shown to be stable against impurity without intervalley scattering [58]. As was later realized and extensively studied [59], these results have been well realized from spinless graphene. On the basis of these two pseudospinors, the Dirac Hamiltonian can be derived in the continuum limit around the two DPs. Within the minimal model, it is given by $H(\mathbf{k}) = v\hbar\tau_3\boldsymbol{\sigma}\mathbf{k}$, with $\hbar\mathbf{k}$ the crystal momentum measured from each DPs, v the Fermi velocity defined by the nearest neighbor lattice site distance and hopping integral, and σ and τ the Pauli matrices acting on the sublattice and the valley spinor, respectively. In this Dirac Hamiltonian, intervalley coupling is absent. This Dirac Hamiltonian carries CS if intervalley is absent, as a result of SLS present in the relevant lattice Hamiltonian [57]. The CS of the Dirac Hamiltonian is termed continuous CS in the following. With λ and ξ denoting the band and valley indices, respectively, the energy of Dirac fermions is given by $E_{\lambda\xi}(\mathbf{k}) = \lambda v\hbar k$. The wave function is given by $\psi_{\lambda\xi}(k) = (1/\sqrt{2})\left(\exp(-i\xi\theta/2), \lambda\xi\exp(i\xi\theta/2)\right)^T$, where $\mathbf{k} = (k, \theta)$ has a phase defined conventionally. It is easy to show that a winding number characterizing the CS class of the Dirac Hamiltonian is given by $w_1 = \lambda\xi$ with respect to the chirality. In the presence of a uniform, perpendicular magnetic field $B\hat{\mathbf{z}}$, continuous CS is still preserved via the minimal coupling. This fact is simply manifested in clean graphene. The Landau level (LL) energy spectrum is obtained as $\epsilon_n = \text{sgn}(n)\hbar\omega_c\sqrt{|n|}$, with $\omega_c = v\sqrt{2eB/\hbar c}$, where the $\sqrt{B}$ function reflects the linear dispersion. The LL energy spectrum is characterized by a zero mode that is constantly pinned to zero energy in a wide range of magnetic field as revealed in the Hofstadter butterfly [60]. The LL wave functions are given by $\Psi(\mathbf{r}) \propto \left(\phi_{|n|}, \text{sgn}(n)\phi_{|n|-1}\right)^T$, with $\phi_{|n|}$ being the nth simple harmonic oscillator wave function, so that the zero-mode LL is pseudospin polarized. Therefore, the zero-mode LL is still chiral and half-filled with respect to the DPs. It is robust under the protection of the continuous CS [57,61,62]. It is remarkable that the chiral zero-mode LL can serve as a mark of 2D massless Dirac fermions existing in a noninteracting fermionic lattice with SLS and the resulting continuous CS. With the zero-mode LL, the half-integer quantized Hall conductivity $\sigma_{xy} = 4(n + 1/2)e^2/h$ has been inferred in the context of spinless graphene in the quantum Hall limit [63,64], which was realized soon later [65,66]. It has been understood that the half-integer QHE is dictated by the chiral, half-filled zero-mode LL [61,67], which is ascribed to continuous CS of 2D massless Dirac fermions. Hence, disorder that degrades continuous CS would wash away such a characteristic and makes the half-integer QHE impossible. It should be noted that the once upon a transition between Hall plateaus, a topological QPT takes place as the chemical potential passes through an LL where the criticality should be characterized [48,68,69]. A significant characteristic of continuous CS is that it anomalously dominates the criticality at the zero-mode LL [47,67,70,71], which is preserved by random happoins, irrespective of that the Hall plateaus are assured by existing localization states due to disorder in ordinary integer QHE [72].

The present concern is about the topological characteristics of 2D massless Dirac fermions that are hosted in the vicinity of nodes in 3D layered systems. Generally speaking, it is easy to figure out that there are only three types of nodes with massless Dirac fermions in 3D noninteracting lattice systems [73]. In the first type shown in Fig. 11.1(a), 3D massless Dirac fermions are hosted around isolated DPs in the bulk, where the Dirac cones exhibit linear dispersion in all the three dimensions. The notable Weyl semimetals [74–78] and Dirac semimetals [79, 80] are of this type and are excluded. Intuitively, the possibility for 2D massless Dirac fermions existing in 3D is rendered by the rest two types. In the second type, 2D massless Dirac fermions exist in codimension $p = 2$ around Dirac nodal lines (DNLs) in the bulk of a 3D

(a)

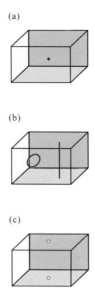

(b)

(c)

FIGURE 11.1 Types of nodes in the 3D BZ. (a) Nodal points inside the bulk. (b) Two types of nodal lines inside the bulk. The line can be an interior loop or a periodic line across the BZ boundaries. (c) Nodal points on the 2D surface BZ.

BZ, as shown in Fig. 11.1(b). There are infinitely many Dirac cones along a DNL, each dispersing linearly in the 2D plane of its own. The stable existence of DNLs indicates a nodal-line TSM. On each of the surface layers, the surface state manifests itself by a drumhead flat band corresponding to each bulk DNL. The simplest model is, among others, the rhombohedral lattice, which is the rhombohedral stack of honeycomb-lattice layers. The physical realization has been found from spinless rhombohedral graphite (RG) [81–86]. The first and second types comprise TSMs inclusively [87–89]. On the other hand, the third type includes those having fully gapped bands in the bulk and corresponding surface Dirac cones, where 2D massless Dirac fermions are hosted on each surface, as shown in Fig. 11.1(c). This type indicates 3D time reversal invariant TIs [39, 40, 90, 91].

At final, the special character of 3D layered systems is remarked. When identical layers of 2D noninteracting fermionic lattice are stacked into 3D, the tight binding interlayer hoppings, whether strong or weak, would bring the stack into a topological class that is either the same as or different from a single layer. People usually tend to use a model in the 2D limit for a 3D layered system provided that the interlayer coupling is weak compared to certain attributes of the 2D lattice. However, the quasi-2D model does not always work. The most famous example is the spinless 2D graphite sheet (nowadays, graphene), which were for modeling graphite [92]. It has been known, from the magneto-electronic and magneto-optic properties [83, 84, 93–95], that this quasi-2D model fails for Bernal graphite but reveals mimic results for RG. Other examples have been found in some organic conducting salt, like α-(BEDT-TTF)$_2$I$_3$, or strained graphene [96], in which interlayer magneto-resistance plays a role though the quasi-2D model is frequently employed. The success of the quasi-2D model for spinless RG suggests the topological characteristic inherited. When a 2D layer of lattice possesses node points in the bulk, whether it being realistic material or being an idealized model, one might ask of what if the layers are stacked to 3D in some ways. For systems having a quasi-2D used for granted, such as d-wave cuprate superconductor with negligible interlayer coupling [97], one would naively derive vertical nodal lines from the nodal points of each layer [50]. However, in stacks of graphene layers, where the interlayer hoppings are significant, the results could disperse in the stacking dimension in various manners depending on the stacking configurations [98]. Bernal (AB stacked) graphite has a nonsymmorphic lattice [99]. In such a system, nexuses can show up where nodal lines merge [85, 100]. By contrast, in spite of that DPs are known to exist in the π-flux square lattice, chiral TIs and Weyl semimetals can be obtained by stacking layers of the lattice into a specific π-flux cubic lattice [74]. All these are examples of the dimensional crossover from 2D to 3D [14]. The (co)dimensional-periodic table in the *tenfold way* classification dictates that the topological class would or would not alter after the dimensional crossover. RG is a 3D layered system consisting of graphene layers stacked in rhombohedral (ABC) configuration. From graphene to RG, the topological phase changes from a 2D TSM with DPs to a 3D TSM with DNLs in a dimensional crossover [82, 86], where the codimension $p = 2$ holds. As having illustrated [Fig. 11.1(b)], the DNLs in spinless RG fall into the second type of nodes, around which 2D massless Dirac fermions are hosted as well as in spinless graphene. In the presence of a perpendicular magnetic field, a 3D layered system has the easiest way to satisfy the Diophantine equation for LL gaps opened in the 3D bulk, as is required for the 3D integer QHE [101]. It is interesting to compare the 2D integer QHE of a single layer to the 3D integer QHE of the layered system [102]. A comparison between graphene and RG is of particular interest, where 2D massless Dirac fermions hosted in both systems are responsible for the existence of a robust zero-mode LL with respect to the Dirac nodes, which, in turn, is a necessary condition of the half-integer QHE. The 3D half-integer QHE held has been experimentally confirmed [103, 104]. A schematic plot of 3D integer QHE in a 3D layered system is provided in Fig. 11.2.

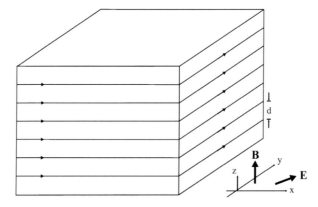

FIGURE 11.2   Schematic plot for 3D QH effect with spin current along the chiral edges or for 3D weak TI with spin current along the helical edges.

11.2   CHIRAL SYMMETRY PROTECTED NODAL-LINE TOPOLOGICAL SEMIMETALS

Here, RG as a 3D layered system consisting of graphene layers is introduced, with the assumption of spinless condition or $SU(2)$ symmetry. This system has been well analyzed but less topologically characterized in spite of the well understanding of graphene. Spinless graphene is not only deemed a realistic system with negligible spin orbital coupling (SOC) or a toy model, but also anticipated to be an artifact on the honeycomb lattice [105] as being realizable by means of cold atoms in optical lattices [106], where the tight-binding hoppings are included to adapt to the model. The honeycomb lattice is symmorphic [99], for which the relevant point-group symmetries are set forth as follows [107], referring to Fig. 11.3(b). There is a C_3 rotation (by $2\pi/3$) symmetry at each of the A (B) sublattice sites. Also, there is a mirror reflection symmetry (MRS) with the reflection axis put between A and B; yet, another one, between A (B) and A (B), is present as well but does not play an important role for protecting the DPs. It is the C_3 rotation symmetry together with certain symmetries, if being preserved in the model Hamiltonian, that makes the two DPs fixed at the K

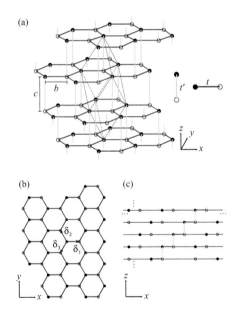

FIGURE 11.3 (a) Rhombohedral lattice of 3D stack of 2D layers of honeycomb lattice, where the rhombohedron (red) is the biparticle primitive unit cell. The present minimal model is described in terms of the intralayer hopping t and interlayer hopping t'. The two sublattices are respectively given by solid and open dots. (b) Honeycomb lattice of each 2D layer. The nearest-neighbor sites associated with hopping t are connected by three vectors δ_m. The two sublattices are respectively given by solid and open dots. (c) Schematic of the 3D extension of SSH model constructed by a rhombohedral stack of 2D honeycomb-lattice layers. One representative chain is shown by linked thick sticks where intralayer hopping t (blue) and interlayer hopping t' (yellow) take place.

and K' points of the BZ [49, 107]. This *essential* band crossing is only limited, not required for the existence and stability of DPs at more generic momenta with *accidental* band crossing [108]. For the model including up to the next-nearest neighbor hoppings, TRS together with MRS can be present and protects the DPs [50, 105], which are located on the reflection axis. This SPT semimetallic phase belongs to the symmetry class AI with MRS. The associated topological invariant is given by the integer $Z_{R^\pm}$ ascribed to the eigenvalues of MRS, where TRS is respected in both the eigenspaces [50]. This model could also render DPs each protected by space-time inversion symmetry (PTS), i.e., the combination of TRS and space inversion symmetry (IS), which commutes with the Hamiltonian, i.e., $[H, \mathcal{TI}] = 0$ [109, 110]. The relevant symmetry class also is AI, but now the topological invariant is given by the binary number Z_2 for each DP because $\mathcal{TI}$ transforms a DP into itself in the BZ. This fact can be described in spatial dimension $d = 7$ in the periodic table in the *tenfold way* classification, as if the dimension of the submanifold (S^1 here) around each DP were -1 [111, 112]. It is noted that PTS can protect DPs at generic momenta off the reflection axis in the case of generally anisotropic hoppings due to random hopping, say, while TRS together with MRS provides the protection up to the case of the hoppings do not break MRS [108].

If the model for spinless graphene is reduced to merely include the nearest neighbor hoppings, there is SLS additionally, which coexists with PTS. In this model, SLS is preserved under the presence of IS as a necessary condition. Hence, SLS should be respected in each eigenspace of PTS [49]. As mentioned above, SLS behaves as CS so that continuous CS in S^1 around each DP is present in the absence of intervalley coupling. The SLS operator anticommutes with the Hamiltonian, i.e., $SHS^{-1} = -H$, and transforms an eigenstate to its conjugate that has inversed eigenenergy and an eigenfunction having one of the two components inversed on the two sublattices, i.e., $S\psi_E = \psi_{-E}$ with $\psi_E = (a, b)^T$ and $\psi_{-E} = (a, -b)^T$ [21, 57]. This minimal model also belongs to an additional symmetry class BDI, where the present SLS comes out to be the combination of TRS and PHS. The energy symmetry between particles and holes is held by SLS as well as PHS. In the momentum representation, the topological invariant derived from the Berry phase is now identical to the winding number $w \in Z$ [21, 49], which is defined in each S^1 submanifold [113]. The bulk-edge correspondence can be deduced from a knowledge of the winding number, manifesting itself by flat bands pinned to the Fermi energy $E_F = 0$ on the zigzag edges [30, 32]. Moreover, as a result of SLS, the chiral zero-mode LL is protected by continuous CS and pinned to $E_F = 0$ as well [57, 62]. Beyond the minimal model, the DPs are protected by PTS only. Inclusion of the next-nearest neighbor hoppings destroys SLS and introduces significant effects. In this model, the DPs shift away from the Fermi energy and the surface bands change from being flat to being linear; besides, the Dirac cones become tilted where tilted 2D massless Dirac fermions are hosted, similar to certain organic conducting salt [96]. Nevertheless, it is remarkable that tilted massless 2D Dirac fermions still have generalized continuous CS so that the zero-mode LL is protected [114, 115].

The lattice of spinless RG, as shown in Figs. 11.3(a) and 11.3(c), is the 3D rhombohedral (ABC) stack (along the z direction) of 2D honeycomb-lattice layers (coordinated by (x, y)). Obviously, the lattice consists of two sublattices as well as the

(a)

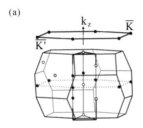

(b)

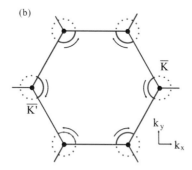

FIGURE 11.4 (a) Projections (red circles) of the spiraling DNLs on the 2D projected BZ (blue hexagon), where the portions of DNLs inside (solid) and outside (dotted) are shown. To sum up, there are two inequivalent DNLs in the 3D rhombohedral BZ. The arrows indicate the spiraling senses in the increase of k_z. (b) 3D rhombohedral BZ (red), with the unfilled dots on the high-symmetry points, in company with the 2D projected BZ.

honeycomb lattice does; also, it is symmorphic [99]. Regarding the point group symmetries, it is easy to know that there is no any rotation symmetry about the hexagonal K or K' lines, referring to 11.4(a) and 11.4(b). Moreover, MRS is absent from this lattice while IS is present. The lattice Hamiltonian including up to the next-nearest neighbor hoppings follows as $\mathcal{H} = \beta_0 \sum_l \sum_{\langle i,j \rangle} [a_{l,i}^\dagger b_{l,j} + \text{h.c.}] + \beta_1 \sum_l \sum_i [b_{l+1,i}^\dagger a_{l,i} + \text{h.c.}] + \beta_4 \sum_l \sum_{\langle i,j \rangle} [a_{l+1,i}^\dagger a_{l,j} + b_{l+1,i}^\dagger b_{l,j} + \text{h.c.}]$, with $a_{l,i}^\dagger$ ($b_{l,i}^\dagger$) create fermions in the sublattice A (B) at site i on layer l, where l labels the layers and $\langle i,j \rangle$ denotes both the nearest neighbor intralayer and the next-nearest neighbor interlayer hoppings. Besides the nearest neighbor intralayer (β_0) and interlayer (β_1) hoppings, the next-nearest neighbor interlayer (β_4) hoppings are included optionally. Under the Fourier transformation $a_{l,i} = N^{-\frac{1}{2}} \sum_{\mathbf{k}} e^{-i\mathbf{k} \cdot \mathbf{r}_{l,i}} a_{\mathbf{k}}$, (similar for $b_{\mathbf{k}}$), the Bloch Hamiltonian is written as $H(\mathbf{k}) = \sum_{i=0,1,2} d_i(\mathbf{k})\sigma_i$, with the Pauli matrices acting on the space spanned by the two sublattices ($a_{\mathbf{k}}, b_{\mathbf{k}}$), where $d_1(\mathbf{k}) = -\beta_0 \sum_m^3 \cos(\mathbf{k}_\parallel \cdot \delta_m) + \beta_1 \cos(k_z c)$, $d_2(\mathbf{k}) = \beta_0 \sum_m^3 \sin(\mathbf{k}_\parallel \cdot \delta_m) - \beta_1 \sin(k_z c)$ and $d_0(\mathbf{k}) = 2\beta_4 \sum_m^3 \cos(\mathbf{k}_\parallel \cdot \delta_m - k_z c)$ are obtained with δ_m the three vectors connecting nearest neighbor lattice sites in a layer and c the layer distance.

The symmetries of the Bloch Hamiltonian $H(\mathbf{k})$ is described as follows. At first, IS is preserved as $\mathcal{I} H(\mathbf{k}) \mathcal{I}^{-1} = H(-\mathbf{k})$, with the operator $\mathcal{I} = \sigma_1$ [109]. Regarding the nonspatial symmetries, TRS is preserved as $\mathcal{T} H(\mathbf{k}) \mathcal{T}^{-1} = H(-\mathbf{k})$, with the operator

$\mathcal{T} = \sigma_0 \mathcal{K}$, where σ_0 is the 2×2 identity matrix and $\mathcal{K}$ is the complex conjugation operator. In the presence of TRS, SLS and PHS are preserved or not in company. For SLS with $S = \sigma_3$, it is preserved as $SH(\mathbf{k})S^{-1} = -H(\mathbf{k})$ if $\beta_4 = 0$ (minimal model) while being broken if $\beta_4 \neq 0$. It is easy to verify the situation of PHS with the operator given by $C = \sigma_3 \mathcal{K}$. Therefore, the model for spinless RG including the next-nearest neighbor hoppings belongs to the symmetry class AI, in terms of the *tenfold way* classification. This class, however, does not allow any nontrivial nodes at accidental band crossing, whether the codimension $p = 2$ or 3 [50]. The protection of nodes in this 3D layered system needs additionally certain spatial symmetries. Indeed, there is a large category of spinless nodal-line TSMs that is protected by PTS, i.e., $[\mathcal{H}, \mathcal{T}\mathcal{I}] = 0$ [116–118], and spinless RG falls into this category. It can be shown that, given PTS, DNLs can stably exist in spinless RG under the protection of nonzero Z_2 topological invariant [12]. When SLS and PHS is recovered to coexists with PTS within the minimal model, the system belongs to an additional class BDI. Now, the winding number $w_1 = (2\pi i)^{-1}\nabla_{\mathbf{q}} \log \sigma_3 H(\mathbf{k})$ can be defined in each S^1 submanifold in momentum representation [82].

To determine the existing nodes in spinless RG, here one has three variables (k_x, k_y, k_z) and two equations ($d_1 = d_2 = 0$) from the Hamiltonian $H(\mathbf{k})$. Hence, the solution generally manifests itself by lines in 3D k space. The very character of the present layered system is disclosed by not only the typical values of the hopping integrals, $\beta_0 \gg \beta_1 \gg \beta_4$, but also the form of $H(\mathbf{k})$. It guides one to find the zeros around the hexagonal K and K' lines with respect to graphene [83, 93]. As expected, a pair of DNLs has been found as shown in 11.4(a) and 11.4(b), which are almost exactly expressed by $\hbar k_{DL} = \beta_0/v_0$ and $\phi_{DL} = \xi k_z c$ in polar coordinates $\mathbf{k} = (k, \phi)$ with respect to the K and K' lines, where $v_0 = 3a\beta_0(2\hbar)^{-1}$ is defined for a graphene layer with a the nearest neighbor site distance and $\xi = \pm 1$ respectively denote the two DNLs. The two DNLs appear to spiral around the K and K' lines respectively in opposite senses across the BZ boundaries from $k_z = -\pi$ to $k_z = \pi$. One can carry out a coordinate transformation in terms of (q, θ) measured from the DNLs at constant k_z plane [83]. Consequently, two tilted Dirac cones stand at the DNLs for constant k_z, given by $\varepsilon(q, \theta) = \lambda[2v_4 \cos(\theta + \xi k_z c) + \xi v_0]\hbar q$, with $v_4 = 3a\beta_4(2\hbar)^{-1}$, where $\lambda = \pm$ are the bands indices. The wave functions of spinless RG have also been known [84], which were shown to be almost chiral as those of spinless graphene, to next higher order of β_4 while being independent of β_1. Thus, there are tilted 2D massless Dirac fermions hosted around the DNLs in spinless RG.

The Dirac cones in spinless RG become normal in the minimal model with β_4 vanishing, so that the stacking dimension k_z completely drops out. It has been proven that, with this, RG mimics graphene in every aspect, including the density of states [86, 94], magneto-electronic properties such as LLs with a zero mode [83, 84, 94] and magneto-optic properties [95]. It is reasonable since spinless RG bearing SLS hosts 2D massless Dirac fermions around the two DNLs. In the more realistic model, the inclusion of β_4 brings out certain modifications in regard to the SLS breaking. That is, the DNLs do not lie at $E_F = 0$ and the topologically corresponding drumhead surface bands would deviate from being exactly flat. Those modifications have been shown in previous experiments to be negligible, not yet topologically characterized, such that almost exactly flat surface bands were shown in epitaxied RG [119] and the 3D half-integer

QHE was shown in natural graphite [103, 104]. In spite that the DNLs for tilted 2D massless Dirac fermions in spinless RG are protected by PTS and characterized by the Z_2 invariant, there remains an issue that is crucial in the topological characterization of LLs. Specifically, it is desirable to determine whether a stable zero-mode LL exists, which would be responsible for the 3D half-integer QHE. As numerical results have shown [83, 84], the LL spectrum (Fig. 11.5) and LL wave functions (Fig. 11.6) for

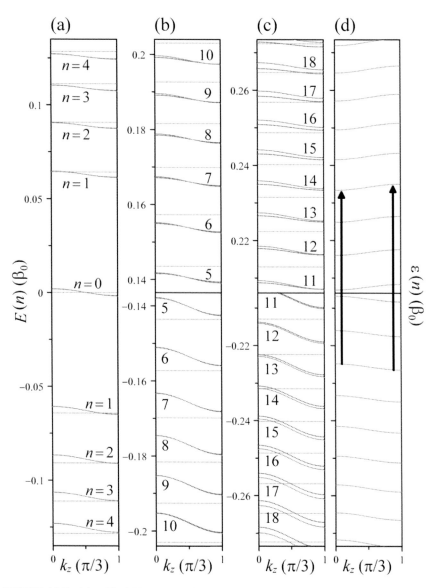

FIGURE 11.5 Magnified LL spectra of RG at $B_0 = 30$ T for (a) $|E(n, B_0, k_z)| \leq 0.12$, (b) $0.195 \leq |E(n, B_0, k_z)| \leq 0.255$ and (c) $0.335 \leq |E(n, B_0, k_z)| \leq 0.365$ (in the unit of t), black: minimal model; red: Onsager quantization; blue: numerical result.

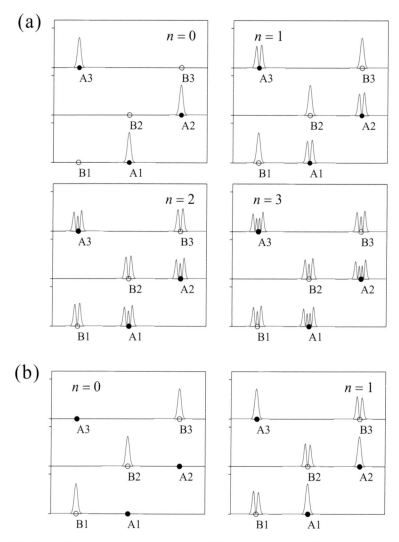

FIGURE 11.6 LL wave functions for $k_z = \pi/6d$ in RG at $B_0 = 20$ T. (a) One of the two degenerate set of LSs, plotted for $n = 0$ and unoccupied $n = 1$, 2, and 3. (b) The other set plotted for $n = 0$ and unoccupied $n = 1$. All the ordinate tick marks are labeled at 0 and 0.005.

spinless RG with $\beta_4 \neq 0$ can be identified to mimic those of spinless graphene except for the k_z dispersion in the LL spectrum. However, the model system lacks continuous CS for protecting the zero-mode LL since SLS is absent due to nonvanishing, though small, β_4. Here, a generalized continuous CS operator anticommuting with the Hamiltonian $H(\mathbf{k})$ is proposed, given by $\Gamma = \rho^{-1}[\sigma_3 - i\eta(\sin(\xi k_z c)\sigma_1 + \cos(\xi k_z c)\sigma_2)]$, with $\rho^{-1} = [1 - (\beta_4/\beta_0)^2]^{1/2}$. In a similar manner to other systems that have tilted 2D massless Dirac fermions hosted, the generalized continuous CS is respected so that the zero-mode LL is protected.

11.3   TIME REVERSAL SYMMETRY PROTECTED 3D STRONG TOPOLOGICAL INSULATORS

Graphene, again! For the distinguished topological phases of time reversal invariant TIs, it is graphene that invoked the first notification [120–122]. The TRS protection makes this kind of TIs differing from those Chern insulators, e.g., integer quantum Hall insulators, whose topological invariant, e.g., Hall conductivity, is odd under time reversal operation. With SOC, SLS and IS are broken. A mass term, though being tiny, is then induced so as to gap the bulk bands of graphene. In the relevant Hamiltonian $H(\mathbf{p}) = vs_3\boldsymbol{\tau}_3\boldsymbol{\sigma}\mathbf{p}$, an additional spinor acting on the electron spin space is included. In comparison to integer quantum Hall insulators, which have chiral edge states carrying electrical current, TRS protected TIs have helical edge states carrying spin current and exhibiting the quantum spin Hall effect (QSHE). This bulk-edge correspondence can be proved solidly through the Kramer degeneracy derived from TRS [31]. Because of TRS, such a nontrivial topological phase is characterized by nonzero Z_2 [123]. The experimental realizations were achieved soon later by means of HgTe/CdTe quantum wells that have much larger SOC gap [124, 125].

Still, the 3D QSHE is possible if 2D layers of quantum spin Hall insulator are stacked to 3D in a specific way [126, 127], while keeping the QSHE on each layer in analog to the 3D integer QHE [101]. The resulting 3D layered system is a kind of 3D weak TI. There exist an even number of Dirac cones on each surface layer. This might be a manifestation of the quasi-2D character of 3D TRS protected weak TIs, in which each layer has fermion doubling. In the *tenfold way* classification [17], these systems belong to the symmetry class AII, to which the Z_2 is specified.

On the other hand, TRS protected 3D strong TIs are found as having been predicted [90, 91]. They also are layered systems as realized by Bi_2Se_3, Bi_2Te_3, etc, which have the stacking units consisting of one quintuple layer sandwiched by Bi_2 and Se_3 (Te_3) in a rhombohedral configuration [128, 129]. These TIs exhibit strong SOC in an inverted band structure as required. Under the protection of TRS, 3D strong TIs belong to the symmetry class AII and are characterized by Z_2. By contrast, there exist a single or an odd number of Dirac cones on each surface [128, 129], corresponding to the gapped bulk bands [31]. The Kramer degeneracy forces the DP be located at the time reversal invariant point on each surface BZ. The existing 2D massless Dirac fermions on the surfaces is attributed to the third type of nodes in 3D, as described in Fig. 11.1(c). Because the two surface layers are practically distant from each other and, hence, their hybridization can be neglected; that is, the intervalley mixing is circumvented. Moreover, the Dirac fermions can penetrate into the bulk hardly. Hence, it is reasonable to consider the single DP on an individual surface. TRS also leads to an exotic helical spin texture, which has been experimentally observed [130]. In this texture, electron spin is always locked to momentum, as shown in Fig. 11.7. Thus, there is spin polarization away from the DP, rather than spin degeneracy as in graphene. To sum up, the TRS protected strong TI host one-fourth massless Dirac fermions with only one spinor in comparison to spinless graphene.

To characterize the single DP on a surface of TRS protected 3D strong TI, one might require an effective Dirac Hamiltonian $H(\mathbf{p}) = v\boldsymbol{\sigma}\mathbf{p}$, with v the Fermi velocity

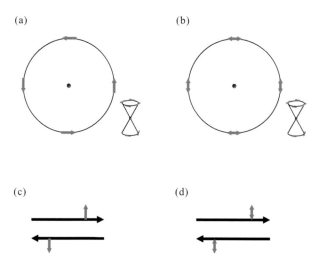

FIGURE 11.7   (a) Spin polarization around the DP on the surface of TRS protected TI, where electron spin is locked to its momentum. (b) Electron spin degeneracy around a DP in the bulk of graphene.

and σ the Pauli matrices acting on the electron spin space. Of course, a Dirac Hamiltonian acquires a continuous CS. This seems to be intriguing since the system lacks SLS. The cause should be attributed to the Kramer degeneracy, again, since there is no symmetry constraint else. However, how to derive such an effective Hamiltonian from the lattice Hamiltonian is crucial. In the conventional methodology, it is proper to get a surface Hamiltonian from the finite lattice Hamiltonian since both the bulk and the surface are non-local to each other. This is why fermion doubling is absent from the surface of the TRS protected 3D strong TI. However, in so doing some perturbation terms would be brought out, which can degrade the continuous CS and, therefore, gap the DP. This problem should be reconciled in order to cast a Dirac Hamiltonian to characterize the 2D massless Dirac fermions on the surfaces of the system [131].

As well realized, the surface Dirac cone is gapped by the exchange field due to magnetic doping or a proximate magnetic material [132, 133]. Referring to Fig. 11.8(c), the LLs lose the characteristic of 2D massless Dirac fermions. QHE in this case is anomalous [Fig. 11.8(d)]. However, all experiments till now have shown that applying a magnetic field perpendicular to the surface can lead to a chiral, half-filled zero-mode LL [Fig. 11.8(a)], which is ascribed to 2D massless Dirac fermions [134, 135]. The half-integer QHE has also been observed [136, 137], referring to [Fig. 11.8(b)]. The LL degeneracy is one fourth of those in graphene. Remember, however, the continuous CS is preserved via the minimal coupling. Thus, one would obtain an anomalous QH effect.

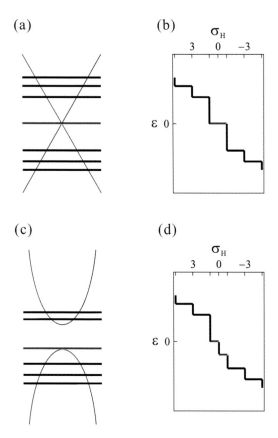

FIGURE 11.8 Schematics for interpreting TRS protected 3D strong TI. (a) LL spectrum when the continuous CS holds. (b) Hall plateaus arising from (a). (c) LL spectrum when the continuous CS is broken due to magnetic doping. (d) Hall plateaus with magnetic doping.

REFERENCES

1. Seifert H and Threlfall W 1980 *A Textbook of Topology*. San Diego: Academic Press.
2. Klitzing K von, Dorda G, and Pepper M 1980 New method for high-accuracy determination of the fine-structure constant based on quantized Hall resistance *Phys. Rev. Lett.* **45** 494.
3. Thouless D J, Kohmoto M, Nightingale M P, and Nijs M den 1982 Quantized Hall conductance in a two-dimensional periodic potential *Phys. Rev. Lett.* **49** 405.
4. Avron J E, Seiler R, and Simon B 1983 Holonomy, the quantum adiabatic theorem, and Berry's phase *Phys. Rev. Lett.* **51** 51.
5. Simon B 1983 Holonomy, the quantum adiabatic theorem, and Berry's phase *Phys. Rev. Lett.* **51** 2167.
6. Kosterlitz J M 2017 Nobel Lecture: Topological defects and phase transitions *Rev. Mod. Phys.* **89** 040501.
7. Haldane F D M 2017 Nobel Lecture: Topological quantum matter *Rev. Mod. Phys.* **89** 0405502.

8. Wen X G 2017 Colloquium: Zoo of quantum-topological phases of matter *Rev. Mod. Phys.* **89** 041004.

9. Weinberg S 1995 *The Quantum Theory of Fields* vol. 1. New York: Oxford University Press.

10. Landau L D and Lifschitz E M 1958 *Course of Theoretical Physics* vol. 5. London: Pergamon.

11. Mermin N D 1979 The topological theory of defects in ordered media *Rev. Mod. Phys.* **54** 591.

12. Volovik G E 2013 The topology of the quantum vacuum *Lect. Notes Phys.* **871** 343.

13. Wen X G 1990 Topological orders in rigid states *Int. J. Mod. Phys.* B **4** 239.

14. Volovik G E 2007 Quantum phase transitions from topology in momentum space *Lect. Notes Phys.* **718** 31.

15. Kitaev A and Preskill J 2006 Topological entanglement entropy *Phys. Rev. Lett.* **96** 110404.

16. Levin M and Wen X G 2006 Detecting topological order in a ground state wave function *Phys. Rev. Lett.* **96** 110405.

17. Schnyder A P, Ryu S, Furusaki A, and Ludwig A, W W 2008 Classification of topological insulators and superconductors in three spatial dimensions *Phys. Rev.* B **78** 195125.

18. Kitaev A 2009 Periodic table for topological insulators and superconductors *AIP Conf. Proc.* **1134** 22.

19. Ryu S, Schnyder A P, Furusaki A, and Ludwig A W W 2010 Topological insulators and superconductors: Tenfold way and dimensional hierarchy *New J. Phys.* **12** 065010.

20. Hořava P 2005 Stability of Fermi surfaces and K theory *Phys. Rev. Lett.* **95** 016405.

21. Zhao Y X and Wang Z D 2013 Topological classification and stability of Fermi surfaces *Phys. Rev. Lett.* **110** 240404.

22. Matsuura S, Chang P Y, Schnyder A P, and Ryu S 2013 Protected boundary states in gapless topological phases *New J. Phys.* **15** 065001.

23. Zhao Y X, Schnyder A P, and Wang Z D 2016 Unified theory of PT and CP invariant metals and nodal superconductors *Phys. Rev. Lett.* **116** 156402.

24. Chiu C K, Teo J C Y, Schnyder A P, and Ryu S 2016 Classification of topological quantum matter with symmetries *Rev. Mod. Phys.* **88** 035005.

25. Tsui D C, Stormer H L, and Gossard A C 1982 Two-dimensional magnetotransport in the extreme quantum limit *Phys. Rev. Lett.* **48** 1559.

26. Bednorz J G and Müeller K A 1986 Possible high T_c superconductivity in the Ba-La-Cu-O system *Z. Phys.* B **64** 189.

27. Berry M V 1984 Quantum phase factors accompanying adiabatic changes *Proc. R. Soc. London, Ser.* A **392** 45.

28. Xiao D, Chang M C, and Niu Q 2010 Berry phase effects on electronic properties *Rev. Mod. Phys.* **82** 1959.

29. Hatsugai Y 1993 Chem mumber and edge states in the integer quantum Hall effect *Phys. Rev. Lett.* **71** 3697.

30. Ryu S and Hatsugai Y 2002 Topological origin of zero-energy edge states in particle-hole symmetric systems *Phys. Rev. Lett.* **89** 077002.

31. Qi X L, Wu Y S, and Zhang S, C 2006 General theorem relating the bulk topological number to edge states in two-dimensional insulators *Phys. Rev.* B **74** 045125.

32. Hatsugai Y 2009 Bulk-surface correspondence in graphene with/without magnetic field: Chiral symmetry, Dirac fermions and edge states *Solid State Commun.* **149** 1061.

33. Mong R S K and Shivamoggi V 2011 Edge states and the bulk-boundary correspondence in Dirac Hamiltonians *Phys. Rev.* B **83** 125109.

34. Demkov Y N and Kurasov P B 2007 von Neumann-Wigner Theorem: Level's repulsion and degenerate eigenvalues *Theor. Math. Phys.* **153** 1407.

35. Young S M, Zaheer S, Teo J C Y, Kane C L, Mele E J, and Rappe A M 2012 Dirac semimetal in three dimensions *Phys. Rev. Lett.* **108** 140405.

36. Young S M and Kane C L 2015 Dirac semimetal in two dimensions *Phys. Rev. Lett.* **115** 126803.

37. Yang B J, Bojesen T A, Morimoto T, and Furusaki A 2017 Topological semimetals protected by off-centered symmetries in nonsymmorphic crystals *Phys. Rev. B* **95** 075135.

38. Herring C 1937 Accidental degeneracy in the energy bands of crystals *Phys. Rev.* **52** 365.

39. Murakami S 2007 Phase transition between the quantum spin Hall and insulator phases in 3D: Emergence of a topological gapless phase *New J. Phys.* **9** 356.

40. Murakami S and Kuga S 2008 Universal phase diagrams for the quantum spin Hall systems *Phys. Rev. B* **78** 165313.

41. Yang B J, Bahramy M S, Arita R, Isobe H, Moon E G, and Nagaosa N 2013 Theory of topological quantum phase transitions in 3D noncentrosymmetric systems *Phys. Rev. Lett.* **110** 086402.

42. Murakami S, Hirayama M, Okugawa R, and Miyake T 2017 Emergence of TSMs in gap closing in semiconductors without inversion symmetry *Sci. Adv.* **3** e1602680.

43. Anderson P W 1958 Absence of diffusion in certain random lattices *Phys. Rev.* **109** 1492.

44. Abrahams E, Anderson P W, Licciardello D C, and Ramakrishnan T W 1979 Scaling theory of localization: Absence of quantum diffusion in two dimensions *Phys. Rev. Lett.* **42** 673.

45. Anderson P W, Thouless D J, Abrahams E, and Fisher D S 1980 New method for a scaling theory of localization *Phys. Rev. B* **22** 3519.

46. Zirnbauer M R 1996 Riemannian symmetric superspaces and their origin in random-matrix theory *J. Math. Phys.* **37** 4986.

47. Altland A and Zirnbauer M R 1997 Nonstandard symmetry classes in mesoscopic normal-superconducting hybrid structures *Phys. Rev. B* **55** 1142.

48. Evers F and Mirlin A D 2008 Anderson transitions *Rev. Mod. Phys.* **80** 1355.

49. Koshino M, Morimoto T, and Sato M 2014 Topological zero modes and Dirac points protected by spatial symmetry and chiral symmetry *Phys. Rev. B* **90** 115207.

50. Chiu C K and Schnyder A P 2014 Classification of reflection-symmetry-protected TSMs and nodal superconductors *Phys. Rev. B* **90** 205136.

51. Nielsen H B and Ninomiya M 1981 Absence of neutrinos on a lattice: (I). Proof by homotopy theory *Nucl. Phys. B* **185** 20.

52. Nielsen H B and Ninomiya M 1981 A no-go theorem for regularizing chiral fermions *Phys. Lett. B* **105** 219.

53. Nielsen H B and Ninomiya M 1981 Absence of neutrinos on a lattice: (II). Intuitive topological proof *Nucl. Phys. B* **193** 173.

54. Vafek O and Vishwanath A 2014 Dirac fermions in solids: from high-T_c cuprates and graphene to topological insulators and Weyl semimetals *Annu. Rev. Condens, Matter Phys.* **5** 83.

55. Dirac P A M 1958 *The Principles of Quantum Mechanics*, 4th edition. New York: Oxford University Press.

56. Semenoff G W 1984 Condensed-matter simulation of a three-dimensional anomaly *Phys. Rev. Lett.* **53** 2449.

57. Hatsugai, Y 2011 Topological aspect of graphene physics *J. Phys. Conf. Ser.* **334** 012004.

58. DiVincenzo D P and Mele E J 1984 Self-consistent effective-mass theory for intralayer screening in graphite intercalation compounds *Phys. Rev. B* **29** 1685.

59. Castro Neto A H, Guinea F, Peres N M R, Novoselov K S, and Geim A K 2009 The electronic properties of graphene *Rev. Mod. Phys.* **81** 109.

60. Hofstadter D R 1976 Energy levels and wave functions of Bloch electrons in rational and irrational magnetic fields *Phys. Rev. B* **14** 2239.

61. Kawarabayashi T, Morimoto T, Hatsugai Y, and Aoki H 2010 Anomalous criticality at the n = 0 quantum Hall transition in graphene: The role of disorder preserving chiral symmetry *Phys. Rev.* B **82** 195426.

62. Kawarabayashi T, Hatsugai Y, and Aoki H 2010 Landau level broadening in graphene with long-range disorder—Robustness of the $n = 0$ level *Physica* E **42** 759.

63. Zheng Y and Ando T 2002 Hall conductivity of a two-dimensional graphite system *Phys. Rev.* B **65** 245420.

64. Gusynin V P and Sharapov S G 2005 Unconventional integer quantum Hall effect in graphene *Phys. Rev. Lett.* **95** 146801.

65. Novoselov K S, Geim A K, Morozov S V, Jiang D, Katsnelson M I, Grigorieva I V, et al. 2005 Room-temperature quantum Hall effect in graphene *Nature* **438** 197.

66. Zhang Y, Tan Y W, Stormer H L, and Geim A K 2005 Room-temperature quantum Hall effect in graphene *Nature* **438** 201.

67. Kawarabayashi T, Hatsugai Y, and Aoki H 2009 Quantum Hall plateau transition in graphene with spatially correlated random hopping *Phys. Rev. Lett.* **103** 156804.

68. Laughlin R B 1981 Quantized Hall conductivity in two dimensions *Phys. Rev.* B **23** 5632.

69. Halperin B I 1982 Quantized Hall conductance, current-carrying edge states, and the existence of extended states in a two-dimensional disordered potential *Phys. Rev.* B **25** 2185.

70. Ludwig A W W, Fisher M P A, Shankar R, and Grinstein G 1994 Integer quantum Hall transition: An alternative approach and exact results *Phys. Rev.* B **50** 7526.

71. Hatsugai Y, Wen X G, and Kohmoto M 1997 Disordered critical wave functions in random-bond models in two dimensions: Random-lattice fermions at $E = 0$ without doubling *Phys. Rev.* B **56** 1061.

72. Ostrovsky P M, Gornyi I V, and Mirlin A D 2008 Theory of anomalous quantum Hall effects in graphene *Phys. Rev.* B **77** 195430.

73. Volovik G E 2003 *Universe in a Helium Droplet*. New York: Oxford University Press.

74. Hosur P, Ryu S, and Vishwanath A 2010 Chiral topological insulators, superconductors, and other competing orders in three dimensions *Phys. Rev.* B **81** 045120.

75. Wan X, Turner A M, Vishwanath A, and Savrasov S Y 2011 TSM and Fermi-arc surface states in the electronic structure of pyrochlore iridates *Phys. Rev.* B **83** 205101.

76. Burkov A A and Balents L 2011 Weyl semimetal in a topological insulator multilayer *Phys. Rev. Lett.* **107** 127205.

77. Halász G B and Balents L 2012 Time-reversal invariant realization of the Weyl semimetal phase *Phys. Rev.* B **85** 035103.

78. Fang C, Gilbert M J, Dai X, and Bernevig B A 2012 Multi-Weyl TSMs stabilized by point group symmetry *Phys. Rev. Lett.* **108** 266802.

79. Wang Z J, Sun Y, Chen X Q, Franchini C, Xu G, Weng H M, et al. 2012 Dirac semimetal and topological phase transitions in A_3Bi (A = Na, K, Rb) *Phys. Rev.* B **85** 195320.

80. Liu Z K, Zhou B, Zhang Y, Wang Z J, Weng H M, Prabhakaran D, et al. 2014 Discovery of a three-dimensional topological Dirac semimetal, Na_3Bi *Science* **343** 864.

81. Xiao R, Tasnádi F, Koepernik K, Venderbos J W F, Richter M, and Taut M 2011 Density functional investigation of rhombohedral stacks of graphene: Topological surface states, nonlinear dielectric response, and bulk limit *Phys. Rev.* B **84** 165404.

82. Heikkilä T T and Volovik G E 2011 Dimensional crossover in topological matter: Evolution of the multiple Dirac point in the layered system to the flat band on the surface *JETP Lett.* **93** 59.

83. Ho C H, Chang C P, Su W P, and Lin M F 2013 Precessing anisotropic Dirac cone and Landau subbands along a nodal spiral *New J. Phys.* **15** 053032.

84. Ho C H, Chang C P, and Lin M F 2014 Landau subband wavefunctions and chirality manifestation in rhombohedral graphite *Solid State Commun.* **197** 11.

85. Heikkilä T T and Volovik G E 2015 Nexus and Dirac lines in topological materials *New J. Phys.* **17** 093019.

86. Ho C H, Chang C P, and Lin M F 2016 Evolution and dimensional crossover from the bulk subbands in ABC-stacked graphene to a three-dimensional Dirac cone structure in rhombohedral graphite *Phys. Rev. B* **93** 075437.

87. Burkov A A, Hook M D, and Balents L 2011 Topological nodal semimetals *Phys. Rev. B* **84** 235126.

88. Yang B J and Nagaosa N 2014 Classification of stable three-dimensional Dirac semimetals with nontrivial topology *Nat. Commun.* **5** 4898.

89. Armitage N P, Mele E J, and Vishwanath A 2018 Weyl and Dirac semimetals in three-dimensional solids *Rev. Mod. Phys.* **90** 015001.

90. Fu L and Kane C L 2007 Topological insulators in three dimensions *Phys. Rev. Lett.* **98** 106803.

91. Fu L and Kane C L 2007 Topological insulators with inversion symmetry *Phys. Rev. B* **76** 045302.

92. Wallace P R 1947 The band theory for graphite *Phys. Rev.* **71** 622.

93. McClure J W 1969 Electron energy band structure and electronic properties of rhombohedral graphite *Carbon* **7** 425.

94. Guinea F, Castro Neto A H and Peres N M R 2006 Electronic states and Landau levels in graphene stacks *Phys. Rev. B* **73** 245426.

95. Ho C H, Chang C P, and Lin M F 2015 Optical magnetoplasmons in rhombohedral graphite with a three-dimensional Dirac cone structure *J. Phys.: Condens. Matter* **27** 125602.

96. Goerbig M O, Fuch J N, Montambaux G, and Piéchon F 2008 Tilted anisotropic Dirac cones in quinoid-type graphene and α-(BEDT-TTF)$_2$I$_3$ *Phys. Rev. B* **78** 045415.

97. Anderson P W 1987 The resonating valence bond state in La$_2$CuO$_4$ and superconductivity *Science* **235** 1196.

98. Dresselhaus M S and G Dresselhaus 2002 Intercalation compounds of graphite *Adv. Phys.* **51** 1.

99. Dresselhaus M S, Dresselhaus G, and Jorio A 2008 *Group Theory: Application to the Physics of Condensed Matter*. Berlin, Heidelberg: Springer.

100. T. Hyart and T. T. Heikkilä 2016 Momentum-space structure of surface states in a TSM with a nexus point of Dirac lines *Phys. Rev. B* **93** 235147.

101. Kohmoto M, Halperin B I, and Wu Y S 2006 Diophantine equation for the three-dimensional quantum Hall effect *Phys. Rev. B* **45** 13488.

102. Balents L and Fisher M P A 1996 Chiral surface states in the bulk quantum Hall effect *Phys. Rev. Lett.* **76** 2782.

103. Kopelevich Y, Torres J H S, da Silva R R, Mrowka F, Kempa H, and Esquinazi P 2003 Reentrant metallic behavior of graphite in the quantum limit *Phys. Rev. Lett.* **90** 156402.

104. Kempa H, Esquinazi P, and Kopelevich Y 2006 Integer quantum Hall effect in graphite *Solid State Commun.* **138** 118.

105. Wunsch B, Guinea F, and Sols F 2008 Dirac-point engineering and topological phase transitions in honeycomb optical lattices *New J. Phys.* **10** 103027.

106. Chen Y H, Wu W, Liu G C, Tao H S, and Liu W M 2012 Quantum phase transition of cold atoms trapped in optical lattices *Front. Phys.* **73** 223.

107. Asano K and Hotta C 2011 Designing Dirac points in two-dimensional lattices *Phys. Rev. B* **83** 245125.

108. Hasegawa Y, Konno R, Nakano H, and Kohmoto M 2006 Zero modes of tight-binding electrons on the honeycomb lattice *Phys. Rev. B* **74** 033413.

109. Mañes J L, Guinea F, and Vozmediano M A H 2007 Existence and topological stability of Fermi points in multilayered graphene *Phys. Rev. B* **75** 155424.

110. Raghu S and Haldane F D M 2008 Analogs of quantum-Hall-effect edge states in photonic crystals *Phys. Rev.* A **78** 033834.

111. Teo J C Y and Kane C L 2010 Topological defects and gapless modes in insulators and superconductors *Phys. Rev.* B **82** 115120.

112. Morimoto T and Furusaki A 2014 Weyl and Dirac semimetals with Z_2 topological charge *Phys. Rev.* B **89** 235127.

113. Wen X G and Zee A 1989 Winding numbers family index theorem, and electron hopping in a magnetic field *Nucl. Phys.* B **316** 641.

114. Kawarabayashi T, Hatsugai Y, Morimoto T, and Aoki H 2011 Generalized chiral symmetry and stability of zero modes for tilted Dirac cones *Phys. Rev.* B **83** 153414.

115. Kawarabayashi T, Aoki H, and Hatsugai Y 2016 Lattice realization of the generalized chiral symmetry in two dimensions *Phys. Rev.* B **94** 235307.

116. Kim Y, Wieder B J, Kane C L, and Rappe A M 2015 Dirac line nodes in inversion-symmetric crystals *Phys. Rev. Lett.* **115** 036806.

117. Fang C, Chen Y, Kee H Y, and Fu L 2015 Topological nodal line semimetals with and without spin-orbital coupling *Phys. Rev.* B **92** 081201.

118. Chan Y H, Chiu C K, Chou M Y, and Schnyder A P 2016 Ca_3P_2 and other topological semimetals with line nodes and drumhead surface states *Phys. Rev.* B **93** 205132.

119. Pierucci D, Sediri H, Hajlaoui M, Girard J C, Brumme T, and Calandra M 2015 Evidence for flat bands near the Fermi level in epitaxial rhombohedral multilayer graphene *ACS Nano* **9** 5432.

120. Kane C L and Mele E J 2005 Z_2 Topological order and the quantum spin Hall effect *Phys. Rev. Lett.* **95** 146802.

121. Kane C L and Mele E J 2005 Quantum spin Hall effect in graphene *Phys. Rev. Lett.* **95** 226801.

122. Hasan M Z and Kane C L 2010 Colloquium: Topological insulators *Rev. Mod. Phys.* **82** 3045.

123. Moore J E and Balents L 2007 Topological invariants of time-reversal-invariant band structures *Phys. Rev.* B **75** 121306.

124. Bernevig B A, Hughes T A, and Chang S C 2006 Quantum spin Hall effect and topological phase transition in HgTe quantum wells *Science* **314** 1757.

125. König M, Wiedmann S, Brüne C, Roth A, Buhmann H, and Molenkamp L W 2007 Quantum sspin Hall insulator state in HgTe quantum wells *Science* **318** 766.

126. Ran Y, Zhang Y, and Vishwanath A 2009 One-dimensional topologically protected modes in topological insulators with lattice dislocations *Nat. Phys.* **5** 298.

127. Slager R J, Mesaros A, Juričić V, and Zaanen J 2014 Interplay between electronic topology and crystal symmetry: Dislocation-line modes in topological band insulators *Phys. Rev.* B **90** 241403.

128. Hsieh D, Qian D, Wray L, Xia Y, Hor Y S, Cava R J, et al. 2008 A topological Dirac insulator in a quantum spin Hall phase *Nature* **452** 970.

129. Chen Y L, Analytis J G, Chu J H, Liu Z K, Mo S K, Qi X L, et al. 2009 Experimental realization of a three-dimensional topological insulator *Science* **325** 178.

130. Kung H H, Maiti S, Wang X, Cheong S W, Maslov D L, and Blumberg G 2017 Chiral spin mode on the surface of a topological insulator *Phys. Rev. Lett.* **119** 1346802.

131. de Resende B M, de Lima F C, Miwa R H, Vernek E, and Ferreira G J 2017 Confinement and fermion doubling problem in Dirac-like Hamiltonians *Phys. Rev.* B **96** 161113.

132. Eremeev S V, Men'shov V N, Tugushev V V, Echenique P M, and Chulkov E V 2013 Magnetic proximity effect at the three-dimensional topological insulator/magnetic insulator interface *Phys. Rev.* B **88** 144430.

133. Yoshimi R, Yasuda K, Tsukazaki A, Takahashi K S, Nagaosa N, Kawasaki M, et al. 2015 Quantum Hall states stabilized in semi-magnetic bilayers of topological insulators *Nat. Commun.* **6** 8530.

134. Cheng P, Song C, Zhang T, Zhang Y, Wang Y, and Jia J F 2010 Landau quantization of topological surface states in Bi_2Se_3 *Phys. Rev. Lett.* **105** 076801.

135. Hanaguri T, Igarashi K, Kawamura M, Takagi H, and Sasagawa T 2010 Momentum-resolved Landau-level spectroscopy of Dirac surface state in Bi_2Se_3 *Phys. Rev.* B **75** 121306.

136. Xu Y, Miotkowski I, Liu C, Tian J, Nam H, Alidoust N, et al. 2014 Observation of topological surface state quantum Hall effect in an intrinsic three-dimensional topological insulator *Nat. Phys.* **10** 956.

137. Yoshimi R, Tsukazaki A, Kozuka Y, Falson J, Takahashi K S, Checkelsky J G, et al. 2015 Quantum Hall effect on top and bottom surface states of topological insulator $(Bi_{1-x}Sb_x)_2Te_3$ films *Nat. Commun.* **6** 6627.

12 Concluding Remarks

Shih-Yang Lin,[e] *Thi-Nga Do,*[c,d] *Chiun-Yan Lin,*[a]
Jhao-Ying Wu,[b] *Po-Hsin Shih,*[a] *Ching-Hong Ho,*[b]
Ming-Fa Lin[a,f,g]

[a] Department of Physics, National Cheng Kung University,
Tainan 701, Taiwan
[b] Center of General Studies, National Kaohsiung University of
Science and Technology, Kaohsiung 811, Taiwan
[c] Laboratory of Magnetism and Magnetic Materials, Advanced
Institute of Materials Science, Ton Duc Thang University,
Ho Chi Minh City, Vietnam
[d] Faculty of Applied Sciences, Ton Duc Thang University,
Ho Chi Minh City, Vietnam
[e] Department of Physics, National Chung Cheng University,
Chiayi 621, Taiwan
[f] Quantum Topology Center, National Cheng Kung University,
Tainan 701, Taiwan
[g] Hierarchical Green-Energy Materials Research Center,
National Cheng Kung University, Tainan, Taiwan

CONTENTS

Apparently, this current book presents the delicately theoretical frameworks about the diverse quantization phenomena, especially for those due to a uniform perpendicular magnetic field and lattice symmetries in layered condensed-matter systems. The generalized tight-binding model, the dynamic Kubo formula, the static one, and the sublattice- and layer-dependent random-phase approximation are developed/modified to thoroughly explore the electronic properties, optical absorption spectra, quantum transports, and Coulomb excitations under the magnetic quantization, respectively. Furthermore, the first method is closely combined with the second/third/fourth one. The existence of many-particle excitonic effects in absorption spectra of layered materials and the modified magneto-optical theory needs to be thoroughly clarified in the near-future works. All the critical mechanisms, the planar or buckling structures, the uniform or nonuniform geometries, the different layer numbers, the distinct stacking configurations, the atom-induced ionization potentials, the strong electron/hole dopings, the significant guest-atom substitutions, the important intralayer and interlayer hopping integrals, the various spin-orbital couplings, the intrinsic intralayer and interlayer Coulomb interactions, and the electric and magnetic fields, are included in the calculations simultaneously using the very efficient manners. Part of calculated results are consistent with the high-resolution measurements Zhang;156801, e.g., the low-lying Hall conductivities of the monolayer, AB- and ABC-graphene systems [1], and the magneto-optical selection rules of few-layer AB-stacked graphene systems. However, most of them require a series of experimental examinations, such as the spin-orbital-diversified Landau levels and magneto-optical absorption spectra in bilayer silicene of AB and AA stackings, four kinds of Landau levels and selection rules in Si-substituted graphene, and the doping- and field-modulated plasmon modes in monolayer germanene. The above-mentioned models are very useful in understanding the essential physical properties of emergent materials, such as few-layer Si [20, 21], Ge [22, 23], Sn [24], Pb [2], GaAs [3], P [26–28], Sb [4], Bi [5], MoS_2 [25], and topological insulators [6].

The analytic derivations and numerical evaluations are available in fully exploring the diverse quantization phenomena. We have successfully established the generalized tight-binding model which is based on the layer- and sublattice-dependent subenvelope functions. The developed theoretical framework is suitable for the various geometric structures, the uniform/nonuniform external fields, the multi-orbital hybridizations in chemical bonds, and the environment-induced spin-orbital couplings. It can analytically deal with the dramatic changes in the extra phases and magnitude of the neighboring hopping integrals. The geometry- and field-created unit cell is directly reflected in the Hamiltonian matrix as a giant Hermitian one, being efficiently solved by the exact diagonalization method. The necessary calculation parameters in the calculations are examined from the well fitting of the low-lying energy bands at the high-symmetry points with the first-principles results. The main features of electronic properties can be utilized under the very effective ways to comprehend the absorption spectra, quantum Hall conductivities, and electronic Coulomb excitations, such as the same contributions from the highly degenerate Landau levels, the localized probability distributions of the magnetic wave functions in largely reducing the computer time, the specific oscillation modes and strong couplings of the initial and final states for the well decision of magneto-optical selection rules, and the Coulomb matrix elements due to the subenvelope functions for determining the bare and screened response behaviors.

The twisted bilayer graphenes show the diversified fundamental properties for the Moire superlattices with the nonuniform physical environments; furthermore, such systems are quite different from the sliding ones or the normal stackings. Most importantly, the generalized tight-binding model can well characterize the complex subgroups of the highly degenerate Landau levels, further illustrating that this model is very suitbale for exploring the diverse magnetization phenomena even under the various geometric structures and intrinsic interactions. The electronic energy spectra, van Hove singularities, optical transition spectra, and magnetic subenvelop functions strongly depend on the multi-combined effects arising from the stacking symmetries/twisted angles, gate voltages, and magnetic fields. There are many pairs of $2p_z$-induced valence and conduction energy subbands, mainly owing to the zone folding effect. However, two pairs appear in the bilayer AA and AB stackings. Among of them, the lowest two pairs, which cover the Fermi level, present the monolayer-like

Dirac-cone structures being initiated from the K/K′ valley. All the $\theta \neq 0°$ and 60° twisted bilayer graphenes are zero-gap semiconductors because of the vanishing density of states at E_F. Specifically, the energy ranges of Dirac-cone structures quickly decline as θ decreases from 30′ to 0° (θ increases from 30° to 60′). The degenerate Dirac cones become split and thus create the finite Fermi momenta (free electrons and holes simultaneously), when gate voltage starts from zero; that is, the semiconductor-semimetal transition occurs during the variation of V_z. With the increasing energy of $|E^{c,v}|$, the stable valleys could form near the M, Γ, and K points. In addition to the extreme band-edge states, the M-point valleys frequently come to exist as the saddle forms. A lot of critical points in the energy-wave-vector space are created by the zone-folding effect of the Moire superlattice and the layer-dependent Coulomb potentials, in which three kinds of van Hove singularities, V-shape/a plateau plus two temple-like cusp structures, prominent symmetric peaks and shoulders (measurements from the Fermi level), respectively, correspond to the K/K′-point, M-point, and M-, Γ-, and K-related valleys. Apparently, a lot of optical absorption structures created the vertical excitations of the rich band-edge states. According to the ranges of optical excitation frequencies, there exist (I) the gapless, featureless, and linear-ω spectral functions (the K-valley Dirac cone), (II) three symmetric absorption peaks (the saddle points near the M ones), and (III) the shoulder and peak structures (near the K, M, and Γ points). The gate voltages lead to an optical gap and the dramatic changes in the number, intensity, and frequency of absorption spectra. Very interestingly, the magneto-electronic properties of the twisted bilayer graphene systems, being based on many subenvelopes functions of multi-sublattices, present the hybridized phenomena arising from monolayer graphene, and AA & AB bilayer stackings. They directly reflect the robust relations between the neighboring A and B lattice sites on the same layer, and for the projections of them on the upper/lower layer. All the oscillation modes in the layer-dependent distinct sublatices are well characterized through the delicate analysis, and they belong to the perturbed ones in the presence of complicated inter-layer hopping integrals. These results suggest that the effective-mass approximation might be not suitable for the current topic. Moreover, the significant differences between the twisted and sliding bilayer graphenes lie in the main features of electronic structures, absorption spectra, magneto-electronic properties and quantum transports [discussed later in Sec. 9.1], mainly owing to the various stacking configurations in the Moire superlattices of the former. On the experimental side, the high-resolution measurements from angle-resolved photoemission spectroscopy [7], scanning tunneling spectroscopy [8], and Hall-bar equipment [9] have confirmed the low-lying linear and parabolic valence subbands, the V-shape structure and the θ-dependent prominent symmetric peaks (the K and M points), and the quasi-monolayer Hall conductivity. The optical examinations seem to be absent up to now. How to examine the gate-voltage-enriched essential properties require the further experiments, especially for those related to the diverse magnetic quantizations.

Obviously, the stacking modulation, gate voltage, and magnetic field greatly diversify the physical phenomena of bilayer graphene. Furthermore, their essential properties are very different from those in twisted [10] and sliding [11] bilayer systems, especially for the important differences in the 1D and 2D characteristics. The geometric modulation along the x-direction induces a periodical boundary condition and thus the dramatic change from 2D into 1D behaviors. The semimetallic systems present a plenty of 1D valence and conduction subbands, accompanied by the splitting of state degeneracy, the creation of non-monotonous energy dispersions, more band-edge states, and the distorted or irregular standing waves. Specifically, the layer-dependent Coulomb potential destroys a specific relation between the (A^1, A^2) and (B^1, B^2) subenvelope functions, even changes the zero-point number, and create the a lot of van Hove singularities in density of states covering the double- and single-peak in the square-root divergent forms, a pair of very prominent peaks near E_F, and a V_g-induced plateau across the Fermi level. The threshold absorption frequency/optical gap is vanishing in any systems. A pristine bilayer AB stacking exhibits featureless optical spectrum at lower frequencies. Concerning the geometry-modulated systems, the observable absorption peaks, corresponding to the vertical transitions of band-edge states, appear only under the destruction of the symmetric/anti-symmetric linear superposition associated with the (A^1, A^2) and (B^1, B^2) sublattices, leading to the absence of optical selection rules. A simple dependence of absorption structures on the domain-wall width is absent. However, the reduced intensity and the enhanced number are revealed during the increase of gate voltage. The frequency, number, and form intensity of optical absorption peaks strongly depend on the modulation period and electric-field strength. In the presence of magnetic field, its strong competition with the stacking modulation leads to the complicated and non-homogenous Peierls phases and thus the rich magneto-electronic properties. The highly degenerate Landau levels of a normal stacking becomes many quasi-1D Landau subbands, in which the latter have the partially flat and oscillatory energy dispersions and present the frequent crossing and anti-crossing phenomena in the k_y-dependent magneto-electronic energy spectra. The greatly reduced state degeneracy and extra band-edge states create more symmetric and asymmetric peaks in the wider energy ranges of density of states. Three kinds of magnetic wave functions, the well-behaved, perturbed, and seriously distorted ones, come to exist within the specific regions related to the stacking configurations; that is, each Landau-subband state might be the superposition of the various Landau-level ones. Based on the detailed and delicate analyses, the close relations among the geometric structures, the band-edge states, and the spatial probability distributions account for the main features of Landau subbands. In addition to the stacking modulation, the quantum confinement of 1D graphene nanoribbons [12], the significant interlayer hopping integrals in 3D graphites [13], and the composite external fields [14] create the Landau subbands, being quite different from one another. All the theoretical predictions on the essential properties require the high-resolution experimental verifications. For the stacking-, gate-voltage-, and magnetic-field-manipulated bilayer graphene systems, the above-predicted essential properties, valence bands, van Hove singularities of electronic/magneto-electronic energy spectra and band property near the Fermi level, spatial distributions of subenvelope functions, and absorption spectra of vertical excitations, could be, respectively, verified from the angle-resolved photoemission spectroscopy (ARPES), V_z-dependent scanning tunneling spectroscopy (STS), energy-fixed and various optical spectroscopies.

A AA-bt bilayer silicene exhibits the rich and unique electronic and optical properties under the specific buckling and stacking configuration. The M, K, and Γ valleys, as measured from the Fermi level, are responsible for the diversified phenomena.

The significant differences with monolayer silicene and few-layer graphene, the main features of the low-lying Landau levels, lie in the absence of stable K-valleys near the Fermi level. Specifically, the first pair of energy bands in this system presents a concave-downward/cancave-upward K-point valleys at the higher/deeper energies for conduction/valence states, being accompanied with the similar and independent Γ-point valley; that is, both parabolic valleys are well established and separated. It should be noticed that the unstable M-point valley belongs to the middle structure of the parabolic K-point one. The first and other two band-edge states, which correspond to the saddle and extreme points, respectively, reveal the prominent symmetric peaks and shoulder structures as the 2D van Hove singularities. Furthermore, the constant-energy valence and conduction loops cross $E_F = 0$ with a very narrow spacing of ~ 8 meV and thus create a temple-like cusp structures. The band gap is opened and gradually enhanced along the KΓ direction as the gate voltage grows from zero. Also, it is characterized by a pair of prominent peaks in the square-root divergent form. However, the spin splitting never comes to exist, owing to the absence of the combined effects arising from the spin-orbital interaction and the sublattice-dependent Coulomb potentials. The above-mentioned unusual valley structure in energy spectrum can create the rich magnetic quantizations and optical absorption spectra. The magneto-electronic states are initiated from the K and Γ valleys simultaneously, thus leading to the direct superposition of two Landau-level subgroups with the frequent crossing behaviors. Their Landau levels, respectively, have four- and two-fold degenerate localization centers [1/6, 2/6, 4/6, 5/6] and [1/2, 2/2]. Any magneto-electronic states, which are magnetically quantized from the first pair of energy bands, are dominated by the B^l sublattices; that is, the equivalence of A^l and B^l sublattices on the same layer is destroyed by the critical interlayer hopping integral [Figs. 6.1(a) and 6.1(b)]. They display the almost monotonous Landau-level energy spacings and B_z-dependences. With the energy ranges covering the saddle M point and the Fermi level, the very large quantum numbers (oscillation modes) come to exist, especially for the Γ-dependent ones covering the larger carrier density. On the other hand, the magneto-electronic states, being magnetically quantized from the second pair of conduction and valence bands, exhibit the normal behavior. As to the vertical optical excitations, only the valence and conduction constant-emery loops display the observable absorption structure, the square-root asymmetric peak, while the Γ-, K-, and M-dependent band-edge states hardly make any contributions. Such unusual optical property mainly comes from the vanishing dipole matrix elements (the wave-vector-independent interlayer hopping integrals and the symmetric/anti-symmetric superposition of the tight-binding functions for valence/conduction states). The spectral characteristics strongly depend on the variation of gate voltage, especially for the enhanced threshold peak intensity and optical gap. Apparently, there exist certain important differences between the bilayer AA-bt and monolayer silicene systems in electronic and optical properties, such as the well-separated parabolic K- and Γ-point valleys or the merged K-point Dirac cone and parabolic Γ-point valley, the independent valence/conduction constant-energy loops across the Fermi level or the roughly isotropic Dirac-cone structure, and the composite two Landau-level subgroups or the dramatic transformation of localization center and state degeneracy within the same group during the variation from the K-point valley/the Fermi level and then

M- and Γ-point valleys, the form, number, frequency, and intensity of optical absorption structures spectra, and their strong dependences on the external fields. The high-resolution experimental measurements from ARPES, STS and optical spectroscopies are available in examining the valley-enriched fundamental physical properties due to the quite large interlayer hopping integrals.

The diversified essential properties are obviously revealed in AB-bt bilayer silicene, compared with AB-stacked bilayer graphene. The buckled honeycomb lattice, the large multi-interlayer hopping integrals, and the layer-dependent significant spin-orbital couplings are responsible for the electric-field-enriched phenomena. An indirect semiconducting system presents the specific semiconductor-semimetal transition, the drastic change of energy dispersions near the K and T valleys, and the dramatic formation of the distorted conduction and valence Dirac cones during the variation of electric-field strength. These are directly reflected in the main features of van Hove singularities and optical absorption spectra, such as the frequency, form, number, and intensity of special structures. Very importantly, the threshold absorption frequency is higher than the band gap, and its declining behavior is very sensitive to the ranges of electric field. Apparently, the magnetic quantization, which originates from a magnetic field or a composite electric and magnetic one, is greatly enriched by the critical mechanisms. The main characteristics cover the double degeneracy for each (k_x, k_y) state in the reduced first Brillouin zone, the significant K and T valleys with the totally different behaviors, the spin- and sublattice-dominated magneto-electronic properties, four Landau-level subgroups quantized from the first pair of valence and conduction bands (nearest to the Fermi level), the coexistence of the well-behaved and abnormal sublattice-related subenvelope functions, the frequent anti-crossing/crossing/non-crossing of two Landau levels in the magnetic-/electric-field-dependent energy spectra, and the dramatic transformation in the well separated energy ranges, spatial oscillation modes and occupation number of valence/conduction states during the variation of gate voltage. These clearly indicate the destruction in the equivalence of A^i and B^i on the same layer (A^i/B^i and A^j/B^j on the distinct layers), being absent in the normally stacked graphene systems. The unusual magneto-electronic states are further revealed in the magneto-absorption spectra with a lot of nonuniform delta-function-like peaks. Furthermore, they present the rich and unique magneto-optical properties: the 16 categories for the available magneto-excitation channels/a plenty of absorption structures within a very narrow frequency range, the single-, double- & twin-peak symmetric structures, the dramatic changes through the gate voltage (e.g., the great decrease about the number of excitation categories and the optical gap), and the type-I and type-II absorption peaks, respectively, with & without the magneto-optical selection rules. The above-mentioned results that the layered silicene systems could be applied to novel designs of Si-based nanoelectronics [15] and nanodevices with enhanced mobilities [16]. Their experimental examinations require the high-resolution measurements from the ARPES [17], STS [18], and optical spectroscopies [19]. Both silicene and graphene are in sharp contrast with each other in the essential physical properties, directly reflecting the significant differences in geometric symmetries and intrinsic interactions.

The Si-substituted graphene systems, which belong to the emergent 2D binary materials, have been thoroughly investigated for their electronic and optical properties in the absence/presence of magnetic quantization. They create the various chemical & physical environments by changing the Si-distribution configuration and concentration, leading to the diverse phenomena. There exist the nonuniform site energies and hopping integrals; therefore, the A_i and B_i sublattices in the substitution-enlarged unit cell might be highly non-equivalent or quasi-equivalent. The former present the well separated parabolic valence and conduction bands and belong to the finite-gap semiconductors, e.g., the Si-A_i-dressed configurations. However, the latter have the anisotropic and distorted Dirac-cone structures and exhibit the zero-gap semiconducting behaviors (the vanishing density of states at the Fermi level), such as the Si-$[A_i, B_i]$-decorated systems. Their van Hove singularities are revealed as the shoulder structures and the asymmetric V-shape forms. The zero-field wave functions clearly show that the low-lying conduction and valence states near the K valleys are, respectively, dominated by the Si-guest and C-host atoms. Furthermore, the certain-sublattice envelope functions are negligible there and thus the localized states come to exist. Apparently, the optical spectra of the finite- and zero-gap systems, respectively, show the shoulder and featureless absorption structures. Only a pristine system displays a simple linear relation between the spectral intensity and excitation frequency, while the nonlinear behaviors appear in the other Si-substituted systems. Such results are principally determined by whether the dominating dipole matrix elements depend on wave vectors/energies under the joint density of states proportional to frequency. There are four kinds of Landau levels, according to the spatial probability distributions & oscillation modes on the distinct sublattices, and the concise relations between A_i and B_i sublattices. The typical magneto-electronic states cover (I) the significant B_i sublattices of valence Landau levels & A_i sublattices of conduction Landau levels with the same mode, and the similar relation between the B_i and A_i sublattices of the conduction Landau levels; (II) the important/observable $[A_i, B_i]$ sublattices with a mode difference of ± 1, the serious deviations of localization centers & the highly asymmetric distributions composed of the main and side modes; (III) the same modes for valence B_i and conduction A_i sublattices, the vanishing or ± 1 zero-point differences between conduction A_i and B_i sublattices & the perturbed multi-modes in most of conduction and valence B_i; (IV) the oscillator-like oscillation modes with the equivalent A and B sublattices. Such Landau levels lead to the unusual magneto-optical selection rules: the dominating $\Delta n = 0$ (the first kind of Landau levels), the coexistent $\Delta n = 1$ & 0 with strong competitions (the second and third kinds), and the specific $\Delta n = 1$ (the fourth kind). Obviously, the above-mentioned features, respectively, correspond to the specific concentration and distribution configuration: the Si-A_i-dressed graphene with an enough high concentration, the $[A_i, B_i]$-decorated graphene, the low-concentration A_i-doped system, and the pristine one. Moreover, the nonuniform bond lengths, site energies & hopping integrals, and the effects due to the external fields are simultaneously taken into account within the generalized theoretical framework without the perturbation forms. This method is expected to be very useful in fully understanding the diversified fundamental properties of the adatom-adsorbed graphene systems [details in Chap. 13], e.g., the magnetic quantization phenomena in graphene oxides [Refs.], hydrogenated/halogenated/alkalized graphene systems [Refs.].

The direct combination of generalized tight-binding model and the static Kubo formula, being based on the same framework, is successfully developed to explore to investigate the unusual magneto-transport properties in high- and low-symmetry bilayer and trilayer graphenes, especially for the detailed identifications of the selection rules during the Landau-level scatterings. Such delicate method is very suitable for fully exploring the rich Hall conductivities of the emergent 2D materials, such as, silicene [20,21], germanene [22,23], tinene [24], MoS_2 [25], and phosphorene [26–28] [discussed later in Chap. 13]. The Fermi-energy- and magnetic-field-dependent quantum Hall conductivities are greatly diversified under the various stacking configurations, since they could create the intragroup and intergroup Landau-level transitions with/without the monolayer-like and other selection rules. Generally speaking, three kinds of Landau levels, the non-crossing/crossing/anti-crossing behaviors, and the split magneto-electronic energy spectra account for the diverse quantum transport phenomena, e.g., the integer and non-integer conductivities at zero temperature, the splitting-created reduction and complexity of quantum conductivity, the vanishing or non-zero conductivities at the neutrality point, the well-like, staircase and composite plateau structures, the uniform/distinct step heights, and the simple or irregular B_z-dependences of step widths. Concerning the quantum Hall conductivities of the sliding bilayer graphenes, the AA, AB, and AA' stackings are dominated by two categories of intragroup transition channels under the normal selection rule of $\Delta n = \pm 1$, while the extra intergroup ones and selection rules ($\Delta n = 0$ & ± 2) are clearly revealed in the other low-symmetry systems. The non-integer quantum conductivities only present in the AA' stacking with the energy-dependent localization center, covering the step heights of $3.8e^2/h$ and $4.2e^2/h$. Most of sliding bilayer systems exhibit the monolayer-like step height of $4e^2/h$ even for the undefined Landau levels, such as AA, AB, $\delta = 6b_0/8$, and $\delta = 11b_0/8$ stackings. Specifically, the exclusive step heights of e^2/h, $3e^2/h$, $5e^2/h$, & $7e^2/h$ are observed in the $\delta = 1/8$ stacking. Moreover a non-negligible conductivity appears near the neutrality point, while there exist Landau levels very close to zero energy, e.g., those of $\delta = b/8$, $6b/8$ and $8b/8$. Only the AA stacking shows four types of magnetic-field-dependent plateau structures (the wells, monolayer-like & non-monotonous staircases, and composite ones), mainly owing to the significant overlap of two Dirac-cone structures in the Landau-level energy spectra. On the other side, the AAB-stacked trilayer graphene exhibits the complex plateau structures with $2e^2/h$ height within a wide energy range and the irregular B_z-dependence, directly reflecting the splitting and anti-crossing energy spectra and the localization-split modes. Apparently, such features are absent in AAA-, ABA-, and ABC-stacked trilayer systems. Only the ABA system presents the insulating behavior near zero Fermi energy within a very narrow window. The numbers of intragroup and integroup inter-Landau-level channels, corresponding to the AAA-, ABA-, ABC-, and AAB-stacked graphene systems, are [3, 0], [3, 2], [3, 6], and [3, 6]. Also, the selection rules of static scatterings, respectively, cover $\Delta n = \pm 1$, $\Delta n = \pm 1$ (0 & ± 2) for few anti-crossings, $\Delta n = \pm 1, 0$ & ± 2, and $\Delta n = \pm 1, 0, \pm 2$, ± 3 & ± 4. In general, the above-mentioned theoretical predictions, being excepted for the AB stackings, could be verified from the high-resolution quantum transport measurements [details in Sec. 2.3]. Such examinations are very useful in identifying the diverse quantum transport phenomena, as well as in clarifying the close relations/the one-to-one correspondence between the zero-field band structure and the main features of the highly-degenerate Landau levels.

The thermal properties of monolayer graphene clearly illustrate another magnetic quantization phenomenon. The specific heat strongly depends on the temperature, magnetic-field strength, Zeeman effect and doping density of conduction electrons/valence holes. An unusual T^2 dependence is revealed at zero B_z, directly reflecting the linear and isotropic Dirac-coe structure. Such behavior is in sharp contrast with the different T-dependences of AAA-, ABA-, and ABC-stacked graphites [29], mainly owing to the distinct dimensions and stacking configurations/interlayer hopping integrals. Magnetic fields dramatically change the temperature dependence, where $C(T, B_z)$ presents a composite form of $1/T^2$ and exponential function. An obvious peak, which is induced by the unoccupied spin-up and occupied spin-down Landau levels across the chemical potential, comes to exist at the critical temperature. Also, the B_z-dependent specific heat shows a single peak at the critical field strength (B_{zc}). By the delicate calculations, a simple linear relation is identified to exist between T_c and B_z (B_{zc} and T). Under doping cases, the specific heat might display an extra shoulder at higher temperature. This structure mainly comes from the second lowest unoccupied Landau levels and the second highest occupied ones. The different filling cases determine whether it would survive or not. There are certain important differences between graphene and single-walled carbon nanotubes regarding specific heat as a result of the dimension-dominated van Hove singularities in density of states. The magneto-electronic specific heats of 1D carbon nanotubes demonstrate four kinds of temperature dependences which corresponds to three metallic and one semiconducting band structures. The main features of thermal properties are closely related to the low-lying band structures; therefore, the high-resolution calorimeter measurements could be utilized to test them, e.g., the ratio of g/m^* in determining the Zeeman splitting energy.

As for monolayer germanene, the many-body Coulomb excitations have been thoroughly explored by the direct association of the modified random-phase approximation and the generalized tight-binding model [formulas in Sec. 3.4], especially for the bare and screened response functions due to dynamic charge screenings. Apparently, the critical factors, the hopping integrals, the spin-orbital couplings, the directions & magnitudes of transferred momenta, gate voltages, and magnetic fields, are responsible for the diverse single-particle and collective excitations. The low-frequency excitation properties directly reflect the characteristics of the low-lying bands, the strong wave-vector dependence, the anisotropic behavior, the SOC-induced separation of Dirac points, the doping levels, and the V_z-created destruction of spin-configuration degeneracy. There exists a forbidden excitation region between the intraband and interband electron-hole boundaries, being ascribed to the Fermi-momentum and band-edge states. The undamped acoustic plasmons could survive within this region with a prominent peak intensity. All the plasmon modes, being purely due to free conduction electrons, belong to 2D acoustic modes at small transferred momenta, as observed in an electron gas [Refs.]. With the increasing momentum, they might experience the interband damping and become another kind of undamped plasmons, change into the seriously suppressed modes in the heavy intraband electron-hole damping, remain the same undamped plasmons, or gradually vanish during the enhanced interband damping, i.e., there exist four kinds of plasmon modes after free carrier dopings. Specifically, the first kind only appears in germanene with the stronger spin-orbital couplings. The fourth kind of plasmon modes in monolayer germanene are dominated by gate voltage, while they are frequently revealed in few-layer extrinsic graphenes without external fields [30]. Moreover, even at the undoped case, the magneto-electronic Coulomb excitation spectra could exhibit several interband magnetoplasmon modes with the observable intensities under low transferred momenta. Most importantly, their collective oscillation frequencies possess the non-monotonous momentum-dependences, in which the first and second critical momenta represent the dramatic variation in the propagating direction of magneto-plasma wave and the disappearance due to the very strong Landau dopings. The main reason is the strong competition between the longitudinal Coulomb interactions and the transverse cyclotron forces. The free-carrier dopings can lead to a very prominent intraband magnetoplasmon (a 2D-EGS-like mode); furthermore, they have strong effects on the reduced frequencies and intensities of interband magnetoplasmons. Also, the former presents the discontinuous dependence in the increase of the magnetic-field strength, mainly owing to the quick enhancement in Landau-level energies. Generally speaking, the significant differences and similarities among germanene, silicene, and graphene lie in the existence of plasmon/magnetoplasmon modes and excitation gaps. The high-resolution experimental examinations for the reflection electron energy loss spectroscopy (EELS) and the inelastic x-ray scatterings are very useful in providing the the significant intralayer hopping integrals and spin-orbital couplings. Similar studies could be extended to other few-layer IV-group monatomic systems, e.g., bilayer Si [31] and single-layer Sn [32] & Pb [33] materials.

The topological characterization of Landau levels (LLs) in 3D layered systems hosting 2D massless Dirac fermions has been elaborated and discussed in Chap. 11. The focus was put on the chiral zero-mode LL, which is robust under the protection ascribed to 2D massless Dirac fermions. It is this characteristic that dictates the half-integer quantum Hall effect (QHE). Two distinct categories of systems that bear Dirac nodes were considered; that is, spinless nodal-line topological semimetals (TSMs), exemplified by spinless rhombohedral graphite (RG), and spinful time reversal invariant 3D strong topological insulators (TIs). RG is the stack of spinless graphene layers in ABC configuration. In this system, 2D massless Dirac fermions are hosted along the nodal lines in the bulk. This topological phase is protected by spacetime inversion symmetry somewhat generally. Within the minimal model, the protection is supplemented by sublattice symmetry (SLS). For the protection of the zero-mode LL, the continuous CS owing to the present SLS is responsible for it. If SLS is broken by certain tight binding hoppings and, as a result, the 2D massless Dirac fermions would be tilted, a generalized continuous CS is derived for the protection. A discussion on the second category has also been held. Time reversal invariant strong TIs, which can be layered systems such as Bi_2Se_3 and Bi_2Te_3. These systems are insulating in the bulk, while possessing an odd number of Dirac cones on the surface, where 2D massless Dirac fermions are hosted. Owing to the presence of those fermions, a chiral zero-mode LL and the dictated half-integer QHE have been expected and experimentally realized. However, how to derive an effective Dirac Hamiltonian from the finite lattice Hamiltonian is a crucial problem. The points have been introduced and discussed in Chap. 11.

REFERENCES

1. Zhang F, Jung J, Fiete G A, Niu Q, and MacDonald A H 2011 Spontaneous quantum Hall states in chirally stacked few-layer graphene systems *Phys. Rev. Lett.* **106** 156801.
2. Brunt T A, Rayment T, O'shea S J, and Welland M E 1996 Measuring the surface stresses in an electrochemically deposited metal monolayer: Pb on Au (111) *Langmuir* **12** 5942.
3. Akiyama M, Kawarada Y, and Kaminishi K 1984 Growth of single domain GaAs layer on (100)-oriented Si substrate by MOCVD. *Jpn. J. Appl. Phys.* **23** L843.
4. Singh D, Gupta S K, Sonvane Y, and Lukacevic I 2016 Antimonene: a monolayer material for ultraviolet optical nanodevices *J. Mater. Chem. C* **4** 6386.
5. Chen C H, Kepler K D, Gewirth A A, Ocko B M, and Wang J 1993 Electrodeposited bismuth monolayers on gold (111) electrodes: comparison of surface x-ray scattering, scanning tunneling microscopy, and atomic force microscopy lattice structures *J. Phys. Chem.* **97** 7290.
6. Luo Z, Huang Y, Weng J, Cheng H, Lin Z, Xu B, et al. 2013 1.06 μm Q-switched ytterbium-doped fiber laser using few-layer topological insulator Bi_2Se_3 as a saturable absorber *Opt. Express* **21** 29516.
7. Kim K S, Walter A L, Moreschini L, Seyller T, Horn K, Rotenberg E, et al. 2013 Coexisting massive and massless Dirac fermions in symmetry-broken bilayer graphene *Nat. Mater.* **12** 887.
8. Brihuega I, Mallet P, Gonzalez-Herrero H, Trambly de Laissardiere G, Ugeda M M, Magaud L, et al. 2012 Unraveling the intrinsic and robust nature of van Hove singularities in twisted bilayer graphene by scanning tunneling microscopy and theoretical analysis *Phys. Rev. Lett.* **109** 196802.
9. Wang Z F, Liu F, and Chou M Y 2012 Fractal Landau-level spectra in twisted bilayer graphene *Nano Lett.* **12** 3833.
10. Mele E J 2010 Commensuration and interlayer coherence in twisted bilayer graphene *Phys. Rev. B* **81** 161405.
11. Son Y W, Choi S M, Hong Y P, Woo S, and Jhi S H 2011 Electronic topological transition in sliding bilayer graphene *Phys. Rev. B*, **84** 155410.
12. Lin Y M, Perebeinos V, Chen Z, and Avouris P 2008 Electrical observation of subband formation in graphene nanoribbons *Phys. Rev. B* **78** 161409.
13. Ohta T, Bostwick A, McChesney J L, Seyller T, Horn K, and Rotenberg E 2007 Interlayer interaction and electronic screening in multilayer graphene investigated with angle-resolved photoemission spectroscopy *Phys. Rev. Lett.* **98** 206802.
14. Castro E V, Novoselov K S, Morozov S V, Peres N M R, Dos Santos J L, Nilsson J, et al. 2007 Biased bilayer graphene: semiconductor with a gap tunable by the electric field effect *Phys. Rev. Lett.* **99** 216802.
15. Zhao J, Liu H, Yu Z, Quhe R, Zhou S, Wang Y, et al. 2016 Rise of silicene: a competitive 2D material *Prog. Mater. Sci.* **83** 24.
16. Akinwande D, Petrone N, and Hone J 2014 Two-dimensional flexible nanoelectronics *Nat. Commun.* **5** 5678.
17. Vogt P, De Padova P, Quaresima C, Avila J, Frantzeskakis E, Asensio M C, et al. 2012 Silicene: compelling experimental evidence for graphenelike two-dimensional silicon *Phys. Rev. Lett.* **108** 155501.
18. Feng B, Ding Z, Meng S, Yao Y, He X, Cheng P, et al. 2012 Evidence of silicene in honeycomb structures of silicon on Ag (111) *Nano Lett.* **12** 3507.

19. Zhuang J, Xu X, Du Y, Wu K, Chen L, Hao W, et al. 2015 Investigation of electron-phonon coupling in epitaxial silicene by in situ Raman spectroscopy *Phys. Rev. B* **91** 161409.

20. Tahir M and Schwingenschlogl U 2013 Valley polarized quantum Hall effect and topological insulator phase transitions in Silicene *Sci. Rep.* **3** 1075.

21. Kane C L and Mele E J 2005 Quantum spin Hall effect in graphene *Phys. Rev. Lett.* **95** 226801.

22. Tabert C J and Nicol E J 2013 AC/DC spin and valley Hall effects in silicene and germanene *Phys. Rev. B* **87** 235426.

23. Kim Y, Choi K, and Ihm J 2014 Topological domain walls and quantum valley Hall effects in silicone *Phys. Rev. B* **89** 085429.

24. Mu Y S, Xue Y, Zhou T, and Yang Z Q 2019 Quantum anomalous Hall effects and various topological mechanisms in functionalized Sn monolayers *New J. Phys.* **21** 023010.

25. Li X, Zhang F, and Niu Q 2013 Unconventional quantum Hall effect and tunable spin Hall effect in Dirac materials: application to an isolated MoS_2 trilayer *Phys. Rev. Lett.* **110** 066803.

26. Ghazaryan A and Chakraborty T 2015 Aspects of anisotropic fractional quantum Hall effect in phosphorene *Phys. Rev. B* **92** 165409.

27. Ma R, Liu S W, Yang W Q, and Deng M X 2017 Quantum Hall effect in monolayer and bilayer black phosphorus *Europhys. Lett.* **119** 37005.

28. Yang J, Tran S, Wu J, Che S, Stepanov P, Taniguchi T, et al. 2018 Integer and fractional quantum Hall effect in ultrahigh quality few-layer black phosphorus transistors *Nano Lett.* **18** 229.

29. Lin S Y, Ho Y H, Shyu F L, and Lin M F 2013 Electronic thermal property of graphite *J. Phys. Soc. Jpn.* **82** 074603.

30. Shih P S, Chiu Y H, Wu J Y, Shyu F L, and Lin M F 2017 Coulomb excitations of monolayer germanene *Sci. Rep.* **7** 40600.

31. Do T N, Shih P H, Gumbs G, Huang D, Chiu C W, and Lin M F 2018 Diverse magnetic quantization in bilayer silicene *Phys. Rev. B* **97** 125416.

32. Chen S C, Wu C L, Wu J Y, and Lin M F 2016 Magnetic quantization of sp3 bonding in monolayer gray tin *Phys. Rev. B* **94** 045410.

33. Kaloni T P, Modarresi M, Tahir M, Roknabadi M R, Schreckenbach G, and Freund M S 2015 Electrically engineered band gap in two-dimensional Ge, Sn, and Pb: a first-principles and tight-binding approach *J. Phys. Chem. C* **119** 11896.

13 Future Perspectives and Open Issues

Ching-Hong Ho,[b] *Po-Hsin Shih,*[a] *Chiun-Yan Lin,*[a] *Thi-Nga Do,*[c,d] *Jhao-Ying Wu,*[b] *Shih-Yang Lin,*[e] *Ming-Fa Lin*[a,f,g]

[a] Department of Physics, National Cheng Kung University, Tainan 701, Taiwan
[b] Center of General Studies, National Kaohsiung University of Science and Technology, Kaohsiung 811, Taiwan
[c] Laboratory of Magnetism and Magnetic Materials, Advanced Institute of Materials Science, Ton Duc Thang University, Ho Chi Minh City, Vietnam
[d] Faculty of Applied Sciences, Ton Duc Thang University, Ho Chi Minh City, Vietnam
[e] Department of Physics, National Chung Cheng University, Chiayi 621, Taiwan
[f] Quantum Topology Center, National Cheng Kung University, Tainan 701, Taiwan
[g] Hierarchical Green-Energy Materials Research Center, National Cheng Kung University, Tainan, Taiwan

CONTENTS

Very interestingly, the current book proposes future perspectives and issues, being very useful in exploring the emergent condensed-matter systems, especially for the layered materials. The main-stream open topics cover the diversified essential properties in the physical, chemical, and material sciences. They are closely related to the dimension crossover (3D→0D), the stacking modulation, the composite uniform-nonuniform fields, the deformed/curved surfaces, the chemical absorptions/substitutions, the various defects, the amorphous structures, the significant combinations of multi-orbital hybridizations and spin-orbital interactions, the magneto-excitonic effects, and the many-particle e-e Coulomb interactions. Furthermore, the theoretical models would be modified to include the many-body effects in the complex scattering processes of optical absorptions, quantum transports and Coulomb decay rates.

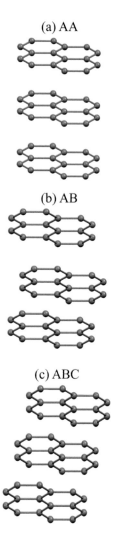

FIGURE 13.1 Geometric structures of (a) AA-, (b) AB-, and ABC-stacked graphite systems.

The dimension crossover phenomena will become one of the main-stream topics since the first discovery of few-layer graphene systems by the mechanical exfoliation in 2004 [1]. The 2D layered materials can dramatically change into the 3D bulk systems, the 1D nanomaterials, or even the 0D quantum dots. Up to now, the hexagonal honeycomb lattices can form 3D graphites [AA-, AB-, and ABC-stacked ones; Fig. 13.1], 2D layered graphenes, 1D graphene nanoribbons [Fig. 13.2], 1D carbon nanotubes [Fig. 13.3], and 0D graphene disks, in which the sp^2 bondings dominate the optimal geometric structures (the planar/curved/folded/cylindrical geometries). In general, the π bondings, being due to the carbon-$2p_z$ orbitals, are sufficiently in explaining the low-energy physical properties except for the highly deformed surfaces with the strong hybridization of $(2s, 2p_x, 2p_y, 2p_z)$ orbitals, e.g., those in the large- and

(a) Armchair nanoribbon

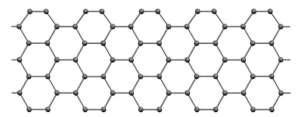

(b) Zigzag nanoribbon

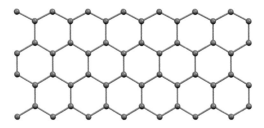

FIGURE 13.2 (a) Armchair and (b) zigzag graphene nanoribbons.

small-radius carbon nanotubes. Apparently, only three nearest neighbors are revealed in these systems, being quite different from the sp^3 bondings in the well-known diamond. Very interestingly, the 3D layered structures are never successfully synthesized in experimental laboratories using the other group-IV elements. This clearly indicates that the dramatic structure transformations might come to exist during the growth of Si/Ge/Sn/Pb layers. Among the significant mechanisms, the buckled geometries, with the very strong competitions between the sp^2 and sp^3 bondings, are expected to play the critical factors. The $2D \rightarrow 3D$ structural changes can only be done through the first-principles methods [discussed later in Chap. 13]. There are more complicated geometric structures even under the few-layer cases, compared with those of graphene systems. For example, the bilayer silicene systems are predicted to present four kinds of typical geometries, covering the AA-bb, AA-bt, AB-BB, and AB-bt ones. Obviously, the distinct geometric symmetries can greatly diversify the essential physical properties. How to get the structure-complicated site energies, intralayer & interlayer hopping integrals, and spin-orbital couplings is the key evaluation in the further developments of the phenomenological models. On the other side, the $2D \rightarrow 1D$ and $2D \rightarrow 0D$ transformations have been achieved in graphene-related systems, 1D/0D graphene [2, 3], silicene [4], and germanene nanoribbons/quantum dots [5, 6]. The quantum confinement and open/terminated edge structures lead to very close cooperations with the other intrinsic interactions. The various fundamental properties will exhibit the unusual behaviors, being worthy of systematic investigations.

There are five typical methods in creating the nonuniform magnetic quantization environments, covering the stacking modulation [7, 8], the reduced or enhanced dimensions [8, 9], the composite external fields [7, 10], the deformed/curved

(a) Armchair nanotube

(b) Zigzag nanotube

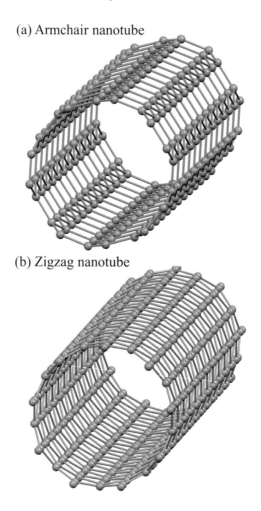

FIGURE 13.3   (a) Armchair and (b) zigzag carbon nanotubes.

surfaces [11, 12], and the hybridized structures [13]. In general, the quasi-1D magneto-electronic states have strong energy dispersions, at least, along one direction of wave vector, in which such relations are responsible for the unusual transport conductivities and absorption spectra. For example, the previous theoretical predictions show that it is difficult to observe the quantum Hall effects in Bernal graphite, as a result of the significant overlaps of various Landau subbands [14]. Furthermore, the ABC-stacked graphite presents the well-separated Landau subbands [15], directly indicating the possibilities in observing the quantum Hall effects. In the stacking-modulated bilayer graphene, the Landau subbands possess the lower state degeneracy [details in Chap. 5], and the k_y-dependent electronic states will make different contributions to the magnetic-field- and carrier-density-dependent transport properties. Equation (3.3.1) needs to be accurately calculated for each k_y state. This fact not only greatly enhances the numerical barriers, but also largely induces the analytic

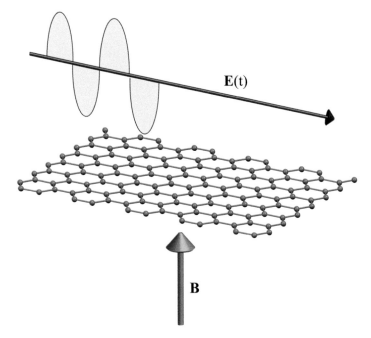

FIGURE 13.4 Monolayer graphene in a composite field; a uniform perpendicular magnetic field mixed with a spatially modulated electric field/Coulomb potential.

difficulties in proposing the concise and full scattering mechanisms. The existence of selection rule, which appears during the static scatterings between the occupied and unoccupied Landau-subband states, is the near-future studying focus. The quantized conductivities are expected to exhibit the fractal structures or be thoroughly absent under the specific geometry-modulated configurations. The similar researches could be conducted on the rich magneto-optical properties [9].

The composite external fields, which consists of a uniform magnetic field and a spatially modulated electric/magnetic field [Fig. 13.4], can create the unusual fundamental properties. The previous theoretical studies on monolayer graphene [7–9] show that their strong effects on magneto-electronic (optical properties) cover the one-dimensional Landau subbands with the significant energy dispersions, many band-edge states strong anisotropy, lower state degeneracy, seriously distorted/perturbed/well-behaved spatial probability distributions; a transformation of oscillation modes, and a lot of square-root-form asymmetric peaks in density of states (many asymmetric absorption peaks, blue or red shifts, the reduced spectral intensities, and the specific rule of $\Delta n = 1$ and the extra ones of $\Delta n = 0, 2, 3$). Up to now, only few studies are conducted on few-layer graphene systems, and the similar works on monolayer silicene and germanene are absent. The complex effects, being due to the buckled structures, the intralayer & interlayer hopping integrals, the spin-orbital interactions, and the composite fields, are deduced to create the unique physical phenomena, especially for those from the two latter factors. For example, the spin- and wave-vector-complicated magnetic wave functions might be composed of more oscillation modes and present the highly active behaviors in optical transitions, Coulomb excitations, and transport scatterings.

The various curved structures have become one the main stream-main-stream topics, e.g., the folded graphene nanoribbons [18, 22], curved ones [17, 23], carbon nanoscrolls [19, 24], carbon toroids [20, 25], and carbon onions [21, 26] purely due to the dominant sp^2 chemical bondings. For the diverse magnetic quantization phenomena, carbon allotropes are very ideal condensed-matter systems in fully exploring the curvature effects (the significant multi-orbital hybridizations of C-$(2s, 2p_x, 2p_y, 2p_z)$ orbitals), the critical interactions of the nonuniform interlayer hopping integrals [16], the strong competition between the quantum confinement and the magnetic localization, and the diamagnetism, paramagnetism, ferromagnetism, and anti-ferromagnetism [arising from the circulating charges and spin distributions; Refs. [17–21]]. Apparently, the systematic investigated will be made on whether the highly degenerate Landau levels, Landau subbands, or quantum discrete states could survive on the closed/open surfaces. They could be done by the generalized tight-binding model which is developed in this book. The important mechanisms should cover the scale of surface area, the degree of curvature, and the strength of magnetic field. Also, the geometry-enriched layered materials are very suitable for observing the diversified magneto-optical selection rule, abnormal quantum Hall effects, and the momentum-dependent inter-Landau-level magnetoplasmon modes.

The hybridized condensed-matter materials, which are composed of the lower-dimensional systems with the same element, have been successfully synthesized in experimental laboratories by the various methods, such as, a 1D multi-walled carbon nanotube [a superposition of some single-walled ones; [27]], a 3D multi-shell carbon onion [due to the various fullerens, a 3D carbon nanotube bundle [28], a 1D carbon nanotube-graphene nanoribbon hybrid [a CNT-GNR system; Ref. [29]]; a 3D compound of carbon nanotubes and graphitic layers. Up to now, the theoretical predictions on the magneto-electronic and optical properties are only done for the specific CNT-GNR hybridized materials [30], clearly indicating that the geometric symmetries/asymmetries dominate the formation of the quasi-Landau subbands. That is to say, the composite effects, arising from the 1D quantum confinement, the transverse cyclotron motion, and the complex interlayer atomic interactions, are responsible for the nonunifrom magnetic quantization phenomena. The main features of Landau subbands are very sensitive to the width of graphene nanoribbon, the radius and position of carbon nanotube, and the strength of magnetic field. Apparently, the diverse magnetic properties are expected to be easily observed/predicted in the other hybridized carbon-related systems. For example, the almost identical single-walled carbon nanotubes could be intercalated into the bulk Bernal graphite (the few-layer graphene systems) because of the weak van Der Waals interactions between two neighboring graphene layers. Such systems are very suitable for the model studies of the 3D or 2D Landau subbands. How to derive the independent magnetic Hamiltonian matrix elements through the generalized tight-binding model, would become an interesting work, since they strongly depend on the nanotube radius, the intertube distance, and the relative position between carbon nanotube and graphene.

The developed methods for efficiently solving the various Hamiltonians are one of the basic challenges in theoretical physics. The engineering of an energy gap, being

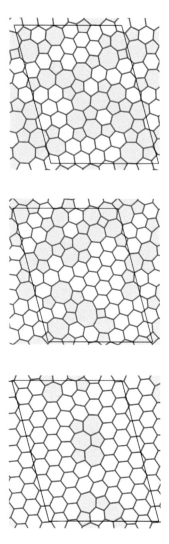

FIGURE 13.5 A defective graphene with hexagons, pentagons and heptagons [Gou, Hi-Gem book; Fig. 3].

closely related to semiconductor applications of layered material, can diversify the main features of magneto-electronic properties. A very effective way is to create the various defects. Different defects are frequently produced during the experimental syntheses, e.g., the vacancies [the studying focus; Ref. [31]], modified lattices [e.g., hexagons replaced by pentagons and heptagons in Fig. 13.5; Ref. [32]], and adatom impurities [33,34]. There are only a few theoretical predictions on the defect-enriched essential properties for the layered group-IV systems [35]. The defect-related silicene, germanene, and graphene will be very suitable for a model study. For such systems, the defects can induce the destruction of lattice symmetry, the distinct site energies (the change of the ionization potential) and the nonuniform atomic interactions (the position-dependent hopping integrals). The dependences of electronic properties on the type, distribution, and concentration of defects are worthy of a systematic investigation. For example, whether the semiconductor-semimetal transition, the localization of wave functions, and the valley- and spin-split states come to exist under the various defect configurations could be explored thoroughly. The complicated cooperative/competitive relations among the combined effects are proposed to comprehend the diverse electronic properties. Such effects are greatly enhanced by the external electric and magnetic fields, e.g., the drastic changes on the magnetic quantization in terms of the distortions and mixing of the Landau-level wave functions, the changes of the localization ranges, the reduced state degeneracy, and the irregular field-dependent energy spectra covering the crossing and anti-crossing behaviors.

In addition to the guest-atom substitutions [discussed in Chap. 8], the chemical absorptions are very efficient in drastically changing the fundamental physical properties, as thoroughly investigated in the previous two books using the first-principles method for adatom-adsorbed graphene-related systems [e.g., the hydrogenated, oxidized, and alkalized graphene systems in Figs. 13.6(a), 13.6(b), and 13.6(c), respectively; Ref. [36]]. They might provide another effective ways to greatly diversify the magnetic quantization. According to the detailed analyses, the critical factors, which determine the diverse phenomena, lie in the optimal geometric structures and the host-guest atomic interactions. Apparently, the guest-adatom positions in the quasi-sublattices also play an important role in the significant contributions of the magnetic wave functions. The corresponding adatom lattice needs to be taken into consideration simultaneously. Whether guest adatoms could exhibit the localized spatial distributions under a uniform perpendicular magnetic field is one of the studying focuses. The distribution configuration and concentration are expected to play important roles in any physical quantities, mainly owing to the strong effects on the single-orbital or multi-orbital hybridizations in adatom-host, adatom-adatom, host-host chemical bonds, and the significant competition between the guest-adatom and magnetic periods. Up to date, the magneto-electronic energy spectra and wave functions can only be solved by the generalized tight-binding model, but not the effective-mass approximation or the first-principles calculations [discussed in Chap. 3]. The various non-negligible hopping integrals and the atom-ionization-induced site energies could be obtained only by fitting the low-energy valence and conduction bands from the first-principles results. For many adatom chemisorptions on graphene, the key orbital-orbital interactions have been successfully identified from the atom-dominated energy bands, the spatial charge distributions before and after chemical modifications, and atom- and orbital-projected density of states. However, the delicate fittings and the very complicated magnetic Peierls phases will come to exist and become the high barriers in fully exploring the diversified quantization phenomena. In general, the magnetic Hamiltonian is a giant Hermitian matrix with many independent and imaginary elements. The critical roles of the chemically adsorbed adatoms on the main features of highly degenerate Landau levels will be very interesting, e.g., their effects on the magnetically quantized energy spectra, the magneto-state degeneracy, the destruction of the well-behaved oscillation modes, and the very close relations between the original and adsorption-created lattices. Apparently, whether the Landau wave functions could built by the guest adatoms is one of the studying focuses;

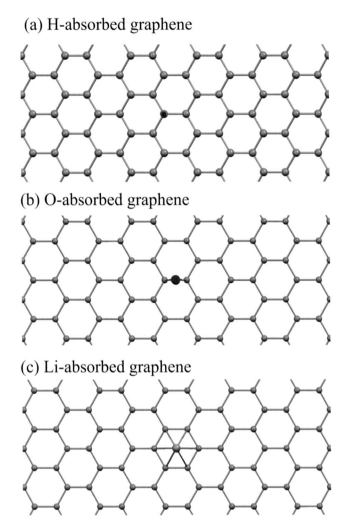

(a) H-absorbed graphene

(b) O-absorbed graphene

(c) Li-absorbed graphene

FIGURE 13.6 The chemical absorptions on graphene surface through the (a) hydrogen, (b) oxygen, and (c) alkali adatoms.

that is, two kinds of magnetic wave functions, which are, respectively, due to host atoms and guest adatoms, might coexist in different layers.

The emergent layered materials are expected to exhibit the rich magnetic quantizations in optical properties under the single-particle scheme characteristics of magneto-optical excitations, such as, the stacking-modulated bilayer graphenes, the twisted ones, the AA- and AB-stacked bilayer silicene/germanene/stanene/bismuthene, and so on [37–41]. To fully comprehend the key characteristics, the very close associations between the generalized tight-binding mode and the linear Kubo formula are necessary in overcoming the rather high barriers induced by the numerical calculations. For example, the Landau subbands, which mainly originate from the stacking modulations, possess the wave-vector-dependent energy spectra and magnetic wave functions, but not the highly degenerate and well-behaved Landau levels. The numerical calculations need to evaluate for the continuous wave vectors; therefore, the giant and **k**-dependent magnetic Hamiltonian matrix, being accompanied with the more complicated dipole moments. Apparently, the completely numerical calculations will be almost impossible. How to greatly enhance the efficiency of computation should be the near-future studying focus. As for the twisted bilayer graphene systems, a lot of Landau-level subgroups, being induced by the Moire superlattice, might show the unique magneto-optical properties, e.g., more inter-Landau-level vertical transition categories and extra selection rules [9]. The unusual phenomena are under the current investigations, since the computational problems could be solved for the highly degenerate magnet-electronic states.

The optical and magneto-optical properties, with the many-body effects (the excitonic effects), are very interesting topics in the fundamental researches and potential applications. The excitations, which are created during the optical excitation processes, belong to the strongly bound or metastable states of electron-hole pairs. They are formed by the attractive Coulomb forces between the negative and positive charges. Such quasiparticles could be condensed together and clearly reveal the macroscopic phenomena in some semiconductors at very low temperatures [42, 43], being quite different from the Cooper pairs in all the superconductors [44]. Any semiconductors might exhibit the excitonic effects under the suitable conditions, e.g., the red shift of threshold absorption frequency and the enhanced spectral intensity [45]. Apparently, the full theoretical framework, which covers the screened Coulomb interactions in the nonlinear scattering events (the standard Kubo formula), is required to display the basic many-particle characteristics in various optical spectra. Up to now, some previous theoretical works are suitable for the undoped/doped 3D semiconductors [46, 47], while those for 2D emergent materials seem to be absent. The difficulties in the theoretical calculations lie in the non-parabolic energy dispersions in the whole first Brilloun zone [48], the optical vertical excitations due to several pairs of low-lying valence and conduction bands [49], the dynamic charge scatterings in the electron-hole pairs [the random-phase approximation; Ref. [50]], and the greatly enhanced complexity arising from the external fields [51]. Apparently, that only the lowest energy of bound-state excitations is regarded as the threshold excitation frequency through the Schrodinger-like equation under the effective-mass approximation is not suitable in explaining the full excitonic effects. If the magnetic quantization could be included in the new theoretical model, the magneto-exciton effects are worthy of the systematic studies. Obviously, the exchange self-energies of Landau levels are not sufficient in calculating the many-body excitation frequencies [9]. It should be noticed that only few experimental cases on 2D few-layer systems might show the weak evidences of magneto-excitonic effects [7–9]. The specific excitation frequencies due to the prominent inter-Landau-level transitions are available in optical detectors [52], sensors [53], and attenuators [54]. In short, how to combine the generalized tight-binding model [7–9], the modified random-phase approximation [55–57], and the nonlinear dynamic Kubo formula [58, 59] are the key points of view in constructing the exact and reliable theoretical framework.

Obviously, the layered condensed systems will show the diverse quantum Hall conductivities, and the modified static Kubo formula might be necessary in thoroughly understanding the delicate transport phenomena due to the various scatterings/excitations (e.g., the electron-electron, electron-phonon and electron-impurity interactions, and the thermal effects). For example, both twisted bilayer graphenes and AB- and AA-stacked silicene/germanene [details in Chaps. 4, 6, & 7] present the rich and unique magneto-electronic energy spectra and wave functions, clearly illustrating the existence of unusual magneto-transport properties. The splittings of K/K′ valleys and/or spin configurations and the complex crossing & anti-crossing behaviors, and the irregular magnetic wave functions with the multi-oscillation modes could play the critical roles in greatly diversifying the quantum transport phenomena, such as, the abnormal integer conductivities, the non-integer ones, and the special or random plateau structures in the Fermi-energy- and magnetic-field-dependences. Furthermore, the distinct selection rules for the static inter-Landau-level transitions are worthy of the full explorations through the detailed analysis. On the other side, only few theoretical investigations are conducted on the complex effects arising from the inelastic scatterings [60,61]. Specifically, it is very difficult to directly cover the many-particle interactions in the transport formula; that is, a complete theoretical model should be further developed to identify the novel results from the high-resolution measurements [details in Sec. 2.3]. A possible algorithm is the close associations of the generalized tight-binging mode, the modified random-phase approximation, and the Kubo formula.

The electron-electron Coulomb interactions are one of the main-stream topics in condensed-matter physics, since they play critical roles in the essential physical properties. Apparently, the many-body phenomena, with the rich and unique behaviors, are expected to be revealed in the emergent layered materials. To fully explore the diversified single-particle and collective excitations, a lot of open issues need to be overcome. The theoretical framework, which covers the close association of the generalized tight-binding model, the modified random-phase approximation, and the screened exchange self-energy, might be suitable in thoroughly investigating the quasiparticle states. In general, the intralayer & interlayer hopping integrals in the band-structure calculations are obtained from the numerical fittings with the first-principles results near the high-symmetry **k**-points. However, it would become very difficult to the get reliable parameters if the buckled structures, stacking configurations, multi-orbital hybridizations, interlayer atomic interactions, and spin-orbital couplings in few-layer systems are taken into account simultaneously. For example, a bilayer AB-bt silicene is predicted to present the layer-dependent spin-orbital interactions and the largest interlayer hopping integral for the first pair of valence and conduction bands with the unusual energy dispersions [details in Ref. [62]]. The previous study clearly shows that a set of good parameters for two pairs of energy bands is absent up to now. The similar cases appear in other bilayer systems, e.g., germanene [63], stanene [64], antimonene [65], and bismuthene [66] without the full parameters in the published papers. When the random-phase approximation is used to include the band-structure effects, one must solve the Coulomb matrix elements [Eq. (3.23)] arising from the orbital-dominated tight-binding function. How to get rid of the significant calculation errors in such elements will be the studying focus. Also, the delicate Coulomb excitation spectra can provide the full information in the further topics, such as, the dominance on the quasiparticle energies and lifetimes [67], the time-dependent propagations of plasma waves in the **r**-space, the screened Coulomb potentials and charge density distributions in planar structures, the simultaneous combinations of electron-electron and electron-phonon interactions. Furthermore, the analytic formulas for distinct many-body properties have to be modified/derived for the highly efficient calculations.

In addition to the above-mentioned phenomenological models, the first-principles methods using the Vienna ab initio simulation packages are very reliable for certain essential physical properties, e.g., the optimal geometric structures [68, 69], electronic properties [69, 70], spin-related magnetic configurations [71], and phonon spectra [72]. Very importantly, they can provide the suitable parameters (site energies, hoppig integrals, and spin-orbital couplings) for the generalized tight-binding model. Apparently, such methods are useless in the magnetic quantization phenomena as a result of the very large unit cell due to the vector potential. The excellent point of view, being based on the basic chemistry and physics, is proposed through the developed theoretical framework: the critical multi-hybridizations are accurately examined from the host-atom- & guest-atom-dominated energy bands, the spatial charge density & the difference after chemisorption/bonding, and atom-, orbital-, & spin-projected density of states; furthermore, the diverse magnetic configurations are delicately identified from the strong competition between the boundary host atoms and adatoms, the spin-split/spin-degenerate electronic structures, the spin-induced net magnetic moment, the spatial spin distributions around two kinds of atoms, and the spin-split density of states. This has been successfully developed to fully explore and comprehend the geometric, electronic, and magnetic properties for the 2D graphene-reflated systems [the 1D graphene-nanoribbon-related ones; details in Ref. [36]], such as, the essential properties in the AA-, AB-, ABC-, & AAB-stacked pristine graphenes, sliding graphenes, rippled graphenes, graphene oxides, hydrogenated graphenes, halogen-, alkali-, Al-, & Bi-doped graphene compounds (armchair and zigzag graphene nanoribbons without/with hydrogen terminations, curved and zipped graphene nanoribbons, folded graphene nanoribbons, carbon nanoscrolls, bilayer graphene nanoribbons, edge-decorated graphene nanoribbons, alkali-, halogen-, Al-, Ti-, and Bi-absorbed graphene nanoribbons). The diverse physical phenomena, including the semiconductors, semimetals, and metals the absence or recovery of the Dirac-cone structures [73], the linear, parabolic, and oscillatory energy dispersions [74], the critical points in the energy-wave-vector space [the minima, maxima, and saddle points; Ref. [36]], the semiconductor-metal transition [75, 76], the creation of band gap [77], and non-magnetic, ferromagnetic, or anti-ferromagnetic spin configurations [78], are clearly explained by the concise physical and chemical pictures. The close associations of the first-principles calculations and the generalized tight-binding model will be very useful for the emergent 2D materials in exploring the various magnetic quantization phenomena. For example, the diverse magneto-electronic properties, magneto-optical spectra, quantum Hall transports, and magneto-Coulomb excitations can be thoroughly investigated only after the reliable fitting for the intrinsic interactions. A set of suitable parameters is strongly required for few-layer 2D materials, being purely composed by group-IV and group-V atoms. These systems are expected to reveal the rich and unique physical phenomena.

The distinct magnetisms, which are induced by the host atoms/molecules, guest ones, and the external magnetic fields, are closely related to the emergent spintronics and can provide the full informations on potential applications. They cover the paramagnetism, diamagnetism, non-magnetism ferromagnetism, and anti-ferromagnetism. The former two phenomena mainly originate from the orbital currents of valence and conduction electrons, being created by a uniform magnetic field. According to the previous theoretical calculations, the diverse magnetic properties are predicted to survive in 1D carbon nanotubes [79], 0D carbon tori [80], 0D C_{60}-related fullerenes [81], 2D graphenes, and 3D Bernal graphite [82]. For example, the metallic/narrow-gap (one-third) and finite-gap (two-thirds) carbon nanotubes, respectively, exhibit the paramagnetic and diamagnetic phenomena [83], when they are present in a uniform magnetic field parallel to the tubular axis. Also, the specific persistent currents exist in other low-dimension carbon-related systems with the sp^2-dominated surface structures [84]. Generally speaking, the magnetic response is very weak, so that the experimental examinations are required to be finished at very low temperature for a almost homogenous system [85]. To get the magnetization field/the magnetic susceptibility, a very accurate derivative of the total energy versus the magnetic-field strength is required in the numerical calculations. However, the variation of the former might be rather small within a finite difference of the latter, especially for 2D and 3D condensed-matter systems. A lot of magneto-electronic states are taken into account simultaneously. Furthermore, the theoretical models are not so reliable in the higher-energy ones, such as, the effective-mass approximation even for monolayer graphene/silicene/germanene [86–91]. As a result, the delicate investigations on the diverse magnetism of layered group-IV and group-V materials would be a significant challenge. The different magnetic behaviors, but the similar difficulties are expected to appear in the ferromagnetic and anti-ferromagnetic materials. The strong cooperations or competitions among the host-atom-, guest-atom-induced spin moments, and magnetic field will be fully explored through the combination of the generalized tight-binding model and the Hubbard model. For example, how to simulate the magnetic-field effects on the anti-ferromagnetic spin configurations at two open boundaries of zigzag graphene nanoribbons is one of studying focuses.

Besides the 3D layered systems considered elsewhere, the open possibility of dimension crossover from 2D to 3D should be noted since the topological classification of symmetry protected topological phases is according to a system's symmetry and intrinsic spatial dimension. For the presented systems, rhombohedral graphite (RG), as a representative for chiral symmetry (CS) protected spinless nodal-line topological semimetals (TSMs), is a result from 2D TSM-to-3D TSM crossover from graphene to RG. Besides, Bi_2Se_3 or the similar, for time reversal symmetry (TRS) protected spinful strong topological insulators (TIs), results from 2D trivial insulator to 3D TI. In general, other possibility is open. For example, there exist evidences of 2D TSM-to-3D TI and 2D TI-to-3D TSM crossovers with various symmetry protection. On the other hand, topological quantum phase transitions are of fundamental importance, where the perturbation due to disorder for metal-insulator transition has deep meaning. Research on these regards has been growing. By the way, since spinless graphene has been artificially produced by means of cold atoms in optical lattice, artificial spinless RG is expected. Hence, it is not impractical to study topological matter with spinless condition.

REFERENCES

1. Novoselov K S, Geim A K, Morozov S V, Jiang D, Zhang Y, Dubonos S V, et al. 2004 Electric field effect in atomically thin carbon films *science* **306** 666.
2. Ruffieux P, Cai J, Plumb N C, Patthey L, Prezzi D, Ferretti A, et al. 2012 Electronic structure of atomically precise graphene nanoribbons *ACS Nano* **6** 6930.
3. Miccoli I, Aprojanz J, Baringhaus J, Lichtenstein T, Galves L A, Lopes J M J, et al 2017 Quasi-free-standing bilayer graphene nanoribbons probed by electronic transport *Appl. Phys. Lett.* **110** 051601.
4. Kara A, Enriquez H, Seitsonen A P, Voon L C L Y, Vizzini S, Aufrayg B, et al. 2012 A review on silicene-New candidate for electronics *Surf. Sci. Rep.* **67** 1.
5. Salimiana F and Dideba D 2019 Comparative study of nanoribbon field effect transistors based on silicene and graphene *Mat. Sci. Semicon. Proc.* **93** 92.
6. Abhinav E M, Sundararaj A, Gopalakrishnan C, Raja S V K, and Chokhra S 2017 Impact of strain on electronic and transport properties of 6 nm hydrogenated germanane nano-ribbon channel double gate field effect transistor *Mater. Res. Express* **4** 114005.
7. Chen S C, Wu J Y, Lin C Y, and Lin M F 2017 Theory of Magnetoelectric Properties of 2D Systems *IOP Concise Physics.* San Raefel, CA, USA: Morgan & Claypool Publishers.
8. Lin C Y, Chen R B, Ho Y H, and Lin M F 2018 *Electronic and Optical Properties of Graphite-Related Systems.* Boca Raton, Florida: CRC Press.
9. Lin C Y, Do T N, Huang Y K, and Lin M F 2017 Electronic and optical properties of graphene in magnetic and electric fields *IOP Concise Physics.* San Raefel, CA, USA: Morgan & Claypool Publishers.
10. Ou Y C, Sheu J K, Chiu Y H, Chen R B, and Lin M F 2011 Influence of modulated fields on the Landau level properties of graphene *Phys. Rev. B* **83** 195405.
11. Chung H C, Chang C P, Lin C Y, and Lin M F 2016 Electronic and optical properties of graphene nanoribbons in external fields *Phys. Chem. Chem. Phys.* **18** 7573.
12. Lin C Y, Chen S C, Wu J Y, and Lin M F 2012 Curvature effects on magnetoelectronic properties of nanographene ribbons *J. Phys. Soc. Jpn.* **81** 064719.
13. Zhu L, Wang J, Zhang T, Ma L, Lim C W, Ding F, et al. 2010 Mechanically robust tri-wing graphene nanoribbons with tunable electronic and magnetic properties *Nano letters* **10** 494.
14. Zhang Y, Tan Y W, Stormer H L, and Kim P 2005 Experimental observation of the quantum Hall effect and Berry's phase in graphene *Nature* **438** 201.
15. Zhang L, Zhang Y, Camacho J, Khodas M, and Zaliznyak I 2011 The experimental observation of quantum Hall effect of l = 3 chiral quasiparticles in trilayer graphene *Nature* **7** 953.
16. Yan W, He W Y, Chu Z D, Liu M, Meng L, Dou R F, et al. 2013 Strain and curvature induced evolution of electronic band structures in twisted graphene bilayer *Nature communications* **4** 2159.
17. Kim K, Sussman A, Zettl A. 2010 Graphene nanoribbons obtained by electrically unwrapping carbon nanotubes *ACS Nano* **4** 1362.
18. Vo T H, Shekhirev M, Kunkel D A, Morton M D, Berglund E, Kong L, et al. 2014 Large-scale solution synthesis of narrow graphene nanoribbons *Nat. Comm.* **5** 3189.
19. Patra N, Wang B, and Kral P 2009 Nanodroplet activated and guided folding of graphene nanostructures *Nano Lett.* **9** 3766.
20. Tsai C C, Shyu F L, Chiu C W, Chang C P, Chen R B, and Lin M F 2004 Magnetization of armchair carbon tori *Phys. Rev. B* **70** 075411.

21. Tomita S, Fujii M, Hayashi S, and Yamamoto K 1999 Electron energy-loss spectroscopy of carbon onions *Chem. Phys. Lett.* **305** 225.

22. Prada E, San-Jose P, and Brey L 2010 Zero Landau level in folded graphene nanoribbons *Physical Review Letters* **105** 106802.

23. Martins B V C and Galvao D S 2010 Curved graphene nanoribbons: Structure and dynamics of carbon nanobelts *Nanotechnology* **21** 075710.

24. Viculis L M, Mack J J, and Kaner R B 2003 A chemical route to carbon nanoscrolls *Science* **299** 1361.

25. Haddon R C 1997 Electronic properties of carbon toroids *Nature* **388** 31.

26. Banhart F and Ajayan P M 1996 Carbon onions as nanoscopic pressure cells for diamond formation *Nature* **382** 433.

27. Monteiro-Riviere N A, Nemanich R J, Inman A O, Wang Y Y, and Riviere J E 2005 Multi-walled carbon nanotube interactions with human epidermal keratinocytes *Toxicology Letters* **155** 377.

28. Schaper A K, Hou H, Greiner A, Schneider R, and Phillipp F 2004 Copper nanoparticles encapsulated in multi-shell carbon cages *Applied Physics A* **78** 73.

29. Yang Z, Liu M, Zhang C, Tjiu W W, Liu T, and Peng H 2013 Carbon nanotubes bridged with graphene nanoribbons and their use in high-efficiency dye-sensitized solar cells. *Angewandte Chemie International Edition* **52** 3996.

30. Khoeini F and Shokri A A 2011 Modeling of transport in a glider-like composite of GNR/CNT/GNR junctions *Journal of Computational and Theoretical Nanoscience* **8** 1315.

31. Ma J, Alfe D, Michaelides A, and Wang E 2009 Stone-Wales defects in graphene and other planar sp^2-bonded materials *Phys. Rev. B* **80** 033407.

32. Terrones M, Botello-Mendez A R, Campos-Delgado J, Lopez-Urias F, Vega-Cantu Y I, Rodriguez-Macias F J, et al. 2010 Graphene and graphite nanoribbons: Morphology, properties, synthesis, defects and applications *Nano Today* **5** 351.

33. Sahin H, Topsakal M, Ciraci S 2011 Structures of fluorinated graphene and their signatures *Physical Review B* **83** 115432.

34. He Z, He K, Robertson A W, Kirkland A I, Kim D, Ihm J, et al. 2014 Atomic structure and dynamics of metal dopant pairs in graphene *Nano Letters* **14** 3766.

35. Li S, Wu Y, Tu Y, Wang Y, Jiang T, Liu W, et al. 2015 Defects in silicene: Vacancy clusters, extended line defects, and di-adatoms *Scientific Reports* **5** 7881.

36. Lin S Y, Tran N T T, Chang S L, Su W P, and Lin M F 2018 *Structure- and adatom-enriched essential properties of graphene nanoribbons* Boca Raton, Florida: CRC Press.

37. Ho Y H, Chiu Y H, Lin D H, Chang C P, and Lin M F 2010 Magneto-optical selection rules in bilayer Bernal grapheme *ACS Nano* **4** 1465.

38. Huang Y K, Chen S C, Ho Y H, Lin C Y, and Lin M F 2014 Feature-rich magnetic quantization in sliding bilayer graphenes *Sci. Rep.* **4** 7509.

39. Do T N, Gumbs G, Shih P H, Huang D, Chiu C W, Chen C Y, et al 2019 Peculiar optical properties of bilayer silicene under the influence of external electric and magnetic fields *Sci. Rep.* **9** 624.

40. Chen R B, Jang D J, Lin M C, and Lin M F 2018 Optical properties of monolayer bismuthene in electric fields *Opt. Lett.* **43** 6089.

41. Shyu F L 2019 Field-induced spin polarized electronic and optical properties of armchair stanene nanoribbons *Phys. Lett. A* **383** 68.

42. O'neil M, Marohn J, McLendon G 1990 Dynamics of electron-hole pair recombination in semiconductor clusters *Journal of Physical Chemistry* **94** 4356.

43. Brinkman W F and Rice T M 1973 Electron-hole liquids in semiconductors *Physical Review B* **7** 1508.

44. Nakamura Y, Pashkin Y A, and Tsai J S 1999 Coherent control of macroscopic quantum states in a single-Cooper-pair box *Nature* **398** 786.

45. Havener R W, Liang Y, Brown L, Yang L, and Park J 2014 Van Hove singularities and excitonic effects in the optical conductivity of twisted bilayer graphene *Nano Lett.* **14** 3353.

46. Van de Walle C G and Martin R M 1987 Theoretical study of band offsets at semiconductor interfaces *Physical Review B* **35** 8154.

47. Dong J, Sankey O F, and Myles C W 2001 Theoretical study of the lattice thermal conductivity in Ge framework semiconductors *Physical Review Letters* **86** 2361.

48. Lewiner C and Bastard G 1980 Indirect exchange interaction in extremely non-parabolic zero-gap semiconductors *Journal of Physics C: Solid State Physics* **13** 2347.

49. Huang B L, Chuu C P, and Lin M F 2019 Asymmetry-enriched electronic and optical properties of bilayer graphene *Scientific Reports* **9** 859.

50. Sarma S D and Hwang E H 1996 Dynamical response of a one-dimensional quantum-wire electron system *Physical Review B* **54** 1936.

51. Gusynin V P, Sharapov S G, and Carbotte J P 2007 Anomalous absorption line in the magneto-optical response of graphene *Physical Review Letters* **98** 157402.

52. Futia G, Schlup P, Winters D G, and Bartels R A 2011 Spatially-chirped modulation imaging of absorbtion and fluorescent objects on single-element optical detector *Optics express* **19** 1626.

53. Kimura M and Toshima K 1998 Vibration sensor using optical-fiber cantilever with bulb-lens *Sensors and Actuators A: Physical* **66** 178.

54. Li H, Tadesse S A, Liu Q, and Li M 2015 Nanophotonic cavity optomechanics with propagating acoustic waves at frequencies up to 12 GHz *Optica* **2** 826.

55. Shung K W K 1986 Lifetime effects in low-stage intercalated graphite systems *Phys. Rev. B* **34** 2.

56. Ho J H, Lu C L, Hwang C C, Chang C P, and Lin M F 2006 Coulomb excitations in AA- and AB-stacked bilayer graphites *Phys. Rev. B* **74** 085406.

57. Ho J H, Chang C P, and Lin M F 2006 Electronic excitations of the multilayered graphite *Phys. Lett. A* **352** 446.

58. Podobedov V B, Miller C C, and Nadal M E 2012 Performance of the NIST gonio-colorimeter with a broad-band source and multichannel charged coupled device based spectrometer *Rev. Sci. Instrum.* **83** 093108.

59. Martinez L F L, Garcia R C , Navarro R E B, and Martinez A L 2009 Microreflectance difference spectrometer based on a charge coupled device camera: Surface distribution of polishing-related linear defect density in GaAs (001) *Appl. Opt.* **48** 5713.

60. Novikov D S 2007 Elastic scattering theory and transport in graphene *Physical Review B* **76** 245435.

61. Mak K F, Ju L, Wang F, and Heinz T F 2012 Optical spectroscopy of graphene: From the far infrared to the ultraviolet *Solid State Communications* **152** 1341.

62. Yaokawa R, Ohsuna T, Morishita T, Hayasaka Y, Spencer M J S, and Nakano H 2016 Monolayer-to-bilayer transformation of silicenes and their structural analysis *Nat. Comm.* **7** 10657.

63. Shih P H, Chiu Y H, Wu J Y, Shyu F L, and Lin M F 2017 Coulomb excitations of monolayer germanene *Sci. Rep.* **7** 40600.

64. Chen S C, Wu C L, Wu J Y, and Lin M F 2016 Magnetic quantization of sp^3 bonding in monolayer gray tin *Phys. Rev. B* **94** 045410.

65. Yu J, Katsnelson M I, and Yuan S 2018 Tunable electronic and magneto-optical properties of monolayer arsenene: From GW0 approximation to large-scale tight-binding propagation simulations *Phys. Rev. B* **98** 115117.

66. Chen S C, Wu J Y, and Lin M F 2018 Feature-rich magneto-electronic properties of bismuthene *New Jour. Phys. Fast Track Communication* **20** 062001.

67. Shih P H, Chiu C W, Wu J Y, Do T N, and Lin M F 2018 Coulomb scattering rates of excited states in monolayer electron-doped germanene *Phys. Rev. B* **97** 195302.

68. Kostelnik P, Seriani N, Kresse G, Mikkelsen A, Lundgren E, Blum V, et al. 2007 The Pd (100)-(5×5) R27°–O surface oxide: A LEED, DFT and STM study *Surface Science* **601** 1574.

69. Hafner J 2007 Materials simulations using VASP—a quantum perspective to materials science *Computer Physics Communications* **177** 6.

70. Neaton J B, Muller D A, Ashcroft N W 2000 Electronic properties of the Si/SiO_2 interface from first principles *Physical Review Letters* **85** 1298.

71. Eelbo T, Wasniowska M, Thakur P, Gyamfi M, Sachs B, Wehling T O, et al. 2013 Adatoms and clusters of 3D transition metals on graphene: Electronic and magnetic configurations *Physical Review Letters* **110** 136804.

72. Barabash S V, Ozolins V, and Wolverton C 2008 First-principles theory of competing order types, phase separation, and phonon spectra in thermoelectric AgPb m SbTe m+ 2 alloys *Physical Review Letters* **101** 155704.

73. Lin S Y, Chang S L, Shyu F L, Lu J M, Lin M F 2015 Feature-rich electronic properties in graphene ripples *Carbon* **86** 207.

74. Tran N T T, Lin S Y, Glukhova O E, and Lin M F 2015 Configuration-induced rich electronic properties of bilayer graphene *The Journal of Physical Chemistry C* **119** 10623.

75. Lin S Y, Chang S L, Tran N T T, Yang P H, and Lin M F 2015 H–Si bonding-induced unusual electronic properties of silicene: A method to identify hydrogen concentration *Physical Chemistry Chemical Physics* **17** 26443.

76. Lin S Y, Lin Y T, Tran N T T, Su W P, and Lin M F 2017 Feature-rich electronic properties of aluminum-adsorbed graphenes *Carbon* **120** 209.

77. Huang H C, Lin S Y, Wu C L, and Lin M F 2016 Configuration- and concentration-dependent electronic properties of hydrogenated graphene *Carbon* **103** 84.

78. Lin Y T, Lin S Y, Chiu Y H, and Lin M F 2017 Alkali-created rich properties in grapheme nanoribbons: Chemical bondings *Scientific Reports* **7** 1722.

79. Lu J P 1995 Novel magnetic properties of carbon nanotubes *Physical Review Letters* **74** 1123.

80. Tsai C C, Shyu F L, Chiu C W, Chang C P, Chen R B, and Lin M F 2004 Magnetization of armchair carbon tori *Physical Review B* **70** 075411.

81. Jonsson D, Norman P, Ruud K, Agren H, and Helgaker T 1998 Electric and magnetic properties of fullerenes *The Journal of Chemical Physics* **109** 572.

82. Moran D, Stahl F, Bettinger H F, Schaefer H F, and Schleyer P V R 2003 Towards graphite: Magnetic properties of large polybenzenoid hydrocarbons *Journal of the American Chemical Society* **125** 6746.

83. Searles T A, Imanaka Y, Takamasu T, Ajiki H, Fagan J A, Hobbie E K, et al. 2010 Large anisotropy in the magnetic susceptibility of metallic carbon nanotubes *Physical Review Letters* **105** 017403.

84. Kibis O V 2002 Electronic phenomena in chiral carbon nanotubes in the presence of a magnetic field *Physica E: Low-dimensional Systems and Nanostructures* **12** 741.

85. Murmu T, McCarthy M A, and Adhikari S 2012 Vibration response of double-walled carbon nanotubes subjected to an externally applied longitudinal magnetic field: A nonlocal elasticity approach *Journal of Sound and Vibration* **331** 5069.

86. Castro Neto A H, Guinea F, Peres N M R, Novoselov K S, and Geim A K 2009 The electronic properties of graphene *Rev. Mod. Phys.* **81** 109–162.

87. McCann E and Koshino M 2013 The electronic properties of bilayer graphene *Rep. Prog. Phys.* **76** 056503.

88. Koshino M and McCann E 2009 Trigonal warping and Berry's phase N pi in ABC-stacked multilayer graphene *Phy. Rev. B* **80** 165409.

89. Koshino M and Ando T 2008 Magneto-optical properties of multilayer graphene *Phy. Rev. B* **77** 115313.

90. Tabert C J and Nicol E J 2013 Magneto-optical conductivity of silicene and other buckled honeycomb lattices *Phys. Rev. B* **88** 085434.

91. Matthes L, Pulci O, and Bechstedt F 2013 Massive Dirac quasiparticles in the optical absorbance of graphene, silicene, germanene, and tinene *Journal of Physics: Condensed Matter* **25** 395305.

Index

Note: Page numbers in italics refer to figures.